T4-AJS-576

INTEGRATED WATERSHED MANAGEMENT *in the* GLOBAL ECOSYSTEM

INTEGRATED WATERSHED MANAGEMENT *in the* GLOBAL ECOSYSTEM

Edited by
Rattan Lal

CRC Press
Boca Raton London New York Washington, D.C.

Contact Editor: John Sulzycki
Project Editor: Sara Seltzer
Cover design: Dawn Boyd

Library of Congress Cataloging-in-Publication Data

Integrated watershed management in the global ecosystem / Rattan Lal, editor.
p. cm.
Based on the proceedings of the international conference held at Toronto Canada in July 1997, sponsored by the International Affairs Committee of the Soil and Water Conservation Society.
Includes bibliographical references and index.
ISBN 0-8493-0702-3 (alk. paper)
1. Watershed management Congresses. 2. Ecosystem management Congresses. I. Lal, R. II. Soil and Water Conservation Society (U.S.). International Affairs Committee.
TC409.I585 1999
333.73'15--dc21

99-21035
CIP

International Standard Book Number 0-8493-0702-3
Library of Congress Card Number 99-21035
Printed in the United States of America 1 2 3 4 5 6 7 8 9 0
Printed on acid-free paper

This volume is based on the proceedings of the international conference on "Global Challenges in Ecosystem Management in a Watershed Context" held in Toronto, Canada in July 1997. The conference was sponsored by the International Affairs Committee of the Soil and Water Conservation Society and funded by the Natural Resources Conservation Service and the Rockefeller Foundation. The conference organizing committee consisted of Rattan Lal (chair), David Cressman, Samir El-Swaify, and J. Hammond.

Foreword

At the international symposium, "Global Challenges in Ecosystem Management: In a Watershed Context" some of the leading world experts and practitioners in soil and water conservation explored how "ecosystem-based" management approaches might lead to more in arresting the persistent ravages of global land and water degradation. Consistent with an ecosystem-based approach is the possibility of enhancing progress by working with affected populations within a watershed context, where relationships between land-use practices and the condition of soil water resources can be better recognized and managed.

Despite decades of research, education, and development assistance, and more recently an emphasis on concepts of sustainable natural resource management and development, conventional approaches are clearly not meeting the need. Land and water degradation and associated stresses on human populations continue at a startling pace. Whether progress lags because of too few socially relevant technologies and management systems, inadequate information dissemination techniques, lack of donor and in-county agency awareness, distracted political will, or whatever, the need to improve our understanding and our ability to respond more effectively is undeniable.

Notwithstanding all recent urgings to move toward more sustainable forms of development activity, there has been a significant decline in donor agency and national government support for initiatives focused on the very foundation of healthy and productive societies — secure land and water resources. Vast quantities of scarce topsoil resources annually leave fields and forests and fill oceans and lakes, while international development agencies largely attend to other issues.

The particular challenge given to the speakers at this symposium was to share, from their pan-global perspectives, their views generally on the reasons for lack of progress and, in selected case studies, the reasons for success. Readers of this volume will quickly appreciate why continued inattention to the issues of land and water degradation is not a viable option. For those who believe that past failures somehow commit us to continued failure, observe the examples of good physical science interacting with good social science to yield very positive results. See, too, the examples of scientists and practitioners discovering through participatory research with local landowners/users how indigenous knowledge can help shape better methods and systems for sustainable resource use.

As cosponsors of this symposium, the Soil and Water Conservation Society, through its International Affairs Committee, and the World Association of Soil and Water Conservation hope the chapters contained in this volume will be of practical value to those who seek better understanding of the issues, perspectives on improved management approaches, and a sense of the practical challenges that lie ahead.

David R. Cressman, P.Ag
Chair, International Affairs Committee, SWCS, 1997

Preface

The 1996 per capita agricultural production index, relative to that of 1989–91 equals 100, was 103.5 for the world, 99.9 for Africa, 106.2 for North-Central America, 108.2 for South America, 117.2 for Asia, and 103.0 for Oceania. There are several regions of the world where the increase in food production has severely lagged behind the rate of population growth. The problem is particularly severe in several countries of sub-Saharan Africa, e.g., Botswana with a per capita food production index of 92.5, Burundi (83.9), Central African Republic (93.9), Gabon (89.3), Gambia (58.8), Kenya (87.9), Madagascar (86.9), Mauritania (91.5), Sierra Leone (88.6), Swaziland (80.2), Tanzania (83.7), and Zaire (80.6). In addition to socioeconomic and political constraints, there are three principal biophysical factors contributing to declining per capita food production: (1) drought stress, (2) low soil fertility, and (3) soil degradation. Drought stress, lack of adequate soil water to meet the physiological demand for optimum crop growth and yield, is a principal cause of low crop yields in many countries of arid and semiarid regions. While drought stress is a common constraint in dry regions, land misuse and soil mismanagement can lead to drought stress even in the humid regions. Low inherent soil fertility, and rapid depletion of soil nutrient pool, is another major cause of low yields, especially in old and highly weathered soils of the tropics and subtropics. It is estimated that more than 500 million ha (25% of the total land area) of soils in Africa may have a problem of P deficiency. Deficiencies of N and some micronutrients are severe constraints throughout the tropics and subtropics. Drought stress and nutrient deficiencies are exacerbated by a widespread problem of soil degradation and desertification. Global extent of extreme and strong soil degradation is estimated at about 300 million ha.

The regions prone to perpetual food shortages are also characterized by some environmental issues of major global concern. Decline in water quality and emissions of radiatively active gases from natural and managed ecosystems to the atmosphere are major environmental concerns. Decline in water quality is due to non-point-source pollution leading to contamination and eutrophication of surface waters and pollution of groundwater. Land misuse and soil mismanagement can lead to emissions of CO_2, CH_4, N_2O, and NO_x from soil to the atmosphere.

The answer to the problems of food insecurity and poor environment quality lies in adopting a strategy leading to sustainable use of natural resources. Watershed management is a strategy that can conserve water, enhance water use efficiency, and improve water quality. Adoption of a judicious land use and appropriate soil and water management practices, which enhance nutrient cycling and decrease losses, can improve soil fertility. Soil degradation, caused by land misuse and soil mismanagement, is a physical problem driven by socioeconomic and political factors. Watershed management techniques, specific to landscape units depending on their pedological and hydrological features, can reverse the degradative trends and lead to soil quality improvements.

Watershed management involves conservation and development of water resources in the upstream regions to (1) maintain or increase water quantity, (2) maintain or enhance water quality, (3) minimize risks of soil erosion and degradation, and (4) improve soil quality. Conservation and development of water resources involve a range of techniques involving vegetative management, runoff management, and soil management.

Whereas techniques of watershed management are well known, their adoption has met with only limited success. The lack of or slow adoption of improved technology is due to a combination of several factors, including socioeconomic factors and participating approaches to problem identification and technology implementation.

It was with this background that an international symposium was organized in Toronto on 25-26 July 1997. The symposium was held in conjunction with the 52nd annual conference of the Soil and Water Conservation Society (SWCS). Because of the global importance of the subject, the symposium was organized under the auspices of the International Affairs Committee of the SWCS. The Organizing Committee comprised Rattan Lal (chair), David Cressman, Samir El-Swaify, and J. Hammond. The symposium was sponsored by the Natural Resources Conservation Service (NRCS), and its World Soils Program, and the Rockefeller Foundation.

This volume is based on the invited and contributory papers presented at the Toronto symposium. The symposium was successful because of the enthusiasm and interest of all participants. Most authors produced high-quality manuscripts that included state-of-the-art knowledge on the subject. The support and cooperation received from all authors and members of the Organizing Committee are much appreciated.

The staff of the SWCS was highly professional and supportive in organizing the symposium. In this regard, special thanks are due to Jennifer Pemble, Sue Ballentine, Doug Klein, and Charlie Persinger. The tedious task of correspondence, typing, and providing logistic support was done by Ms. Brenda Swank of the School of Natural Resources, the Ohio State University.

Rattan Lal

Editor

Dr. Rattan Lal is a professor of Soil Science in the School of Natural Resources at the Ohio State University. Prior to joining Ohio State in 1987, he served as a soil scientist for 18 years at the International Institute of Tropical Agriculture, Ibadan, Nigeria. While based in Africa, Professor Lal conducted long-term experiments on soil erosion processes as influenced by rainfall characteristics, soil properties, methods of deforestation, soil tillage, crop residue management, and cropping systems including cover crops and agroforestry and mixed/relay cropping methods. He established critical limits of soil properties in relation to the severity of soil degradation and assessed the effectiveness of different restorative measures. Data from these long-term experiments facilitated identification of indicators of soil quality and development of the concepts of soil resilience. He also assessed the impact of soil erosion on crop yield and related erosion-induced changes in soil properties to crop growth and yield. Since joining the Ohio State University in 1987, he has continued research on erosion-induced changes in soil quality and developed a new project on soils and global warming. He has demonstrated that accelerated soil erosion is a major factor affecting emission of C from soil to the atmosphere. Soil erosion control and adoption of conservation-effective measures can lead to C sequestration and mitigation of the greenhouse effect. The impact of severity of soil erosion on crop yield is evaluated at the landscape scale to quantify the compensatory effects of depositional sites. Erosion-induced changes in soil quality are related to crop growth and yield. The research has helped establish indicators of soil quality and resilience in relation to land use and management practices. Professor Lal is a fellow of the Soil Science Society of America, the American Society of Agronomy, the Third World Academy of Sciences, the American Association for the Advancement of Science, the Soil and Water Conservation Society, and the Indian Academy of Agricultural Sciences. He is the recipient of the International Soil Science Award, the Soil Science Applied Research Award of the Soil Science Society of America, The International Agronomy Award of the American Society of Agronomy, and the Hugh Hammond Bennett Award of the Soil and Water Conservation Society. He is past president of the World Association of Soil and Water Conservation and the International Soil Tillage Research Organization. He is a member of the U.S. National Committee on Soil Science of the National Academy of Sciences. He has served on the Panel on Sustainable Agriculture and the Environment in the Humid Tropics of the National Academy of Sciences. He is a member of the Committee on Sustainable Agriculture in the Developing Countries of the World Resources Institute.

Contributors

Inder P. Abrol
Facilitator
The Rice–Wheat Consortium
International Crops Research Institute for Semi-Arid Tropics
New Delhi, India

Fahmuddin Agus
Researcher
Center for Soil and Agroclimate Research
Bogor, Indonesia

Jock R. Anderson
Senior Economic Advisor
The World Bank
Washington, D.C.

Jacqueline A. Ashby
Rural Sociologist and Director NMR Research
Centro International de Agricultura Tropical
Cali, Colombia

William C. Bell
GIS Specialist
Centro International de Agricultura Tropical
Cali, Colombia

Cyril A. A. Ciesiolka
Department of Natural Resources
Toowoomba, Australia

Eric T. Craswell
International Board to Soil Research and Management
Jatujak
Bangkok, Thailand

David R. Cressman
Ecologistics Ltd.
Waterloo, Ontario, Canada

Pierre Crosson
Senior Fellow and Resident Consultant
Resources for the Future
Washington, D.C.

Samir A. El-Swaify
Professor of Soil and Water Conservation
Chair, Department of Agronomy and Soil Science
Interim Chair, Department of Natural Resources and Environmental Management
University of Hawaii at Manoa
Honolulu, Hawaii

Hari Eswaran
Soil Management Support Services
Washington, D.C.

Dennis P. Garrity
Agronomist and Regional Coordinator
International Centre for Research in Agroforestry, Southeast Asian Regional Programme
Bogor, Indonesia

Francis Gichuki
Department of Agricultural Engineering
University of Nairobi
Nairobi, Kenya

Roy K. Gupta
Project Implementation Unit
NATP
IARI Campus
New Delhi, India

William L. Hargrove
Professor and Director
KCARE
College of Agriculture
Kansas State University
Manhattan, Kansas

Ghulam Mohd Hashim
Malaysian Agricultural Research
Development Institute
Kuala Lampur, Malaysia

Fred Hitzhusen
Professor of Resource Economics and
Environmental Science
The Ohio State University
Columbus, Ohio

Jeffrey Hopkins
Research Associate
Department of Agricultural Economics
The Ohio State University
Columbus, Ohio

Anthony S. R. Juo
Department of Soil Science
Texas A&M University
College Station, Texas

Kurt Christian Kersebaum
Scientist
Institute of Landscape Modelling
Center for Agricultural Landscape and
Land Use Research
Muencheberg, Germany

John M. Kimble
Research Soil Scientist
National Soil Survey Center
National Resources Conservation Service
U.S. Department of Agriculture
Lincoln, Nebraska

E. Bronson Knapp
Soil and Systems Specialist and Project
Manager
Centro Internacional de Agricultura
Tropical
Cali, Colombia

Rattan Lal
Professor of Soil Science
School of Natural Resources
The Ohio State University
Columbus, Ohio

Andrew Manu
Department of Plant and Soil Science
Alabama Agricultural and Mechanical
University
Normal, Alabama

Christoph Merz
Scientist
Institute of Hydrology
Center for Agricultural Landscape and
Land Use Research
Muencheberg, Germany

Constance L. Neely
AGLS
Food and Agriculture Association
Rome, Italy

Chalinee Niamskul
International Board for Soil Research
and Management
Jatujak
Bangkok, Thailand

Helle M. Ravnborg
Rural Sociologist
Centro International de Agricultura
Tropical
Cali, Colombia

Robert E. Rhoades
Professor
Department of Anthropology
The University of Georgia
Athens, Georgia

Edie Salchow
School of Natural Resources
The Ohio State University
Columbus, Ohio

Jagir S. Samra
Central Soil and Water Conservation
Research and Training Institute
Dehra Dun, India

T. Francis Shaxon
Greensbridge
Dorset, United Kingdom

Thomas L. Thurow
Department of Range Science
Texas A&M University
College Station, Texas

Mary Tiffen
Parsonage House
Crewkerne, Somerset, United Kingdom

Luther Tweeten
Anderson Professor of Agricultural
Marketing, Policy, and Trade
Department of Agricultural Economics
The Ohio State University
Columbus, Ohio

Surinder M. Virmani
Principal Scientist (Agroclimatology)
Soils and Agroclimatology Division
International Crops Research Institute
for Semi-Arid Tropics Asia Center
Patancheru
Andhra Pradesh, India

X. X. Wang
Institute of Soil Science
Academia Sinica
Nanjing, Peoples Republic of China

Angelika Wurbs
Scientist
Institute of Land Use Systems and
Landscape Ecology
Center for Agricultural Landscape and
Land Use Research
Muencheberg, Germany

Ibrahim Zanguina
Institut National de Recherches
Agronomiques du Niger
Niamey, Niger

B. Zhang
Institute of Soil Science
Academia Sinica
Nanjing, Peoples Republic of China

T. L. Zhang
Institute of Soil Science
Academia Sinica
Nanjing, Peoples Republic of China

Q. G. Zhao
Institute of Soil Science
Academia Sinica
Nanjing, Peoples Republic of China

Table of Contents

TECHNOLOGICAL OPTIONS FOR WATERSHED MANAGEMENT

ENVIRONMENTAL QUALITY AND WATERSHED MANAGEMENT

Introduction

1 Rationale for Watershed as a Basis for Sustainable Management of Soil and Water Resources

Rattan Lal

CONTENTS

INTRODUCTION

Principal global issues of the 21st century are food security and environment quality. Despite the phenomenal advances made in agricultural technology, there are several regions of the world where food production has either not kept pace with the increase in population (e.g., sub-Saharan Africa) or has barely kept ahead of the ever-increasing demand (e.g., South Asia). Although stagnation or decline in agricultural production can be due to political and social reasons, degradation of soil and water resources and lack of appropriate technology to address the basic issue of resource management may be the primary factors responsible for low agricultural productivity. Important among environmental issues are poor water quality and the accelerated greenhouse effect. Water scarcity and poor water quality are major concerns in numerous countries. Fresh water availability is already a major factor in sustainable use of natural resources. Fresh water use worldwide was about 1500 km^3/yr in 1940

0-8493-0702-3/00/$0.00+$.50

and is projected to be 5000 km^3/yr in the year 2000 (Engelman and LeRoy, 1993). The number of water-scarce countries increased from 7 in 1955 to 20 in 1995, and is projected to increase to 34 in 2025. The water scarcity is accentuated by deteriorating water quality. Potential risks of global warming, due to both industrial and agricultural activities, are major concerns of environmentalists and policy makers. World soils and land resources have an important impact on the potential risks of the enhanced greenhouse effect (Lal et al., 1995a; 1998a).

Both food security and environmental issues are related to watershed management. Watershed is a basic hydrologic unit, and hydrologic and ecologic processes govern the quality of soil and water resources within the watershed. Soil degradative processes are accentuated by anthropogenic factors. It is appropriate, therefore, that issues related to sustainable management of natural resources (e.g., food security and environment quality) are addressed within the context of watershed management.

WATERSHED: THE BASIC HYDROLOGIC UNIT

Watershed is defined as a delineated area with a well-defined topographic boundary and water outlet. A watershed is a geographic region within which hydrological conditions are such that water becomes concentrated within a particular location, e.g., a river or a reservoir, by which the watershed is drained. Within the topographic boundary or water divide, watershed comprises a complex of soils, landforms, vegetation, and land uses. The terms *watershed*, *catchment*, and *basin* are often used interchangeably.

Hydrologic processes (e.g., infiltration, runoff, seepage flow, evapotranspiration) within a watershed are interlinked and appropriately assessed within its confine. Being the basic hydrologic unit, water from outside cannot enter the watershed and that from within leaves it from a well-defined point; properties of the watershed determine the nature and rate of fluvial processes. A watershed may range in size from a few hundreds of square meters to millions of square kilometers. It may be simple one with a first-order stream, or a complex conglomerate comprising a network of drainage channels.

Soil erosional processes (e.g., detachment, transport, deposition) and transport of nutrients and pollutants (chemicals) are determined by lateral and vertical flow of water within the watershed. When the watershed is drained by a single stream and the underlying geological strata provide an impermeable base, measurements of water flow and dissolved and suspended loads in runoff provide an ideal condition to study the mass balance of water and nutrients under specific land-use conditions. Since fluvial processes are governed by watershed characteristics (e.g., slope gradient and length, vegetation cover, soil types and management), controlling erosion and minimizing risks of water pollution require an understanding of hydrologic processes at the watershed scale. Soil degradation, a physical process driven by socioeconomic and political forces, recognizes only the natural water divide rather than social, ethnic, and political boundaries. Therefore, success and failure of erosion control and other processes depend, to a large extent, on whether control measures are implemented at a watershed scale or not.

THE IMPORTANCE OF WATERSHED MANAGEMENT TO MEET GLOBAL CHALLENGE

Prime soil resources of the world are finite, nonrenewable over the human time frame, and prone to degradation through misuse and mismanagement. Rapid increase in global population, especially in several developing countries of Africa and Asia, accentuates stress on finite soil resources and exacerbates risks of soil and environmental degradation. Principal reasons for addressing the issue of sustainable management of soil and water resources in the watershed context include the following.

DECREASE IN PER CAPITA ARABLE LAND AREA

The arable land area of the world has been relatively constant (at about 1.35 billion ha) since mid 1980s. In contrast, however, the world population has increased from about 4 billion in 1977 to 6 billion in 1998, and is expected to level off at about 11 billion by the end of the 21st century. In addition to soil degradation, prime agricultural land is also being converted to industrial, urban, recreational, and other nonagricultural uses. Urbanization is a principal threat to prime agricultural land, especially in densely populated countries (e.g., China, India). The data in Table 1.1 show trends in total and per capita arable land area in the world from 1977 to 2100. The per capita arable land area of 0.33 ha in 1977 declined by 33% to 0.22 ha in 1998, is projected to decrease to 0.16 ha by 2025, and may stabilize at 0.10 ha by the year 2100 (Table 1.1).

In view of the ever-shrinking arable land resources, it is important to identify and implement strategies for restoration of degraded soils and intensification of

TABLE 1.1
Per Capita Arable Land Area of the World

Year	World Population (10^6)	Arable Land Area	Per Capita Land Area (ha/person)
1977	4,000	1311	0.33
1980	4,447	1332	0.30
1985	4,854	1348	0.28
1990	5,282	1357	0.26
1995	5,687	1362	0.24
1998	6,000	1362	0.22
2010	7,000	1335[a]	0.19
2025	8,000	1295[a]	0.16
2050	9,400	1243[a]	0.13
2100	11,000	1168[a]	0.10

[a] Projected arable land area considering 2, 3, 4, and 6% decrease due to soil degradation and conversion to nonagricultural uses.

Source: Calculated from FAO, 1996 and other sources.

TABLE 1.2
Global Extent of Strong and Extreme Soil Degradation

Degradation Process	Land Area under Extreme and Strong Degradation (10^6 ha)	% of the Total Specific Global Degradation[a]
Water	224	20.5
Wind	26	4.7
Chemical	43	18.0
Physical	12	14.5
Total	305	15.5[b]

[a] Specific global degradation refers to the estimated total (slight, moderate, strong, and extreme) degradation by the specific process.
[b] Total global soil degradation = 1965 × 10^6 ha.

Source: Calculated from Oldeman, 1994.

existing prime agricultural land. Watershed management is an appropriate option to implement these strategies.

Natural Resources Degradation

Degradation of soil and water resources is a global threat. Oldeman (1994) reported that out of a total degraded land area of 1965 Mha, over 300 Mha (about the size of India) are strongly degraded on a world scale (Table 1.2). These lands have lost their productive capacity, and restoration can only be achieved through major investments and engineering works involving massive landforming. Land degraded by water erosion (Figure 1.1) and wind erosion (Figure 1.2) constitute 82% (250 out of 305 Mha) of the total strong and extremely degraded land area. On a global basis, high sediment load is carried by rivers draining very densely populated regions of the world, i.e., Yellow River in China, Ganges in India (Table 1.3). The problems of erosion and sedimentation are accentuated by misuse and mismanagement of resources within the watersheds of these river systems. A strategy based on watershed management is essential to effective erosion control and restoration of degraded soils.

Desertification

Desertification refers to a special type of land/soil degradation, where desertlike conditions spread to the areas on the fringe of the desert in semiarid and arid regions. Desertification implies decline in soil quality leading to reduced biological productivity and environmental moderating capacity of land in arid regions. Total land area in arid regions of the world is estimated at 4.85 billion ha (Table 1.4), of which two thirds occur in Africa and Asia. Desertification may happen as a result of natural and anthropogenic or human-induced factors. The global land area prone to human-induced desertification is about 1.02 billion ha or 20% of the total dryland areas of the world (Table 1.5). In addition to climate and soil factors, risks of desertification also depend on land use and farming systems. Rangelands and rainfed croplands are

FIGURE 1.1 Soil degradation by water erosion.

FIGURE 1.2 Soil degradation by wind erosion.

highly susceptible to desertification. Desertification is an especially severe problem in dry regions of sub-Saharan Africa, West Asia and North Africa, Central Asia, parts of northern and western Australia, western U.S., and southwestern regions of South America. An appropriate strategy for desertification control may involve natural resource planning at the watershed level.

TABLE 1.3
Runoff and Sediment Load in Major Rivers of the World

River	Drainage area (10^6 km^2)	Runoff (km^3/yr)	Sediment Load (10^6 tons/yr)
Amazon	6.2	6300	900
Danube	0.8	206	67
Ganges	1.5	971	1670
Irrawaddy	0.4	428	265
Magdalena	0.2	237	220
Mekong	0.8	470	160
Mississippi	3.3	580	210
Niger	1.2	192	40
Ob	2.5	385	16
Orinoco	1.0	1100	210
Yangtze	1.9	900	478
Yellow	0.8	49	1080
Yenisei	2.58	560	13
Zaire	3.8	1250	43

Source: Data from Milliman and Meade, 1983; Van der Leeden et al., 1994.

TABLE 1.4
Land Area in Arid Regions of the World

Region	Land Area (10^6 ha)	% of the Total Arid Area of the World
Africa	1689	34.8
Asia	1596	32.9
Australasia	612	12.6
Europe	116	2.4
North America	504	10.4
South America	335	6.9
Total[a]	4852	100

[a] Total arid area is 37.3% of the Earth's land area.

Source: UNEP, *World Atlas of Desertification,* Arnold, London, 1992. With permission.

Tropical Deforestation

Tropical rain forest (TRF) is an important ecosystem because of its impact on global hydrologic and C cycles, biodiversity, and numerous social, economic, and political issues. However, the TRFs are rapidly dwindling (Table 1.6; Figure 1.3). The rate

TABLE 1.5
Land Area Affected by Human-Induced Soil Degradation in Dry Regions of the World

Land Use	Area (10^6 ha)	% of Total Drylands
Irrigated cropland	43	0.8
Rainfed cropland	216	4.1
Rangeland	757	14.6
Total	1016	19.5

Source: UNEP, 1991.

TABLE 1.6
Tropical Rain Forests and the Rate of Deforestation

Region	Area (10^6 ha)	Rate of Deforestation (10^6 ha/yr)
Central Africa	204.10	1.08
Tropical southern Africa	100.46	0.84
West Africa	55.60	0.53
South Asia	63.90	0.50
Southeast Asia	210.60	2.80
Mexico	45.60	0.59
Central America	19.50	0.36
Brazil	561.10	3.42
Andea region and Paraguay	241.80	2.25
Total	1,505.7	12.37

Note: Faminow (1998) estimated total TRF at 1756.3 Mha and the rate of deforestation at 15.4 Mha/yr.

Source: Modified from WRI, 1996; Southgate, 1998; Faminow, 1998).

of deforestation of TRF ecosystems ranges from 0.36 million ha/yr in central America to 3.42 million ha/yr in Brazil. The global annual rate of deforestation is estimated at 12.37 Mha out of a total remaining area of 1505 Mha or 0.82%/yr.

Once cleared of its protective vegetal cover, soils of the TRF ecoregions are prone to severe degradation by accelerated soil erosion (Figure 1.4) and other degradative processes including soil compaction (Figure 1.5) and decline in soil structure (Figure 1.6). Colonization by imperata (*Imperate cylindrica*) (Figure 1.7) and other undesirable species is also a severe problem following deforestation and conversion to agricultural land uses. Watershed management can play a crucial role in planning for a judicious management of TRF ecosystems, establishing criteria for selecting land that is suitable for conversion to agricultural land uses, and in restoration of degraded soils.

FIGURE 1.3 Rate of tropical deforestation is high in Brazil, Indonesia, and Central Africa.

FIGURE 1.4 Deforestation by heavy machinery can cause severe soil erosion.

Water Scarcity

Renewable fresh water scarcity remains a problem for millions of people around the world, especially those in arid and semiarid regions. In 1995, there were 436 million people living in 29 countries considered water stressed (Table 1.7). It is estimated

FIGURE 1.5 Deforestation leads to soil compaction.

FIGURE 1.6 Decline in soil structure and soil physical quality are accentuated by mechanized systems of deforestation.

that by 2050, for medium population projections, there will be 4 billion people living in 54 countries which will experience some level of water scarcity (Figure 1.8). The problem of water scarcity is accentuated by water quality. Soil degradation and water

FIGURE 1.7 Colonization by *Imperata cylindrica* is a severe problem on land converted to agriculture from TRF ecosystem in Sumatra and elsewhere in the tropics.

TABLE 1.7
Status of Fresh Water Availability

Parameter	1995	2050[a]
Population (billions)	5.7	9.8
Population affected by water scarcity (millions)	166	1700
Population affected by water stress (millions)	270	2300
Countries affected by water scarcity	18	39
Countries affected by water stress	11	15

[a] Medium population projection.

Source: Modified from Gardner-Outlaw and Engelman, 1997.

quality are interrelated issues. Soil degradative processes lead to pollution (Figure 1.9) and eutrophication (Figure 1.10) of natural waters.

IMPORTANCE OF WATERSHED MANAGEMENT TO FOOD SECURITY AND THE GREENHOUSE EFFECT

The declining per capita agricultural production remains a serious issue affecting millions of people around the world. The data in Table 1.8 show that in comparison with 1989–91, agricultural production in several countries declined severely. There were at least six countries in sub-Saharan Africa where agricultural production

FIGURE 1.8 Scarcity of drinking water is a major problem. Families leave their pots and pans in a queue at a municipal water tap which runs only for 2 or 3 h/day.

FIGURE 1.9 Safe drinking water is a severe problem in densely populated countries of South Asia.

declined by 12 to 40%. There were also several countries with similar downward production trends in central Asia (newly independent states), western Asia, and South

FIGURE 1.10 Eutrophication of water due to agricultural activities is a serious threat in developing countries. High nutrient content in the lake causes growth of water weeds.

and Central America (see Table 1.8). The problem of food security needs to be addressed at the global scale.

Another serious issue is that of global warming. The atmospheric concentration of CO_2 and other greenhouse gases is rapidly increasing (Lal et al., 1995ab; 1998ab). Emission of greenhouse gases from soil to the atmosphere is accentuated by soil degradation and desertification, and soil degradation processes are driven by deforestation, land misuse, and soil mismanagement. Therefore, agricultural intensification, adoption of recommended agricultural practices, and restoration of degraded soils are options to sequester C in soil and mitigate the greenhouse effect (Lal et al., 1998c). Adoption of improved watershed management technologies has a potential to sequester C in soil, enhance soil quality, improve productivity, and mitigate the greenhouse effect.

CONCLUSIONS

Food security and environment quality are serious global issues that need to be addressed. Watershed management is not a new concept. Its importance to natural resource management has been recognized since the 1950s. However, the concept needs to be revisited to address the emerging issues of the 21st century. Research and development projects undertaken since the 1950s need to be critically reviewed to determine the causes of success and failure. Needless to say, a vast majority of the projects have met their objective to a limited extent. Therefore, methodologies and strategies need to be reviewed, reassessed, and revised to address the issues of the modern era. An important question is whether or not the watershed management approach can solve the pressing problems of food security and environmental quality. How we can learn

TABLE 1.8
Agricultural Production Index of Some Countries from 1985 to 1996 (1989–91 = 100)

Country	1985	1986	1987	1988	1989	1990	1991	1992	1993	1994	1995	1996
World	99.1	99.0	98.2	98.2	99.8	100.7	99.5	100.1	99.2	101.3	102.2	103.5
Africa	97.7	99.5	96.9	99.1	100.1	98.4	101.6	97.3	98.2	97.5	95.4	99.9
Burundi	104.5	106.6	106.7	106.0	95.1	102.2	102.8	103.6	98.7	83.8	86.0	83.9
Gambia	108.6	125.5	125.3	109.4	125.5	83.9	90.7	67.8	77.7	78.2	73.8	58.8
Kenya	95.3	101.8	98.7	104.0	103.1	99.9	97.0	92.0	86.6	89.8	89.9	87.9
Swaziland	104.9	112.5	104.0	100.4	102.1	96.3	101.5	88.0	83.0	86.5	74.4	80.2
Tanzania	105.8	106.5	103.2	99.8	103.8	99.1	97.1	89.7	88.5	83.6	84.3	83.7
Zaire	101.1	100.6	100.0	100.1	100.3	100.2	99.5	98.3	95.5	86.6	82.4	80.6
Cuba	102.9	103.4	100.0	102.7	103.1	100.2	96.8	81.0	64.8	60.0	58.1	61.3
Haiti	121.7	118.6	112.8	108.5	105.8	98.8	95.4	90.2	87.4	86.4	81.1	80.7
Iraq	131.3	119.9	103.8	105.2	106.7	116.7	76.6	87.9	97.4	91.6	86.8	81.5
Yemen	94.6	106.1	101.4	112.6	113.6	101.5	84.9	94.4	95.5	90.3	87.9	82.3

Source: FAO, 1996.

from the past successes and failures and use these experiences to develop strategies for sustainable management of natural resources should also be asked.

REFERENCES

Engelman, R. and P. LeRoy. 1993. *Sustaining Water: Population and the Future of Renewable Water Supplies,* Population and Environment Program, Population Action International, Washington, D.C. 56 pp.

Faminow, M.D. 1998. *Cattle, Deforestation and Development in the Amazon: An Economic, Agronomic and Environmental Perspective,* CAB International, Wallingford, U.K., 253 pp.

FAO 1996. *Production Yearbook,* Food and Agriculture Organization, Rome, Italy.

Gardner-Outlaw, T. and R. Engelman. 1997. *Sustaining Water: Easing Scarcity: A Second Update,* Population Action International, Washington, D.C.

Lal, R., J.M. Kimble, E. Levine, and B.A. Stewart, Eds. 1995a. *Soils and Global Change,* CRC/Lewis Publishers, Boca Raton, FL, 440 pp.

Lal, R., J.M. Kimble, E. Levine, and B.A. Stewart, Eds. 1995b. *Soil Management for Mitigating the Greenhouse Effect,* CRC/Lewis Publishers, Boca Raton, FL, 385 pp.

Lal, R., J.M. Kimble, R. Follett, and B.A. Stewart, Eds. 1998a. *Soil Processes and the Carbon Cycle,* CRC Press, Boca Raton, FL, 609 pp.

Lal, R., J.M. Kimble, R. Follett, and B.A. Stewart, Eds. 1998b. *Management of Carbon Sequestration,* CRC Press, Boca Raton, FL, 457 pp.

Lal, R., J.M. Kimble, R. Follett, and C.V. Cole. 1998c. *The Potential of U.S. Cropland to Sequester C and Mitigate the Greenhouse Effect,* Ann Arbor Press, Chelsea, MI, 128 pp.

Millman, J.D. and R.H. Meade. 1983. Worldwide delivery of river sediment to the oceans, *J. Geol.* 91: 1–21.

Noin, D. and J.I. Clarke. 1998. Population and environment in arid regions of the world, in J. Clarke and D. Noin, Eds. *Population and Environment in Arid Region,* Parthenon, New York, 1–18.

Oldeman, L.R. 1994. The global extent of soil degradation, in D.J. Greenland and I. Szabolcs, Eds. *Soil Resilience and Sustainable Land Use,* CAB International, Wallingford, U.K., 99–118.

Southgate, D. 1998. *Tropical Forest Conservation: An Economic Assessment of Alternatives in Latin America,* Oxford University Press, New York.

United Nations Environment Program. 1991. Status of Desertification and Implementation of the U.N. Plan of Action to Combat Desertification, Report of the Executive Director to the Governing Council, Third Special Session, Nairobi.

United Nations Environment Program. 1992. *World Atlas of Desertification,* Arnold, London.

Van der Leeden, F., F.L. Troise and D.K. Todd, Eds. 1994. *The Water Encyclopedia,* 2nd ed., Lewis Publishers, Chelsea, MI.

WRI. 1996. *World Resources 1996–97,* World Resources Institute, Washington, D.C.

Watershed Management for Soil Erosion Control

2 Challenges in Ecosystem Management in a Watershed Context in Asia

Jagir S. Samra and Hari Eswaran

CONTENTS

INTRODUCTION

Watersheds are environmental and land management natural units which determine the health of a nation. Poor ecosystem management of watersheds has and will result in the impaired functioning of the watershed, which in fragile environments can lead to ecosystem collapse. Erosion rates of 10 to more than 500 t/ha/yr have been estimated for countries of the region. Sedimentation of Indian subcontinent reservoirs has been analyzed to be 2 to 20 times more than that predicted during the design stage. Resource management problems of the Asian region is a complex of increasing soil and water loss, land degradation, sedimentation, irregular stream flow, and poverty. The magnitude of the economic value of land degradation may be as high as three times the gross national product of the country if the value of ecosystems is included in the traditional estimates. Reducing this deficit is the prime challenge of countries with large areas of stressed ecosystems. Until recently, research and development efforts in Asia were primarily concerned with food or biomass production with resource conservation and sustainability generally of much lesser impor-

0-8493-0702-3/00/$0.00+$.50

tance in the planning process. Adequate infrastructure is not available in Laos, Myanmar, Vietnam, Cambodia, Nepal, and Sri Lanka to initiate a conservation program. There is a need for technology development projects, designed carefully for the viability of large-scale ecosystem management on a watershed basis. Watershed management programs are being implemented increasingly by nongovernmental organizations other than the conventional departments of soil conservation and forestry. This significant shift calls for additional capacity building of human resources development. Multiple linkages among countries of Asia are weak, primarily because of language barriers and a general lack of financial resources for exchange of scientific personnel, literature, and holding of workshops, seminars, etc. The region is rich in biodiversity, and the role of watershed management in preserving and utilizing flora and fauna is inadequately quantified. New paradigms of ecosystem management include soil health, land husbandry, resilience, ecofriendly technologies, integrating with traditional knowledge, equity, ethnic sensitivity, and enlisting community participation in watershed development. Transregional implications of ecosystem management need sincere initiatives as there are many similarities and opportunities between countries which can be harmonized to optimize returns from research and development investments.

The basic premise in the management of any system is the ability to minimize risk (Eswaran et al., 1993). In the context of an ecosystem, one of the important questions is the integrity of the environment and how this integrity is compromised by management (Virmani et al., 1994). With mismanagement in ecosystems, the risk of permanent damage to the environment is very high. From a biodiversity point of view, the risk extends from changes in ecosystem composition that have significant positive or negative impacts on its functions to extinction of species. From a land-use point of view, the risks deal with impacts on productivity of the land and concomitant impacts on aquatic systems within the watershed and at the terminus of the watershed.

The causal chain that leads to these marked changes in ecosystems of watersheds are many and, in several countries, are frequently associated with demands for more land by the increased rural populations. In addition, there are also the impending impacts of global climate change, which, depending on the locality, may be strong. Ecosystem degradation processes are thus strongly affected by population pressures, poverty in many of the developing countries, enhanced demand for ecosystem products, and uncontrolled rates of resource consumption.

The negative effects of ecosystem degradation commence with the imperceptible changes in biodiversity and lead eventually to the process commonly called "desertification" (UNEP, 1992). Cropland quality is slowly reduced through land degradation processes. When crop yields reach their marginal utility value or when it is no longer economically productive to grow a crop, the land is either abandoned (example of shifting cultivation) or used for grazing. In the latter case, a frequent process is overgrazing and an indicator of reduction in land quality is when large ruminants are replaced by small ruminants. The consequence of these is a gradual change in the hydrology of the watershed, resulting in reduced biomass quality and quantity and leading to reduced carbon sequestration and enhanced albedo.

FIGURE 2.1 River bank erosion has exposed a well which today stands out as a tower. Torrent flows, partly initiated by low-grade seismic activities in the foothills of Himalayas, are dynamic factors to be reckoned with in watershed management of this region.

Watersheds are environmental and land management units which determine the health of the nation. Poor ecosystem management has and will result in the impaired functioning of the watershed, which in fragile environments can lead to ecosystem collapse (Eswaran et al., 1995). Figure 2.1 is an example of erosion by torrential river flows in the village of Khujnawar in North India. A former well is exposed through erosion of the floodplains by the river and stands as a tower today. The village located on the banks has moved three times over the last five decades. The quality of life is strongly interlinked with the quality and functioning of the watershed. The importance of soil resources for providing sustainable livelihood gathering and environmental securities is adequately emphasized in the Brundtland Commission (WCED, 1987) and Agenda 21 of the Rio Conference (UNCED, 1992). The world population is expected to increase to 8.5 billion by the year 2030 (50% growth over the year 1994), creating more competition and conflicts. Asian countries with already overexploited and excessively stressed resources are projected to make major contributions to the future growth of population. The latest report on Global Assessment of Soil Degradation (GLASOD; Oldeman et al., 1992) estimated degraded lands (in Mha) of 1200 in Asia, 400 in Africa, 245 in South America, 215 in Europe, 150 in Central and North America, and 112 in Australia. Soil erosion (by water and wind) accounts for 60% of the total land degradation and is the most extensive factor of lowering environmental qualities. According to these estimates, more than 50% of the degraded land of all the continents is situated in Asia. Degradation manifests in a high rate of soil erosion, increased sedimentation, reduced farm production and livestock-carrying capacity, and deforestation with consequent loss of biodiversity.

TABLE 2.1
Average Rate of Soil Erosion in Asian River Basins Indicating Natural Resource Degradation

River	Country	Drainage Basin (10^3 km^2)	Estimated Annual Soil Loss (t/ha/yr)
ChaoPhraya	Thailand	106	21
Mekong	China, Thailand, Laos, Tibet, Vietnam, Myanmar	795	43
Red	China, Vietnam	120	217
Ganges	India, Bangladesh, Nepal, Tibet	1076	270
Kosi	India, Nepal	62	555

Average erosion rate of 100 to 115 t/ha/yr during the last century has been estimated for Sri Lanka (De Alwis and Dimantha, 1981). In India, average soil loss of 16.5 t/ha/yr was estimated by Dhruvanarayana and Ram Babu (1983). Sedimentation of Indian subcontinent reservoirs has been analyzed to be 2 to 20 times more than that predicted during the design stage (Galay and Evans, 1989). The estimated soil loss in Asia reported by Holeman (1968) ranged from 21 to 555 t/ha/yr (Table 2.1). Resource management problems of the Asian region are a complex of increasing soil and water loss, land degradation, sedimentation, irregular stream flow, and poverty (Samra, 1997). The natural geophysical drainage unit of a watershed is being considered a single window, and a comprehensive and integrated area for development program addressed to the problems of resource management and economic prosperity (Magrath and Doolette, 1990). Certain environmental concerns like greenhouse gases, contamination, and loss of biodiversity may require wider, transregional efforts. Similarly, catchment of some great rivers runs across several nations, and harmonization of conservation measures across the borders is required. The new paradigms of sustainable development aim at integration of the complementary components of all kinds of production systems, environmental protection/enhancement, biophysical aspects, socioeconomic policies, and a bottom-up approach of people's empowerment. Globalization, competitiveness, diversification, sustainability, management of common property resources, refinement of indigenous technical knowledge, human resource development, equity, reviving/creation of village-level institutions are other important parameters and concerns of the participatory watershed development program of Asia.

HISTORICAL PERSPECTIVE

From the resource degradation/conservation/aggradation viewpoint, the practices of shifting cultivation, terracing, and water harvesting are very ancient. Slashing, burning, and cropping for a few years and moving to other sites is a traditional practice of hill tribes of India, China, Indonesia, Myanmar, Sri Lanka, Thailand, Vietnam,

TABLE 2.2
Growth of Organized Watershed Management Program in India

Year of Commencement	No. Watersheds or Total Area (ha)	Agency or Project	Investment in U.S. $
1956–	42 watersheds	CSWCRTI, Dehradun	Demonstration
1961–1962 (RVP)	3.3 million ha	29 catchments	230
1974	4 watersheds	CSWCRTI, Dehradun	Research project
1980–1981 (FPR)	0.83 million ha	10 catchments	88
1983	47 watersheds	CSWCRTI and CRIDA	Research project
1987	12,000 ha	PIDOW	10
1991–1995	5 million ha	government of India and international agencies	815

RVP = river valley projects; FPR = flood-prone rivers; PIDOW = Participatory Integrated Development of Watersheds; CSWCRTI = Central Soil and Water Conservation, Research and Training Institute; CRIDA = Central Research Institute for Dryland Agriculture (Hyderabad).

and other Asian nations. This practice is known by the name "Jhum" in India, "Chena" in Sri Lanka, and "Kaingineros" in the Philippines and is a serious concern of ecosystem management. The presently reduced cycle of shifting cultivation does not allow restoration of soil fertility through natural biocycling processes, and overall environmental degradation is attracting the attention of researchers, planners, bureaucrats, and investors. Similarly, resource conservation practices of terracing started about 4000 to 2000 B.C. in China, India, Nepal, and other countries which promoted permanent settlements in hilly and mountain ecosystems. China alone had 13.3 Mha of terraced land by 1950, 13 Mha was terraced after 1950, and another 14 to 15 Mha is under active consideration for terracing. Similarly, rainwater harvesting into ponds or tanks and its use as a common property resource is also an ancient traditional practice in Asia (Samra et al., 1996; Aggarwal and Narain, 1997). Organized research and development efforts initiated in the 1920s by China and India were focused mainly on soil and water conservation programs with a bias toward biophysical measures. The Philippines and Thailand initiated serious efforts in the 1950s, Vietnam in the 1970s, and Nepal about two decades ago also in the 1970s. Soil and water conservation is an important component of present-day watershed management, which is holistic.

Since the 1950s, greater emphasis is placed on soil and water conservation programs in many countries of Asia. India set up 42 research watersheds in 1956 in different agroecological zones, and watershed-based development spread at a tremendous rate after the countrywide drought year of 1987 (Table 2.2). In the IX Plan period of 1997 to 2002 in India, watershed-based ecosystem management is to be realized by creating village-level institutions with the active involvement of nongovernmental organizations (NGOs) and voluntary organizations. This mode of development has also witnessed definite reorientation in China since the socialistic reforms of 1980s and land contract systems of various shades. Similar pursuits are being followed in Vietnam.

ISSUES OF WATERSHED MANAGEMENT IN ASIA

The Asian economy is primarily agrarian and land-based activities provide livelihood securities to 90% of the population in China, 75% in India, and 70% in Thailand. Environmentally fragile, marginalized, as well as inaccessible hills and mountain ecosystems constitute 70% of the geographic area of China, 67% each of Nepal and Vietnam, 54% of the Philippines, and 29% of India. In Myanmar, approximately 23% of the area is affected by shifting cultivation and 7.6% by land degradation, depleting the forests. Bangladesh faces severe erosion and impeded drainage due to overexploitation of forest cover. Studies in China have related higher sediment yield to increased farming of steep slopes, rising population, and decreasing forest cover. Similarly, rainfed agriculture is the backbone of Asian food and livestock production systems. An excessive rate of soil erosion, pollution of natural water bodies, depletion of groundwater in tube-well irrigated tracts, salinization in canal command areas, overexploitation of forest cover, shifting cultivation, uncertainties of rainfed biomass productivity, high density of uneconomical livestock, etc. are the significant features of a typical Asian economy. Pro-people, pro-environment, and integrated strategies of watershed management programs are expected to provide sustainability in the natural resource use in the region.

More than 10,000 small watersheds (500 to 3000 ha) in China and 2500 in India have been developed with comprehensive management during last 10 years. In Indonesia, 39 priority watersheds are being developed out of 81 identified at the national level. Planting forests at the top of mountains, horticulture on midslopes, farmland on lower slopes, and water harvesting into reservoirs, ponds, and dams in foothills for irrigation and aquaculture is being practiced in China. Ducks are also introduced and poultry pens are constructed right above the ponds/tanks so that their droppings are eaten by the fish to realize complete organic cycling and complementary trade-off. Different variants of this model are being used in India and other countries of Asia.

POLICY

Various legislations and decrees have been proclaimed in Asia from time to time to comply with trade-related treaties like that of the World Trade Organization (WTO), market forces (pricing), land tenures, institution building (participation and organization), and support services (research, extension, credit), etc. The Torrent (flash flood) Control Act (1901), the Damodar Valley Corporation Act (1949), several States Soil Conservation Acts after the 1960s, joint forest management policies, and the Panchayati Raj Act (1994), etc. are important landmarks in the watershed development history of India.

In 1950, the State Council of China issued a National Provisional Decree for Soil and Water Conservation Commission. These decrees were subsequently revised in 1982 in light of feedback and lessons learned from 30 years of experience. The concept of watershed management was now defined more precisely after the review. Several watershed management–related acts, such as the Flood Protection Ordinance, 1924; the Soil Conservation Act, 1951; the State Land Ordinance, 1947; the National Environment Act, 1980; the Agrarian Service Act, 1991; and the Irrigation Ordinance, 1994, were enacted in Sri Lanka. An exclusive Memoranda for Watershed

Management (1990) was passed in the Philippines. In 1985, the Thai Cabinet approved a watershed classification program for preventing forest conversion to other uses. The National Environment Act (1980) of Sri Lanka placed a greater emphasis on watershed management.

The land tenure system is highly relevant for making decisions about watershed management in Asia (Molnar, 1990). Changes in the land tenure system of China, Vietnam, Sri Lanka, Thailand, and the Philippines since 1980s for conferring leases, 25-year contracts, and land titles to individual farmers played a significant role for production-oriented resource use and conservation on a watershed basis. A stable land tenure system exists in Java and Taiwan, whereas in India, China, and Myanmar common property resources often have undefined customary rights, and, in closed areas of the Philippines, North Thailand, and Malaysia, tribal or indigenous people have informal, traditional rights. Most of the common lands as well as open access or community use resources in India belong to government departments of revenue and forests and a small fraction to the community. This kind of tenancy system promoted land degradation due to excessive deforestation, overexploitation, and weakening of traditional local institutions. Restoration of degraded watershed is less effective because of lack of privatization-like contracting of their use on the pattern of China and Vietnam. Clearly defined ownership or an equitable sharing system is important for community participation in development program.

The importance of social issues, in addition to the conventional biophysical aspects, for successful implementation of environmental management program is being increasingly realized in India and other Asian countries. Considering the enormity of the task, participatory community efforts have to be encouraged by evolving suitable ecoethics, value addition to products, and incentives. An NGO (MYRADA) took up a watershed management program in Karnataka State of India with the help of the Swiss Development Cooperation in 1987, and since then several policy initiatives have been introduced by donors in Asia to promote the involvement of NGOs in resource conservation and ecorestoration.

Because of compulsions of demand and supply, the ideal land-use options may not be feasible because of conflicting needs. Resource management policies and programs should address the emerging issues of sustainability, equity, food security, gender, and incurring minimum environmental costs. Effective policies of ecosystem management will be converging in nature and lead to overall economic efficiency and national as well as continental well-being.

Equity

Equity, in general, is not a serious land-based issue in Nepal, the Himalayan region of India, and some other countries of Asia, which have a democratic tradition where every villager owns land or has acquired the right to contract common land. In other parts of Asia, the landless are quite common and watershed development is alleged to benefit landowners (or the State) to a relatively greater extent. Various strategies of productive employment creation, income-generating activities through apiculture, sericulture, aquaculture, rabbitaries, livestock production, piggery, poultry, goatery, carpentry, handicraft, and other small production systems are being introduced to cover the landless population of watersheds in India. Organization of self-help groups (SHGs) and resource management societies in watersheds where even the landless have a share in developed water and common property resources is a strongly emerging feature in India.

Gender

In the hills, mountains, and Hindu Kush Himalayan region of Asia, male migration to lowland plains for supplementing family income is quite common. Under these situations, women provide 80% of the labor input to agriculture and are the prime managers of the households. They also bear the brunt of child, elderly, as well as livestock care, and it is difficult to bring them to the institutes for upgrading their skills in ecosystem management. Some initiatives to impart training on watershed management in the village itself have been promoted through NGOs, but it is still a formidable challenge for realizing their potentials.

Participation

Percolation of benefits to the grassroot level, cost effectiveness, transparency, sustainability, equity, and harmonization with the indigenous traditional knowledge are among several benefits now being considered to be realized with the bottom-up approach of people's participation in Asia. In some cases, it is a difficult task since communities of this region are accustomed to subsidies and free fertilizers, seeds, and waiving of loans, etc. Farmer's willingness to participate in watershed development is known to be influenced mainly by the improved economic benefits to individual farmers and land tenure. In some cases, the lack of creditability of implementing agencies may be an impediment. In order to win the confidence of the villagers, funds are provided in the projects of India for entry-point activity to be decided in consultation with the community. In the second stage, villagers are involved in agroecoanalysis during participatory rural appraisal (PRA) exercises. They are expected to contribute 10% of the cost of the project, which they generally do in the form of labor, collection of local materials, etc. According to the 8th Plan Document of Nepal, the community was expected to contribute 10 to 50% of the cost of the watersheds. In China, 20% of watershed investment comes from the central government, 10 to 20% from the local government, and 60 to 70% from the farmers, mostly in the form of labor. After completion of the projects, watershed management is handed over to the locally, but democratically constituted institutions. In the new contract system of China, farmers are free to select the crops and planting pattern after meeting certain conditions (Deyi, 1996). In Nepal and China, watershed management demands are identified by village-level development committees and passed on to the district/prefecture level.

As in other countries, there is a whole range of stake holders who are directly or indirectly involved in sustainability of watersheds. The role of each is summarized in Table 2.3 and a successful watershed management program hinges on their working in harmony. In many countries of Asia, this is still not a reality, with decisions of some of the players doing more harm than good to the goal. Sustainable watershed

TABLE 2.3
Role of Different Stake Holders in Watershed Management

Level of Activity	Stake Holder/s	Planning Horizon	Activities
Farm/household	Farmer land user	1 generation	Income generation, risk aversion, intergenerational equity
Community	Leaders	1–3 years	Help decision making, planning, link with larger political entities, facilitate marketing, respond to national policies
	Elders	1 generation	
	Extension	Career	
Local to national	Politicians	1–3 years	Support services
	Religious, cultural leaders	1 generation	Policy environment, economic incentives
	NGOs	Funding period	Ecocentric concerns
International	Donors	Funding period	Policy initiatives
	Institutes	Funding period	Basic research
	Trade institutions	long term	Profit generation

management only results when land users are motivated in not only managing for income generation but also for contributing to environmental integrity. The latter is still elusive in many communities of Asia.

Institution Building

Institutional deficiency is one of the recognized constraints of ecologically desirable resource management. The institution should be able to address the requirements of resource qualities incorporating the watershed community needs and aspirations. The conventional straight-jacket approach has to give way to a participatory, supportive, interactive, and result-oriented institutional framework.

A paradigm shift becomes essential when a land user has to change conceptual directions from the farm/household level to an ecosystem level. To assist the land user in making this shift, institutional changes should address the following:

- Regulations affecting use and ownership rights;
- Framework for support services;
- Training and awareness activities;
- Access to financial inputs;
- Market opportunities and market transparencies;
- Technologies that are appropriate both from a productivity and an ecology viewpoint; and
- Congruency of local to national polices.

Creation of watershed-level institutions is necessary for continuing watershed activities sustainable after withdrawal of interventions of governmental/NGO/international funding agencies. Management of common property resources like grazing lands, rainwater harvesting and recycling, joint forest operations, fishing, etc. is being done efficiently by people's empowerment in many cases. About 70% of irrigation projects in Nepal are community operated. Maintenance of water harvesting structures and distribution of water and fish production in such systems are being organized by a group of respected elderly men or selected members in a village or group of villages in China and India. Under the joint forest management policy of India, protection of watersheds and harvesting of grass is done by pooling labor of the entire watershed community. Produce for home consumption and monetary income is distributed in proportion to the labor contributed by each family. However, these kinds of ethnic institutions are successful in traditional tribal areas, where local leadership is very much revered. In other situations, for example, to share harvested rainwater for irrigation, one has to become a member of a registered society by paying a prescribed membership fee. A committee is elected by the members to prescribe rates, collect charges, regulate distribution, and for the repair and maintenance of conservation programs.

Research and Development

Earlier research and development efforts in the 1920s were mostly concerned with measurement of soil losses and runoff from small plots in China and India. Research entered into a new era with the development of research and development institutes in both countries in the 1950s. The Central Soil and Water Conservation Research, Demonstration and Training Centers were established in India in 1954 under the Indian Council of Agricultural Research and were the nucleus for much of the conservation work on small watersheds. Some initiatives were also undertaken in the Philippines by the Reforestation Administration and Bureau of Forestry, the Baguio Experiment Station, Luzon, the Cebu and Malaybalay experiment stations, etc. A humble beginning was made in Thailand in 1953 with the establishment of four watershed rehabilitation field stations under the Silviculture Division of Royal Forest Department which were subsequently upgraded to Sections in 1965 and to a Watershed Management Division in 1981.

The period of 1950 to 1980 generally concentrated on soil- and water-loss studies, small watershed hydrology, demonstration and extension of agricultural and silvicultural practices. Adequate information about short- and long-term impact of erosion hazards, effectiveness in retaining and improving soil quality, biological and economic efficiency, off-site effects, and the skills needed were necessary prerequisites for a sound transfer of technology program. About 42 small watersheds for monitoring surface hydrology and vegetation successions with the elimination of biotic interference were set up in India during 1956. Four watersheds under village situations were demonstrated in 1974, and another 47 model watersheds were added to this transfer of technology program in 1983 for covering different agroecosystems of India. After the drought year of 1987, large-scale projects on watershed management were undertaken by the development department and focus on this program has been intensified for the 1997 to 2002 Plan period of India. In China, there are at present 12 universities and colleges and 5 professional middle schools for teaching watershed management and soil and water conservation. A scanty research and development infrastructure has been created by other Asian countries from time to time.

It may be correct to state that most agricultural research institutions in Asia are not geared to or even know how to develop research programs that address sustainability. Client-oriented and demand-driven research is not the basis or the desire of scientists for several reasons, including that these are not glamorous and frequently are not easy to justify for government funding. The first hurdle to overcome is for the research institutions to recognize that shifting the research orientation is a problem; this will be quickly followed by questions related to what and how to develop a meaningful research program.

ECOSYSTEM MANAGEMENT CHALLENGES

Awareness of the need to protect and preserve the quality of the ecosystem is well engraved in the minds of the farming community, particularly the traditional farmers of Asia. This is reflected in their religion, traditions, history, and their commitment to ensuring intergenerational equity. However, in recent decades farmers have seen a scramble for survival, resulting from societal conflicts, confusing messages from their local and national leaders, forced reliance on nonfarm inputs, changing demands of the local, national, and international markets, and greater involvement of these market forces in their daily life. The consequence of these has been slowly to transform an ecosystem-conscious agrarian society into a market-responsive production system with no concern for environmental costs. Today, this same group is being asked to reduce its production and conserve the natural resources — a dilemma that presents some of the following challenges:

Enhanced sustainability concerns. Until recently, research and development efforts of the Asian region were primarily concerned with food or biomass production. Resource conservation and sustainability generally occupied a backseat in the planning process. Demand pressure and land quality are major determinants of a given state of degradation. Land clearing of the natural vegetation brings drastic ecological changes. Adequate research and development infrastructure is not available in many countries, including Vietnam, Cambodia, Nepal, and Sri Lanka, to satisfy the intricacies demanded by sustainable agriculture.

Need for holistic approach. Biophysical components of ecosystem like soil, physiography, precipitation, etc. were given higher priority in the earlier endeavors. Of course, the importance of other aspects, such as ethnicity, socioeconomic status, gender, demography, common property resources, participation, people's empowerment, local-level institution building, integrated farming system approach, and environmental externalities, etc., is being increasingly realized. However, integrating these components and enabling land users to implement this vision need a much higher level of appreciation and flexibility by the farm community.

Inducing community participation. Indigenous knowledge, skills, and capabilities should be built upon and integrated with efforts on watershed management. Greater community participation is essential because mechanical, legislative, and policy measures of the past did not produce the desired results. Learning to work with farmers is itself a challenge for many scientists.

Utilizing modern technology. Application of modern tools and procedures like remote sensing, geographic information system (GIS), etc. for evaluating the ecosystem as a whole is still in its infancy in many Asian countries. Systems research in land resource science is yet to evolve and process models which are new research venues. In some of the countries, applications are confined to research organizations only. In many, absence of computer facilities prevents this important step in integration of information. Lack of data is another major impediment in utilizing these tools.

Consequences of mismanagement. Some larger issues of watershed management like ingress of saline seawater into coastal ecosystems, desertification, and subsistence of land due to excessive pumping of groundwater in India, the Philippines, Taiwan, and Thailand, etc. have been reported. Research, development, and policy protocols for reversing these processes are inadequate as compared with the seriousness of their consequences.

Systems monitoring. A reasonable amount of data about on-site effects of watershed management, such as production, surface hydrology, soil loss, *in situ* moisture conservation, and vegetation successions, etc., are available in a few countries and for some localities. Research efforts on modeling the off-site effects like groundwater investigations, aquifer recharge, flooding, and deposits of alluvium, etc. are inadequate in many respects. There is a need for technology development projects, designed carefully for the viability of large-scale ecosystem management on the watershed basis. Very few countries have traditions of monitoring the biophysical resource base, and, unless governments value the need to address the health of the nation, no funds will be made available for such activities.

Predicting and managing large-scale processes. Some of the mass-wasting phenomena — landslides, mining, torrents, coastal erosion, and gully formation — are serious threats to the environmental qualities especially because of high seismicity of the Hindu Kush Himalayas. Apart from China, where an exclusive institute is devoted to such issues, institutionalized infrastructure is very scanty in most of the region. Each such destructive event is treated as unique, and, although an attempt is made to heal the wound, few additional activities to prevent further wounds are initiated.

Quantifying economic benefits. Capabilities of environmental economics for quantifying nontangible benefits of watershed management are weak and many a time it is difficult to convince the planners and financing agencies to justify large investments. Several problems including fast changes in the economic equation among the watershed community, real estate, and growing economic insecurity, utilization of prime land and water resources for nonfarm purposes, and migration of people have all aggravated pressure on the ecosystem. A recent highlight of ecosystem management involved mobilization of community efforts, largely by voluntary agencies, NGOs,and it focused on water resource and tree species valued for economic benefits and sustenance of human and livestock.

Developing training facilities. Watershed management programs are being implemented increasingly by NGOs other than the conventional departments of soil conservation and forestry. This significant shift calls for additional capacity building of human resource development through training activities of NGO staff.

Transnational issues in shared watersheds. Multiple linkages among countries of Asia are weak primarily due to the language barrier and the complete lack of financial resources for exchange of scientific personnel, literature and holding of workshops, seminars, etc. Transregional implications of ecosystem management need sincere initiatives at a global scale. There are many interregional similarities and opportunities which can be harmonized to optimize returns from research and development investments.

CONCLUSION

The Asian region is rich in biodiversity, and the role of watershed management in preserving and utilizing flora and fauna is inadequately quantified to capture their quality and quantity. It is important to achieve diversification in agriculture for both national heritage concerns and international competitive aims of organizations such as the World Trade Organization. New paradigms of ecosystem management include soil quality, land husbandry, resilience, ecofriendly technologies, integration with traditional knowledge, equity, ethnic sensitivity, and enlisting community participation in watershed development. In general, the challenges faced by developing countries, and Asian countries, in particular, include:

1. Realization that watershed management is not merely managing the "shed" but also the "water";
2. Recognition of the holistic nature of the task;
3. Empowerment of the people with appropriate institutional support;
4. Creating awareness in not only the participating communities but also society as a whole;
5. Ensuring political recognition and support for the ecosystem aspects of sustainable land management;
6. Mobilizing the scientific community nationally, regionally, and internationally to mount integrated programs for methods, standards, data collection, and research networks for assessment and monitoring of watershed conditions and functioning;
7. Developing land-use models that incorporate both natural and human-induced factors which contribute to land degradation and which could be used for ecosystem management on a watershed basis;
8. Developing information systems that link environmental monitoring, accounting, and impact assessment of sustainable land management technology;
9. Developing economic instruments in the assessment of land degradation and watershed integrity to encourage the sustainable use of land resources; and
10. Facilitating transnational collaboration in watershed management.

Swaminathan (1986) summarized the *status quo* in India, which is applicable to most of Asia and developing countries in general:

We have now the technical capability to build enduring national and global nutrition security systems based on sound principles where the short and long-term goals of development are in harmony with each other. What we often lack is the requisite blend of political will, professional skill, and farmer's participation. We live in this world as guests of green plants and of farmers who cultivate them. If farmers are helped to produce more, agriculture will not go wrong. If agriculture goes right, every thing else will have a chance for success.

REFERENCES

Aggarwal, A. and S. Narain. 1997. Dying wisdom. State of India's Environment. Citizens' Report 4. Publ. Center for Science and Environment, 41, Tughlakabad Institutional Area, New Delhi, India.

De Alwis, K.A. and S. Dimantha. 1981. Integrated Rural Development Project, Nawara Eliya, Sri Lanka, 65.

Deyi, Wu. 1996. State of art and status of watershed management in China, in P. N. Sharma and M. P. Wagley, Eds., *The Status of Watershed Management in Asia,* ICIMOB, Kathmandu, Nepal, 7–17.

Dhruvanarayana, V.V. and R. Babu. 1983. Estimation of soil erosion in India, *J. Irrigation Drainage Eng. Am. Soc. Civil Eng.* 109(4):409–434.

Eswaran, H., S.M. Virmani, and L.D. Spivey. 1993. Sustainable agriculture in developing countries: constraints, challenges and choices, in J. Ragland and R. Lal, Ed., *Technologies for Sustainable Agriculture in the Tropics,* ASA Spec. Publ. 56, Madison, WI, 7–24.

Eswaran, H., S.M. Virmani, and I.P. Abrol. 1995. Issues and challenges of dryland agriculture in southern Asia, in A.S.R. Juo and R.D. Freed, Ed., *Agriculture and the Environment: Bridging Food Production and Environment Protection in Developing Countries,* ASA Spec. Publ. 60, ASA, CSSA, and SSSA, Madison, WI.

Galay, V.J. and R. Evans. 1989. Sediment transport modeling, in *Proceedings of the International Symposium,* New Orleans, August 14–18, American Society of Civil Engineers, New York.

Holeman, J.N. 1968. Sediment yield of major rivers of the world, *Water Resour. Res.* 4:739–747.

Magrath, W.B. and J.B. Doolette. 1990. Strategic issues in watershed development, World Bank Technical Paper No. 127, 1–34.

Molnar, A. 1990. Land tenure issues in watershed development, 131-158. in J. B. Doolette and W. B. Magrath, Ed., *Watershed Development in Asia — Strategies and Technologies,* World Bank, Washington, D.C.

Oldeman, L.R., R.T.A. Hakkeling, and W.G. Sombroek. 1992. World Map of the Status of Human-Induced Soil Degradation: An Explanatory Note, International Soil Reference Center, Wageningen, Netherlands, 34 pp.

Samra, J.S. 1997. Status of Research on Watershed Management, Central Soil and Water Conservation Research and Training Institute, Dehradun, India, 44.

Samra, J.S., V.N. Sharda, and A.K. Sikka. 1996. Water Harvesting and Recycling: Indian Experiences, Central Soil and Water Conservation Research and Training Institute, Dehradun, India, 248.

Swaminathan, M.S. 1986. Building national and global nutrition security systems, in M.S. Swaminathan and S.K. Sinha, Eds., *Global Aspects of Food Production,* Tycooly Int., Oxford, England, 417–449.

UNCED, 1992. United Nations Conference on Environment and Development, Rio de Janeiro, Brazil.

UNEP, 1992. *World Atlas of Desertification,* United Nations Environment Program, Nairobi, Kenya, E. Arnold, London, 69 pp.

Virmani, S.M., J.C. Katyal, H. Eswaran, and I. Abrol, Eds. 1994. *Stressed Agroecosystems and Sustainable Agriculture,* Oxford & IBH Publishing Co., New Delhi, India.

WCED (World Commission on Environment and Development). 1987. *Our Common Future.* Oxford University Press, Oxford.

3 Operative Processes for Sediment-Based Watershed Degradation in Small, Tropical Volcanic Island Ecosystems

Samir A. El-Swaify

CONTENTS

INTRODUCTION

Small islands comprise the most distinctive terrestrial ecosystems of the Pacific Basin. These islands function as "whole ecosystems" in which all segments of the landscape from the highest point on land to the sea are joined directly and are intimately interdependent. The islands may be atolls or wholly volcanic in origin. The first are living coral reefs with coarse sandy surface layers (soils), generally overlying volcanic foundations. Unless tectonically uplifted, atolls have elevations that seldom exceed 5 m above sea level. Although such (low) atolls are subject to

0-8493-0702-3/00/$0.00+$.50

certain soil and water salinization problems, watershed-based degradation problems common to tropical uplands are absent or rare. Such problems, in contrast, abound in volcanic islands.

This chapter, therefore, focuses on natural resource degradation problems and watershed management challenges in small tropical volcanic island ecosystems. Such islands are subject to high erosion and sedimentation hazards and so are intrinsically sensitive to the impacts of land use and management. Islands of the Hawaiian Archipelago are considered here as a case study as they typify the "whole island ecosystems" in the Pacific and elsewhere. This chapter will review available quantitative information on sediment-based land and soil degradation and discuss ongoing watershed-based conservation and protection efforts addressing these problems in selected areas of the state.

Watershed degradation issues in Hawaii, the compressed spatial and temporal scale in which on-site and off-site impacts are manifested, the dynamic changes in land use, and the overall implications to the sustainability of natural resource base and environmental quality are relevant not only to island settings but also to large continental areas.

GEOLOGIC, CLIMATIC, AND SOIL SETTING

El-Swaify (1992ab) provided detailed accounts of the setting of the Hawaiian Islands and research activities from the perspectives of soil erosion and conservation. Lying in the mid-Pacific Ocean between 18°N and 23°N latitude and 154° and 161° longitude, the islands form an arc running northwest to southeast over a distance of about 1400 km. The total land area of the islands is just over 1.6 million ha. The islands are of volcanic origin and are progressively younger from north to south; the ages of the five major islands range from 5.6 to 0.8 million years for Kauai and Hawaii, respectively. The rocks are primarily vesicular and permeable basalts or andesites, volcanic ash, and/or cinders. Geologic age, associated weathering, and ruggedness of landscape are well reflected in present-day elevations of mountain peaks which get progressively higher from north to south. The highest peak on Kauai is that of Mount Kawaikui at 1575 m, whereas that on Hawaii is Mauna Kea at over 4000 m. Further details and descriptions of all the islands in the chain have been provided elsewhere (e.g., MacDonald and Abbot, 1970).

Hawaii's climate is generally moderate with three dominant characteristics: mild temperatures, persistent northeast trade winds, and extremely variable rainfall over very short land distances. The mean annual temperature at sea level is 22 to 24°C. Rainfall has a mean annual value on the surrounding open ocean of about 760 mm. However, as for volcanic islands in general, this value is meaningless over land where it ranges from 250 to over 10,000 mm, depending on both the elevation and geographic location with respect to prevailing wind direction. Winter storms occurring between November and April bring widespread heavy rains. The dry leeward lowlands may receive half or more of their annual rainfall from a single storm during this period. Storm rainfalls of 300 mm and intensities of 50 to 75 mm/h are not rare. Rainfall is distributed more evenly throughout the year in areas of high rainfall.

Rainfall patterns are primarily attributed to orthographic showers resulting from moist trade winds ascending the steep, rugged terrain. Consequently, precipitation is greater on the windward and mountain ridges than the leeward lowlands, and it varies drastically over very short distances. Investigations reported elsewhere (Lo, 1982; El-Swaify, 1992a) have confirmed that storm EI_{30} (storm kinetic energy $\times$ maximum 30-min intensity) is a valid index of rainfall erosivity for Hawaii. Up-to-date isoerodent maps have been constructed for deriving rainfall erosivity values and applying the revised universal soil loss equation (RUSLE) for erosion prediction and conservation planning in Hawaii. The extreme rainfall gradients across island landscapes are reflected in the wide range of erosivities which, on Maui, for example, range from less than 100 to 1800 erosivity units, all within a distance of less than 20 km.

The diverse geologic age, parent rocks, topographic settings, and climates contribute to the formation of a wide variety of soils in the islands. All 11 orders of the U.S. Department of Agriculture (USDA) soil classification system (taxonomy) are represented. In general, "tropical soils" are presumed to be resistant to erosion, but this quality applies only to the highly weathered soils (El-Swaify et al., 1982). Soil erodibility values (quantitatively expressed as the K value in the RUSLE) were experimentally determined by El-Swaify and Dangler (1977) as 0.05, 0.09, 0.10, 0.18, 0.19, 0.25, 0.31, 0.45, and 0.72 Mg/ha/metric EI units for Tropohumults, Hydrandepts, Ustropepts, Eutrustox, Torrox, Dystrandepts, Chromusterts, Camborthids, Eutrandepts, respectively. Different soils are associated with various climatic regimes and specific locations and slopes along the landscape. This results in different erosion potential for specific sites as determined by the collective "inherent" attributes of these sites.

WATERSHEDS, HYDROLOGIC UNITS, AND THE "AHUPUA`A"

The islands have no "river basins" and only few perennial streams. Watersheds or hydrographic areas are small by continental standards and each consists of many even smaller subwatersheds. A recent survey delineated 614 watersheds ranging in size from less than 0.05 ha to about 23,000 ha (Hawaii CZMP, 1996). Of these, 566 (or 92%) are less than 1700 ha and drain directly into the ocean. Only the larger watersheds are geologically well developed and are amenable to conventional land uses such as agriculture. As discussed later in this chapter, catchment size is a major distinctive feature that determines sediment-based land degradation and its off-site impacts. It also figures prominently in applying the nonpoint pollution requirements of the federal Coastal Zone Management Act (Section 6217). Because of catchment and continuity of surface water pathways from mountain tips to the ocean, the definition of coastal zones extends to every bit of land in the state.

In the context of Hawaiian history and culture, the *Ahupua`a* is the indigenous ancestor of today's "watershed." It represents a land-use principle that encompasses more than mere partitioning of hydrologic units and their boundaries. The author describes it as a "watershed plus" concept as it integrates the land and ocean lying below, including the shoreline and inshore and offshore ocean areas with deliberately subdivided zones within each. It was originally described as follows (from several citations in Hawaii CZMP, 1996):

> A land [segment] running from the sea to the mountains, thus affording to the chief and his people a fishery residence at the warm seaside, together with the products of the high lands [forest], such as fuel, canoe timber, mountain birds, and the right of way to the same, and all the varied products of the intermediate land as might be suitable to the soil and climate of the different altitudes from the sea soil to mountainside or top.

The *Ahupua`a* concept is instrumental in presenting effectively to the land-user community and policy makers the importance of "holistic" land conservation. It lends itself well as a heritage-respected tool for explaining the need for integrating on-site and off-site elements of soil and water conservation to protect the livelihoods and privileges of land users from the crests of hills to coastal and surrounding water areas. It has been lent scientific credibility by use of water balance principles as well as erosion, sediment delivery, and sedimentation processes.

MAJOR LAND USES AND AGRICULTURAL INDUSTRIES

Patterns of agricultural land use in Hawaii have developed, to a large extent, to suit the topography, climatic regime, availability of water, soil capabilities, and economic factors. The steep slopes and higher elevations on most islands are generally forested, accounting for about 40% of the land area (USDA, 1987). Plantations were, and to some extent remain, the dominant enterprises for sugarcane and pineapple production. Today, sugarcane production is rapidly declining but pineapple remains viable, partly due to the shift from canning to fresh fruit production. Pasture is the third major enterprise and "diversified" agriculture, particularly in replacement of sugarcane, is on the rise. As will be discussed in more detail below, agricultural diversification is predicted to have very profound impacts on the sustainability of Hawaii's natural resource base and environmental quality.

Historically, soil erosion has been a problem in the Hawaiian Islands ever since the introduction of goats and cattle soon after European contact (El-Swaify et al., 1982). Large areas of native forests were converted to alternative land uses. Disturbances by proliferating feral animals and by cattle and sheep overgrazing resulted in accelerated erosion. Reforestation has taken place but has been primarily implemented by use of exotic vegetation which comprises over 50% of the present tree population. Later, the introduction of large-scale sugarcane and pineapple production increased erosion on large areas of land. Since statehood, in 1959, development of land for urban and residential use has also contributed to the problem.

EROSION POTENTIAL, EXTENT, AND SEDIMENT SOURCES IN AGRICULTURAL WATERSHEDS

Erosion potential is defined here as the maximum erosion rate that would hypothetically take place if rainfall, soils, and topography at a defined site are allowed to

TABLE 3.1
Erosion Potential for Different Locations in Hawaii

Soil Group	Local R Value Annual EI_{30} (tonne-m/ha/yr)	Prevailing Slope Steepness, %	Estimated Loss (Mg/ha/yr)
Tropohumult	400	9	29
Hydrandept	1000	12	162
Ustropept	200	4	8
Eutrustox	350	6	42
Dystrandept	450	10	117
Eutrandept (T)	300	8	103
Torrox	220	7	57
Chromustert	190	4	23
Camborthid	80	11	58
Eutrandept (E)	190	5	73

express themselves to cause soil loss, unimpeded by human intervention or control practices. The magnitude of potential erosion is important for setting conservation priorities among possible sediment source areas, planning of specific preventive or corrective measures, and designing programs for monitoring and evaluating the performance of these measures.

Estimates of erosion potential for soils in different regions in the Hawaiian Islands were made for "farm"-scale land areas by combining the above information on soil erodibility and rainfall erosivity with topographic attributes which characterize specific sites. Assuming that a common continuous segment of slope on a field averages about 30 m, estimates of erosion at the source were made by applying the RUSLE (Renard et al., 1996). Soil erodibility values stated above were used to develop the estimates presented in Table 3.1. Estimated potential soil losses range from 8 to 162 Mg/ha/yr. Actual soil losses from croplands can be derived by use of appropriate values for crop and land management parameters (C & P values) of RUSLE. The values for the product of $C \times V$ values range from 0.2 to 0.5 (El-Swaify and Cooley, 1980a; El-Swaify et al., 1982). These values result in an estimated soil loss 2 to 81 Mg/ha/yr for agricultural lands. Available data and alternate estimates of actual erosion (e.g., El-Swaify and Cooley, 1980a) provided values ranging from 5 to 35 Mg/ha/yr. No quantitative values are available for noncroplands; however, predicted estimates for undisturbed forest areas are below 1 Mg/ha/yr.

A survey of sediment sources causing non-point-source pollution was completed by the state in 1978 (TCNSPC, 1978). Estimates of actively eroding land reported in the survey are given in Table 3.2. These values included only areas of "bare" soil in the actively eroding category and, therefore, considerably underestimated the area on which erosion is a problem. The survey also estimated soil movement for each hydrologic district on the major islands using the universal soil loss equation (USLE). Figures ranged from less than 1 Mg/ha for two districts on Hawaii to over 40 Mg/ha for a district on Molokai. Average soil movement by island was 15.6, 6.8, 26.8, 31.3,

TABLE 3.2
Estimates of Actively Eroding Areas for the Hawaiian Islands

Island	Total Area (ha)	Actively Eroding Area (ha)
Kauai	143,300	7,892
Oahu	156,415	3,892
Molokai	67,381	11,659
Lanai	36,463	10,361
Maui	188,458	1,703
Hawaii	1,045,852	8,175
Total	1,637,869	42,906

Source: TCNSPC, 1978.

11.4, and 2.0 Mg/ha for Kauai, Oahu, Molokai, Lanai, Maui, and Hawaii, respectively. The relatively low values for Oahu and Hawaii are due to the large amount of urbanization and large areas of lava lands, respectively. This survey alerted the community that large areas around the state are eroding at rates above the maximum "recognized" tolerance limit of 11 Mg/ha/yr. The Second RCA Appraisal (USDA, 1987) estimated that, in 1982, the average annual erosion rate in rural areas of the state was 9.2 Mg/ha and that on cropland was 14.3 Mg/ha, highest in the United States.

The first quantitative assessment of erosion from agricultural lands in Hawaii was initiated in 1972. Its purpose was to relate runoff and sediment losses to rainfall, soils, land use, and commonly practiced field management. Six small, cropped watersheds were chosen, ranging in size from 0.8 to 2.8 ha. The soils were representative of both residuum and volcanic ash. A summary of that study is in Table 3.3. The Laupahoehoe and Honokaa watersheds (on the island of Hawaii, both on volcanic ash soils) were planted in sugarcane. The other four were on residual soils on Oahu; one was in sugarcane (Waialua S), one in pineapple (Kunia), one in pineapple and was converted to sugarcane (Waialua (P)); and the fourth had a mixed land-use history and, therefore, is not included in this summary (Mililani). All the soils in the study were well drained and considered to have low to moderate erodibilty. Other details on the watersheds and soils were given elsewhere (El-Swaify and Cooley, 1980ab; El-Swaify, 1992). The Waialua (P) and Kunia watersheds, both in pineapple, had similar EI_{30} and soil loss values, although cropping histories were different. In both cases it was clear that plantation roads contributed substantially to the sediment loads. The Kunia watershed appeared to be the most susceptible to erosion, having the highest soil loss per unit rainfall erosivity (EI_{30}), reflecting the high erodibility of Kolekole soil (a semiarid Inceptisol) and somewhat rugged topography.

Data from 4 to 10 years of monitoring of these watersheds are given in Table 3.3. Soil losses were very low from the Laupahoehoe and Honokaa watersheds, both with highly weathered volcanic ash soils. The majority of soil loss occurred when

TABLE 3.3
Sediment Loss Logs for Agricultural Watersheds during the Period Beginning December, 1972[a]

Watershed Name (area, ha)	Approimate Elevation (m)	Soil Name (Taxonomy)	Median Annual Rainfall (mm)	Prevailing Slope, %	Primary Cropping	Recorded Mean Annual EI_{30} (tonne-m/ha/yr)	Monitoring Period, months	Mean Annual Soil Loss (Mg/ha/yr)	Mean Soil Loss (Mg/ha per EI_{30})
Laupahoehoe (0.9)	509	Kaiwiki (Typic Hydrandept)	3556	16	Sugarcane	739	60	1.18	0.16
Honokaa (2.2)	492	Kukaiau (Hydric Dystrandept)	1981	17	Sugarcane	166	78	2.37	1.43
Waialua 1 (2.5)	287	Paaloa (Humoxic Tropohumult)	1448	10	Sugarcane	139	109	2.52	1.81
Waialua 2 (0.8)	308	Wahiawa (Tropeptic Eutrustox)	1092	6	Pineapple	276	53	7.02	2.54
Kunia (2.9)	305	Kolekole (Ostoxic Humitropept)	864	7	Pineapple	180	122	7.13	3.96

[a] Additional details on the characteristics of these watersheds were provided by El-Swaify and Cooley, 1980b.

the soils were left bare between crop plantings. However, these net losses did not reflect the substantial amount of soil movement which occurred within the watersheds. Significant redeposition of sediments from the steeper slopes occurred in the more level areas. These sediments were predominantly of sand size and composed of irreversibly dehydrated aggregates characteristic of these and other volcanic ash soils (Andisols). The low mean soil loss for the Waialua (S) watershed (Ultisol) appears to reflect mainly the low annual rainfall and total erosivity (EI_{30}) values over the monitoring period. During 1975, with a total of over 850 mm of rain, about 14 Mg/ha of soil was lost at this site. This exceeded losses recorded on any other watershed for any year. The poorly distributed rainfall during that year, with no rain falling between February and October, resulted in a sparse growth of newly planted sugarcane and the lack of protective cover during periods of heavy rains (November to February).

Three important conclusions were drawn from this study:

1. Rates of soil loss from watersheds were highly variable from year to year, but were lower than the acknowledged maximum tolerance limits.
2. Only a few storm events that coincide with soil exposure are responsible for the majority of soil losses; therefore, the timing of "disturbance" in field operations and maximum soil cover within certain periods of the year is a very important component of conservation planning.
3. Better planning of plantation roads could reduce sediment losses significantly from pineapple lands.

ON-SITE IMPACTS

Studies on highly weathered Oxisols by Yost et al. (1985) quantified the effect of soil erosion and nutrient restoration on soil productivity. Yield on eroded surfaces could not be brought to the same level as on the noneroded soil by restorative fertilization alone. The loss of only a few centimeters led to significant declines in yield, which were quite crop-specific. These observations lead to the conclusion that the soil loss tolerance level should be set very low for highly weathered and fragile tropical soils such as Oxisols. The studies also showed that restricted rooting resulted in significantly reduced water and nutrient use efficiency by the crops grown on eroded soils. Follow-up studies also showed that the eroded Oxisol underwent significant loss of biological quality, in particular the levels of rhizobia and vesicular-arbuscular mycorrhiza (Yost et al., 1985; Habte et al., 1988).

SEDIMENT DELIVERY AND YIELD

Downstream consequences and off-site impacts of erosional processes are a direct function of sediment yield which, in turn, is determined by the sediment delivery ratio (SDR). SDR quantifies the fraction of soil loss emerging into a sediment yield and thus its ability to move off the eroding catchment area and contribute to non-point-source pollution. Sediment yield is reduced if there is a high tendency for it to be deposited along the downhill path or to be trapped by certain other landscape features. For a given set of environmental conditions, vegetative cover, and defined sediment sources (soils), SDR is determined by the location of the source and the geomorphology of the likely sediment path (Walling, 1994). Included in the latter is the area of the drainage "basin" within which erosion is taking place, its macro- and microrelief characteristics, continuity and length of eroding slopes, and the drainage channel frequency, length, and bifurcation. Walling (1994) summarized the available empirical equations depicting the relationship between SDR and these variables for several different regions. He and El-Swaify et al. (1982) presented diagrams of several generalized relationships showing a particularly strong dependence of SDR on catchment area. These relationships are inverse in nature primarily due to the decreased macrorelief and increased opportunity for sediment deposition in larger areas. The relationships show that SDR falls rapidly from the maximum possible value of 1.00 in very small uniformly sloping areas ($\ll$1 ha) to a value of about 0.60, 0.50, 0.40, 0.30, or 0.20 for areas of 25, 250, 2,500, 25,000, or 250,000 ha, respectively.

Recalling that 97% of the Hawaiian catchments are less than 1700 ha in size and drain directly into the ocean, it is clear that, in addition to the high erosion hazard discussed above, small volcanic islands are also characterized by many of the conditions that favor higher sediment yield than for large continental basins. These facts largely explain the high frequency of sediment movement to shoreline areas following significant rainstorms. They also partly justify why emerging regulations on coastal zone non-point-source pollution (Coastal Zone Management, CZM, Section 6217) define coastal zone boundaries for the purpose of sediment movement as encompassing all land areas in every island (Hawaii CZMP, 1996).

As a consequence, 16 watersheds and hydrologic unit areas are specifically undergoing serious erosion and contributing to downstream sediment-based water pollution in the state. The receiving coastal waters (Table 3.4) have been characterized as "Water Quality Limited Segments" (needing special actions to correct serious impairments) in the official Water Quality Nonpoint Source Pollution Assessment Report (Hawaii DOH, 1990; Hawaii CZM, 1996).

Following detailed investigations, four of these were targeted for priority action using federal funding. These are the watersheds and hydrologic unit areas contributing to the Pearl Harbor Estuary of southern Oahu, Kaiaka/Waialua Bay of northern Oahu, Waimanalo/Maunawili Bay of eastern (windward) Oahu, and the Maunawainui Bay of southern Molokai. The respective watershed areas for these basins are, approximately 39,060, 23,600, 4,460, and 4,050 ha. Land uses in these areas are mixed with pineapple, sugarcane, and intensively managed pasturelands representing the major agricultural uses on the first two, diversified farming on the third, and extensively managed pasture and truck farming on the fourth. Urban lands, military lands, and forest reserves are also important uses of the Oahu sites. Water quality impairments are attributed to sediments, agrichemicals, and wastewater disposal. Erosional sediments have some opportunity for deposition in flat field portions but little or no opportunity to deposit once they have moved off and into the short conveyance streams leading to the receiving bays. Although no comprehensive

TABLE 3.4
Water Quality Limited Segments in the State of Hawaii

Segment	Area (ha)	Contaminant/Sources
Ala Wai Canal, Oahu	5	Urban runoff
Hanapepe Bay, Kauai	125	Sediments, agricultural runoff
Hilo Bay, Hawaii	751	Sediments, agricultural runoff
Honolulu Harbor, Oahu	746	Urban runoff
Kahana Bay, Oahu	104	Sediments, agricultural runoff
Kahului Bay, Maui	101	Sediments, agricultural runoff
Kaneohe Bay, Oahu	5014	Sediments, agricultural runoff
Keehi Lagoon, Oahu	1491	Sediments, urban. runoff
Kihei, Maui	[a]	Sediments, agricultural and urban runoff
Kewalo Basin, Oahu	4	Urban runoff
Nawiliwili Bay, Kauai	140	Sediments, agricultural runoff
Pearl Harbor, Oahu	2778	Sediments, agricultural runoff
South Molokai, Molokai	4795	Sediments, agricultural runoff
Waialua-Kaiaki Bay, Oahu	507	Sediments, agricultural runoff
Waimea Bay, Kauai	510	Sediments, agricultural runoff
West Maui	[a]	Urban and agricultural runoff, sewage

[a] These areas are composites of several catchment areas, extent unspecified.

sediment yield assessments have been made for these or any other location in the state, estimates of sediment delivery ratios may be used by assuming the validity of the SDR/area relationships discussed above (Walling, 1994). For most Hawaii basins, the SDR is estimated at between 0.3 and 0.6; i.e., nearly one third to two thirds of the eroded sediments (note above estimated average of 14.3 Mg/ha/yr for cropland) would likely be destined for surrounding bays and estuaries. These values compare with 0.1 or less for continental basins (El-Swaify et al., 1982).

Of equal importance is the temporal dimension of the problem; sediment movements to low-lying and shoreline areas after erosive storms often take only a few hours or even minutes. In addition to the high quantity, sediment characteristics also induce serious water quality impairments. Turbidity values for sediments derived from highly weathered sesquioxide-rich soils were shown to exceed by far those derived from temperate soils (El-Swaify, 1989).

SPECIFIC CHALLENGES: IMPLICATION OF DIVERSIFICATION OF FORMER PLANTATION CROPLANDS

Replacing the ailing plantation industries of sugarcane and pineapple with "diversified" crops is expected to have profound impacts on the natural resource base and environmental quality. Sediment-based non-point-source pollution is illustrated here as a case study. Our monitoring studies cited above provide a baseline for comparison, having shown that sediment loss ranged from 1.2 to 2.5 Mg/ha/yr and 6.9 to 7.1 Mg/ha/yr for sugarcane and pineapple fields, respectively. While sugarcane culture is clearly more protective than pineapple, all these sediment loss rates have been judged as "tolerable." Managing these two crops for high virtual "perennials," with few periods of soil "disturbance" or exposure, leads to effective soil protection "within planted field areas"; the high proportion of roads contributes most to erosion in pineapple land.

The "diversified" crops that are likely to replace sugarcane and pineapple may be annuals, orchards, or pastures. Their impacts are clearly expected to be as diverse as the specific crops and the management practices applied to them. Since "conservation tillage" is not yet commonly practiced in Hawaii, an important element for the purpose of comparison is the frequency and timing of soil "disturbance" by field preparation and crop harvest operations. Our model-based predictions, using the RUSLE show that annuals grown with "conventional" practices would be considerably less protective against sediment losses than either sugarcane or pineapple. Orchards and short-rotation bioenergy plantings are also vulnerable, particularly during the highly exposed early stages of tree growth, after full canopy development shades out ground cover or "understory" vegetation, and during or following harvest operations. Well-managed pastures represent the most "protective" alternative. Similar arguments may be made for agrochemical contamination hazards as diversified crops require higher amounts and more frequent applications of fertilizers and pesticides.

Biological conservation options have received increasing research and demonstrations in recent years and offer substantial promise for controlling sediment losses and protecting both soil and water quality (El-Swaify et al., 1988, 1996). These authors have shown that the use of intercropping and precropping with low-lying ground cover vegetation have also been demonstrated in pilot areas beyond the research farms. As an example, runoff and erosion rates from maize (*Zea mays*, L.) were substantially reduced by use of kalo (*Lotus corniculatus*, *Arvensis*) and rose clover (*Trifolium hortum*) as ground covers. Similar benefits were gained from intercropping cassava (*Manihot esculenta*, *Crantz*) with stylosanthes (*Stylosanthes hamata*, *Verano*). Considerably more runoff and erosion occurred on monoculture plots of maize and cassava than on plots intercropped with legume ground cover. Cassava in a "typical" tropical small farm grows less prolifically and provides less soil protection than noted here because resource-poor farmers normally grow this crop on marginal soils and can afford only limited fertilizer inputs. Thus, the erosion hazards from such plantings may be considerably higher than concluded from the above data. Consequently, the benefits of intercropping with legumes may be even more striking than determined here. Legume intercrops not only offer protection against runoff and erosion, but may also provide forage, increase food production and/or income, and contribute to yield gains in succeeding crops due to nutritional contributions from their residues. In the above experiment, groundnuts yielded over 2 Mg/ha of yield for "cash flow" and *Stylosanthes* contributed the equivalent of 200 kg/ha of fertilizer nitrogen to the following cassava crop. This residual benefit demonstrates the importance of examining not only simultaneous crop combinations but also alternative cropping sequences in order to assess the long-term conservation effectiveness of a cropping system. Residue recycling remains the single most promising but least utilized technology for soil protection (El-Swaify, 1996).

POLICY, PLANNING, AND COMMUNITY EFFORTS

An important consequence of the small area of such island ecosystems is the close proximity between the users of upland regions (sediment source areas) and occupants or users of sediment-receiving lowlands or bays/estuaries. Opportunities for conflict between these communities, which pursue different means of livelihood (e.g., farming vs. fishing), are many and frequent. All, therefore, must be involved in and supportive of charting and implementing the improvements needed in land use or the enactment of protective laws and strategies. Very often, the first and most important step in securing such support and commitment is to convey to the communities and policy makers a clear understanding of the causes, extent, and detrimental impacts of erosion both on site and off site. Such sensitization and increased awareness can now be amply supported by quantitative data and models addressing these concerns. A strong commitment to the regular monitoring of erosion, sedimentation, and their consequences is necessary for assuring the continued effectiveness of applied conservation measures and stability of the natural resource base. Educational vehicles must be carefully selected to assure that information is disseminated effectively to all of the concerned audience, on site and off site by tailored brochures, information bulletins and other readily accessible vehicles. Addressing primary and secondary school teachers and students directly in the classroom is presumed to be particularly desirable for reaching many families.

A model effort incorporating many of these elements has been launched in Hawaii to deal with non-point-source pollution under the auspices of the Hawaii Association of Soil and Water Conservation Districts (HASWCD). The HASWCD is a community-based, statewide organization which assumes leadership and quasi-governmental authority in promoting and approving conservation-effective land-use planning. This effort, named the Hawaii Interagency Water Quality Action Program (HIWQAP), joins HASWCD with all concerned state and federal agencies and land-user organizations in a comprehensive effort to identify and correct sediment and other water contamination problems in the state (El-Swaify, 1991). The program consists of four components, namely, public education, annual in-service training of agency personnel, research and development, and implementation. The HIWQAP is instrumental in the governance of the four water quality limited segment non-point-source pollution projects described above.

SUMMARY

Small tropical islands of volcanic origin are characterized by steep topography, aggressive rainfall, and a proximity between sediment sources on land and ultimate sediment destinations in the surrounding ocean. The inducive physical setting and small size of such "whole island ecosystems" result in inherently high potential runoff and erosion, large sediment yields and delivery to internal or shoreline water bodies, a compression of the temporal spans and spatial scales between on-site production and off-site impacts of sediment, and high levels of exposure of conflicts between upland farming communities and users of sediment-receiving areas.

Islands of the Hawaiian Archipelago typify such ecosystems. The potential for erosion by water is the highest among the 50 United States and ranges upward of 100 Mg/ha. Large sediment delivery ratios allow more substantial movement of eroded sediment per unit area than do large continental basins and watersheds. Sediments derived from prevailing highly weathered soils have the physical and chemical enrichment attributes which cause more impairment to the quality of receiving water, per unit mass, than do erosional sediments derived from temperate soils. In all, 16 coastal locations around this small state are so seriously affected by sediment-based non-point-source pollution that they have been designated as water quality limited segments.

Land protection plans in such ecosystems should extend beyond the customary "watershed" approach. Conservation plans must incorporate the continuum of landscape segments including sediment sources, pathways, and deposition areas. Furthermore, because of social conflicts between upland and downstream land users and communities, watershed management programs should involve all of them in the conceptualization and implementation of management and conservation plans.

ACKNOWLEDGMENT

Research cited here was supported by Grant No. 91-34135-6177, Section 406, Food for Peace Act (USDA Cooperative State Research Service). Journal series No. 4217 of the University of Hawaii Institute of Tropical Agriculture and Human Resources.

REFERENCES

Dangler, E.W., El-Swaify, S.A., and Lo, A. 1985. Predicting the erodibility of tropical soil, in *Soil Conservation and Productivity,* I. Pla Sentis, Ed., Sociedad Venezolana dela Ciencia Del Suelo, Maracay, Venezuala, 822–837.

DOH (Department of Health, State of Hawaii). 1990. Assessment of non-point-source pollution water quality problems, Department of Health, Honolulu, 8 chapters, 4 appendices.

El-Swaify, S.A. 1977. Susceptibilities of certain tropical soils to erosion by water, in *Soil Conservation and Management in the Humid Tropics,* John Wiley & Sons, New York.

El-Swaify, S.A. 1988. Conservation-effective rain fed farming systems for the tropics, in *Challenges in Dryland Agriculture: A Global Perspective*, Unger, D.W., Sneed, T.V., Jordan, W.R., and Jensen, R. (Eds.). Proc. Intl. Conf. on Dryland Farming, Texas Agric. Exp. Stn., Temple, TX, 134–136.

El-Swaify, S.A. 1989. Monitoring of weather, runoff and soil loss. In Soil Management and Small Holder Development in the Pacific Islands, IBSRAM Proc. No.8 Intl. Board for Soil Res. and Management, Bangkok, Thailand, 163–178.

El-Swaify, S.A. 1991. Hawaii's response to the National Water Quality Initiative, University of Ryukus, Okinawa, Galaxia, 12:145–153.

El-Swaify, S.A. 1992a. Degradation hazard and land protection elements for small volcanic tropical island ecosystems, *Proc. 7th ISCO Conf.,* Vol. 1, ISCO, Sydney, Australia, 281–287.

El-Swaify, S.A. 1992b. Soil conservation research in Hawaii — a case study for the tropics, *Austr. J. Soil Water Cons.* (Spec. Intl. Issue), 9–13.

El-Swaify, S.A. 1996. Factors affecting soil erosion hazards and conservation needs for tropical steeplands, *Soil Tech.*, 7:3–14.

El-Swaify, S.A. and Cooley, K.R. 1980a. Soil losses from sugarcane and pineapple in Hawaii, in *Proceedings of the U.S./Europe Workshop on the Assessment of Soil Erosion,* John Wiley & Sons, New York.

El-Swaify, S.A. and Cooley, K.R. 1980b. Sediment losses from small agricultural watersheds in Hawaii (1972–1977), Science and Education Administration, Agricultural Reviews and Manuals, Western Series, No. 17, September 1980, USDA.

El-Swaify, S.A. and Dangler, E.W. 1977. Erodibility of selected tropical soils in relation to structural and hydrologic parameters, in *Soil Erosion: Prediction and Control,* Soil Conservation Society of America, Ankeny, IA, 105–114.

El-Swaify, S.A., Dangler, E.W., and Armstrong, C.L. 1982. Soil Erosion bv Water in the Tropics. Res. Ext. Series 024. College of Tropical Agriculture and Human Resources, University of Hawaii, Honolulu, 173 pp.

Giambelluca, T.U., Nullet, M.A., and Shroeder, T.A. 1986. Rainfall Atlas of Hawaii, Department of Land and Natural Resources, State of Hawaii, Report R76, 267 pp.

Habte, M., Fox, R.L., Aziz, T., and El-Swaify, S.A. 1988. Interaction of vesicular-arbuscular mycorrhizal fungi with erosion in an Oxisol, *Appl. Environ. Micro.* 54(4):945–950.

Hawaii CZMP (Coastal Zone Management Program). 1996. Hawaii's Nonpoint Pollution Control Program, Vol. 1, Management Plan, Office of State Planning, Honolulu.

Lo, A.K.F. 1982. Estimation of Rainfall Erosivity in Hawaii, Ph.D. dissertation, Department of Agronomy and Soil Science, University of Hawaii, Honolulu.

MacDonald, G.A. and Abbott, A.T. 1970. *Volcanoes in the Sea: The Geology of Hawaii,* University of Hawaii Press, Honolulu, 441 pp.

Renard, K.G., Foster, G.R., Welsies, G.A., McCal, D.K., and Yoders, D.C., (Coordinators). 1996. Predicting Soil Erosion by Water: A Guide for Conservation Planning with the Revised Universal Soil Loss Equation (RUSLE), Agricultural Handbook 703, USDA, ARS, Washington, D.C., 384 pp.

SCS (Soil Conservation Service). 1975. Soil Taxonomy: A Basic System of Soil Classification for Making and Interpreting Soil Surveys, Agricultural Handbook 436, USDA, Washington, D.C.

SCS (Soil Conservation Service). 1972. Soil Survey of the Islands of Kauai, Oahu, Maui, Molokai, and Lanai, State of Hawaii, USDA, Washington, D.C.

SCS (Soil Conservation Service). 1981. Erosion and Sediment Control Guide for Hawaii, USDA, Washington, D.C.

TCNSPC (Technical Committee on Nonpoint Source Pollution Control). 1978. Nonpoint Source Pollution in Hawaii: Assessments and Recommendations, Technical Report No. 2. Department of Health, State of Hawaii, Honolulu.

USDA (U.S. Department of Agriculture). 1987. The Second RCA Appraisal: Soil, Water, and Related Resources on Nonfederal Land in the U.S. Analysis of Conditions and Trends, Public Review Draft, USDA, Washington, D.C.

Walling, D.E. 1994. Measuring sediment yield from river basins, in *Soil Erosion Research Methods*, 2nd ed., R. Lal (Ed.), Cons. Soc., Ankeng, IA, 39–80.

Yost, R.S., El-Swaify, S.A., Dangler, E.W., and Lo, A.K.F. 1985. The influence of simulated erosion and restorative fertilization on maize production on an Oxisol, in *Soil Erosion and Conservation,* S.A. El-Swaify et al., Eds., Soil Conservation Society of America, Ankeny, IA.

4 Soil Degradation in Relation to Land Use and Its Countermeasures in the Red and Yellow Soil Region of Southern China

T.L. Zhang, X.X. Wang, B. Zhang, and Q.G. Zhao

CONTENTS

INTRODUCTION

With its large area, abundant natural resources, and great potentials in agricultural production, the Red and Yellow Soil Region has been the important production basis of paddy rice, edible oils, tropical fruits, and fast-growing forests. However, the land-use structure has been irrational for a long time. Although the region is geo-

0-8493-0702-3/00/$0.00+$.50

morphologically dominated by mountains and hills with the percentage of 38.1 and 38.5%, respectively, the predominant land-use systems of the region were the irrational "mono- and valley-cropping" systems. Consequently, intensive agricultural production was confined to the valleys and basins which account for only 20% of the region's total land, while the widely spread mountains and hills were extensively managed, partially exploited, or untapped. Because of the irrational land-use structure, the degradation of soils and environments, such as seasonal drought, soil erosion, soil acidification, decline in fertility, and depletion in biodiversity, have become more serious. It is, therefore, urgent to adjust and optimize the land-use structure to prevent land from continuing degradating and to restore the degraded ecosystems.

The region of southern China covers most parts of the Yangtze River basin, all the areas south of the basin, and southeastern Tibet, including 15 provinces (Figure 4.1). Because of the highly diversified and abundant water, heat, land, and biological resources, the region has contributed greatly to sustainable development in the agriculture and economy of the country. It produced 42.7% of the grains and 75% of the rice in the national totals, respectively, and supports 43% of the population with only 28% of the national cultivated land. However, the long-term irrational paddy rice–based "monocropping" systems, together with the drastically increasing encroachment of cultivated lands by rural industries and/or industrial development districts and the overgrowth of population, have placed a significant strain on land resources, leading to various soil degradation problems. It is, therefore, of crucial importance to develop viable alternatives which combine the integrated utilization of natural resources with soil conservation and improvement (Zhang et al., 1995ab).

LAND-USE STRUCTURE

Although various land-use systems are found in different ecoregions or watersheds (Table 4.1) owing to the high diversity in terms of soil, water, vegetation, topography, and other environmental resources, the predominant land-use system of the region is the traditional "mono- and valley-cropping" system (Xi, 1990), in which only the lands in valleys and basins (about 20% of the total area of the region) covered mainly by paddy soils were intensively cultivated, with cropping industry in an absolutely leading position and rice production as a focal point, whereas large areas of undulating tablelands and rolling hills (about 38.5% of the total area) along the river valleys were extensively managed and mainly used for dry farming, practically serving as the sources for collecting rural fuel, organic manure, and/or for scattered grazing. The tablelands and rolling hills are predominantly covered with red soils derived from series of sedimentary and metamorphic rocks like Quaternary red clay, with weathered red sandstone and shale interbedded beneath, conglomerates, phyllites, quartzites, and particularly the saprolite of granite. Most of the original evergreen broad-leaf vegetation has been replaced by scattered pines (*Pinus massoniana*) and grasses of low nutritional value. This unreasonable land-use structure was also reflected in the economic structure. Although the forestlands accounted for 35% of the total land surface, forestry production occupied only about 6.17% of the total agricultural output, whereas the cultivated land area and cropping industry output were 12.7 and 51.5%, respectively (Zhang et al., 1995a).

TABLE 4.1
Land-Use Structure and Soil Erosion in Representative Watersheds or Ecoregions of Southern China

Watersheds or Ecoregion	Area (km²)	Principal Soil	Dominant Land Uses (%)	Erosion Module (t km^{-2} yr^{-1})	Reference
Tableland eco-region, Yujiang County, Jiangxi Province	936.9	Paddy soils and red soils derived from quaternary red clay and red sandstones	A typical undulating tableland ecoregion, with 18.1% cultivated lands and 53% degraded and sparse woodlands	860–1404	Wang, M.Z. (1995)
Tangbei watershed, Xingguo County, Jiangxi Province	16.39	Red soil derived from granite	With wastelands about 10%, eroded lands more than 80%	5481.1	Yang (1990)
Xiangshuitan watershed, Zhiyang County, Sichuan Province	4.5	Purple soils	Sloping cultivated lands 80%, bare rock and wasteland 8.5%, forestland 20.0%	935–3127	Zhang et al. (1997)
Mengba River watershed, Zhijing County, Guizhou Province	39.58	Yellow soil	Uplands 51.6%, shrub lands 23.3%, forestlands 14.0%	69.9	Xia and Xu (1997)
Wupohe watershed, Wuhua County, Guangdong Province	23.23	Lateritic red earth	Cultivated lands 10.8%, degraded forestlands 7.3%, waste mountainous lands 80.9%	2327.0	Zhong (1993); OSWC[a] (1993)

[a] OSWC: Office of Soil and Water Conservation of Wuhua County, Guangdong Province.

SOIL DEGRADATION

The long-term irrational utilization of red soil resources on hilly lands of a densely populated (245 persons/km²) agricultural region, together with the harsh climatic conditions of severe seasonal drought and flood, and particularly the hilly topography and highly erosive soil properties, has led to and/or aggravated land degradation, such as soil erosion, nutrient depletion and fertility decline, and acidification. The hilly red soil ecosystems have been severely damaged and are in an unbalanced and fragile state.

Soil Erosion

Soil erosion can cause considerable degradation in soil fertility and decline in crop yield (National Soil Erosion-Soil Productivity Research Planning Committee, 1981); it is one of the most serious land degradation forms (Zhao and Liu, 1990). The eroded area in southern China amounted to 61.53 million ha, accounting for about 25% of the total land area. Taking 10 southeastern provinces of hilly red soils as an example, the total area under erosion amounted to 24.8 million ha or 21.4% of the total area. Of these eroded lands 16.81 and 6.62% were severely and extremely eroded, respectively. Data from the five provinces of Jiangxi, Hunan, Fujian, Guangdong, and Guangxi had shown an increase of the area under erosion: from 5.4 million in the 1950s to 15.3 million in the 1980s to 18.1 million ha in 1993 (Figure 4.2). Soil erosion modules of the Yangtze River, the Zhujiang River, and rivers along the southeastern coast also increased (Figure 4.3).

An annual loss from topsoil amounting to 585 million tons in the five provinces of Zhejiang, Hunan, Fujian, Guangdong, and Guangxi implied that through soil erosion the Ap horizon of 2.1×10^5 ha of cultivated lands with 18 million tons of mineral nutrients had been rushed away annually. As a consequence, Panyang Lake had shrunk by 1000 km^2 over the last 30 years; Dongting Lake was only half as large as it was 160 years ago; and more than 20 reservoirs had being silted up and their storage capacities reduced by 8.5% over the last 20 years.

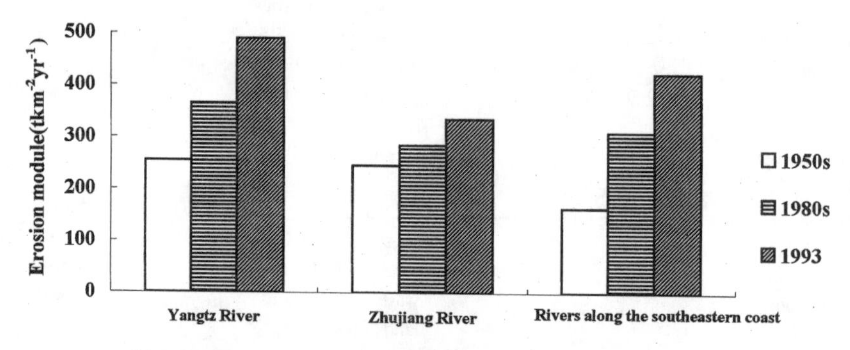

FIGURE 4.2 Dynamics of soil erosion modules in main river.

Decline in Soil Fertility

Decline in soil fertility is another land degradation form in southern China, and it has been threatening agriculture sustainable development of the region (Zhang et al., 1995b). On the one hand, because of the strong weathering and soil leaching processes, most soils of the region have high acidity, low cation exchange capacity, and low reserves of mineral nutrients; on the other hand, long-term irrational utilization and the accelerated soil erosion, especially erosion-induced nutrient deficit in the cycling of the systems, exacerbate the depletion of the nutrients and the decline in fertility of the soils, especially those on the sloping uplands. Moreover, negative balance between application of fertilizer and consumption of plants and unbalanced fertilization are also responsible for nutrient depletion in some soils of this region.

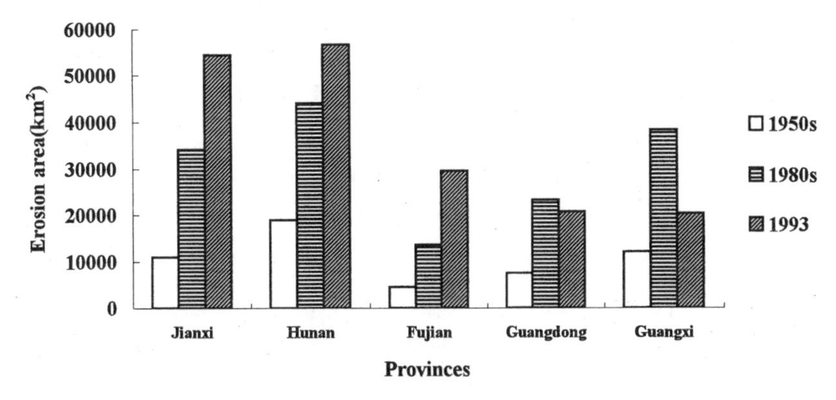

FIGURE 4.3 Dynamics of soil erosion in some provinces.

According to data from the second national soil survey, in southern China, about 68% of the cultivated lands have medium to low yield, and all the cultivated lands are deficient in organic matter and nitrogen, while all uplands and 60% of the paddy fields are deficient in P and 58% of cultivated lands deficient in K, 80% in B, 64% in Mo, 49% in Zn, and 18% in Mg, respectively. Furthermore, an integrated assessment of soil fertility based on 11 different fertility variables in the eastern hilly red soil region has also demonstrated that 25.9, 40.8, and 33.3% of the soils are of the high, medium, and low fertility degradation classes, respectively (Sun et al., 1995).

Another calculation based on the contents of total N, available P, and exchangeable K in soils (Table 4.2) had also shown that most red soils in the hilly region are

TABLE 4.2
Evaluation Standards for Soil Nutrient Depletion

	Nutrient Content		
Depletion Class	**Total N (g/kg)**	**Available P (mg/kg)**	**Exchangeable K (mg/kg)**
Zero	>2.0	>8.7	>83.3
Slight	1.5–2.0	6.6–8.7	66.7–83.3
Moderate	1.0–1.5	4.4–6.6	50.0–66.7
Serious	0.5–1.0	2.2–4.4	33.3–50.0
Extreme	<0.5	<2.2	<33.3

remains of soils subjected to erosion, and their nutrient contents had almost reached the lowest values, which are 0.4 g/kg, 0.44 mg/kg, and 48.3 mg/kg for total N, available P, and available K, respectively. Soils with slight, moderate, and serious nutrient depletion account for 25.1, 49.5, and 29 of the total area, respectively. It was also indicated in Table 4.3 that the upland soils, especially the red soils on sloping waste lands, are more severely degraded than paddy soils. The most depleted nutrient was phosphorus, with moderate degradation amounting to 6.4%, severe and extreme degradation 92.9%; the least depleted nutrient was potassium, with moderate

TABLE 4.3
Degradation Evaluation of Soils on Sloping Uplands

	Total Nitrogen		Available P		Available K	
Degradation Class	Content (g/kg)	Percentage (%)[a]	Content (mg/kg)	Percentage (%)	Content (mg/kg)	Percentage (%)
Zero	>2.0	16.2	>8.7	0	>83.3	34.7
Slight	1.5–2.0	21.3	6.6–8.7	0.7	66.7–83.3	25.4
Moderate	1.0–1.5	43.3	4.4–6.6	6.4	50.0–66.7	26.3
Serious	0.5–1.0	16.2	2.2–4.4	16.7	33.3–50.0	8.6
Extreme	<0.5	2.0	<2.2	76.2	<33.3	5.0

[a] Percentage: percentage of the total uplands area.

degradation amounting to 26.3%, severe and extreme degradation 13.6%; the class of nitrogen degradation fell between potassium and phosphoros, with moderate degradation amounting to 43.3%, severe and extreme degradation 18.2%.

Although the application of chemical fertilizers has apparently increased over the last decade, it was estimated that at present there were still about 30% of the soils in southern China lacking available phosphorus and 70% lacking potassium. Meanwhile, with decreased application of green manure, by 50% from 1978 to 1991, the organic matter content in soils also declined. Furthermore, with the drastically increased intensification of agriculture and irrational fertilization, degradation of trace elements, especially of B have been becoming more and more serious.

TABLE 4.4
Classification of Acid Sensitivity of Red Soils[a]

Sensitivity Index[b]	Class
>1.2	Very sensitive
1.2–0.8	Sensitive
0.8–0.5	Less sensitive
<0.5	Not sensitive

[a] From unpublished report “Red soil degradation and its control” by Zhang Taolin, Lu Rukun, and Ji Guoliang, 1996.
[b] Sensitivity index: changes in pH value caused by adding 10 mmol H_2SO_4/kg soil.

Soil Acidification

Acidification is a natural soil degradation process in the humid subtropics. However, human activities like the irrational application of physiologically acid fertilizers have greatly accelerated this process. Most red soils of the sloping uplands without liming have pH values ranging from 4.6 to 5.3. Moreover, in contrast to acid soils of the temperate region, red soils are extremely vulnerable to acidification, owing to large amounts of variable charges. An assessment based on acid sensitivity index (Table 4.4) showed that most soils in the provinces of Guangdong, Guangxi, and Hainan belong to very sensitive and sensitive class while soils in Province Fujian are mainly in the sensitive and less sensitive class.

INTERACTION OF LAND USE AND SOIL DEGRADATION

In the humid tropical and subtropical areas of southern China, because of the severe seasonal drought, heavy rains, strong weathering, and the specific topography dominated by undulating and rolling hills and mountains, soil erosion, depletion of nutrients, and acidification are all natural processes. However, human activities, especially long-term irrational land use and management, may lead to and/or accelerate the soil degradation process.

Effect of Land Use on Soil Erosion

Various land-use patterns have different effects on soil erosion. As shown in Table 4.5, soil erosion varied with different farming systems after the reclamation

TABLE 4.5
Effect of Land Use in Hilly Red Soil on Soil Erosion (t/km^2)

	A	B	C	D	E
1993	6.50	6.35	0.60	2.23	0.21
1994	1.80	0.45	0.24	0.36	0.13
1995	19.11	21.23	0.53	2.37	0.48
1996	9.38	9.63	0.63	0.56	0.05
Total	36.79	37.66	2.00	5.52	0.87

Note: A: peanut–green manure; B: rape–soybean + maize–buckwheat (conventional tillage); C: rape–soybean + sweet potato (ridge tillage); D: rape–soybean + maize-buckwheat (zero tillage with mulching); E: sparse grassland. – means rotation; + means intercropping

of red soil derived from quaternary red clay on sloping uplands of 7°. Local traditional peanut rotated with green manure (Treatment A) and the designed system of

TABLE 4.6
Soil Erosion (t) under Various Land-Use Systems on Purple Soil in the Watershed of Xiangshuitan

Year	Sloping Cultivated Land, km² 5–10°, 0.796	Sloping Cultivated Land, km² 10–15°, 0.358	Sloping Cultivated Land, km² >15°, 0.185	Bare Land, 0.227	Uncultivated Land, km² Sloping Waste Land, 0.153	Uncultivated Land, km² Sloping Land of Young Forest, 0.587
1989	845.0	549.2	354.3	2197.4	682.4	144.4
1990	1921.0	1826.9	1078.9	3197.3	1055.7	657.4
1991	118.6	407.8	406.4	3135.3	653.3	73.4
1992	230.8	469.0	383.0	2695.9	752.8	363.9
1993	1516.3	2878.3	1345.3	3862.6	997.6	454.3
1994	307.3	339.7	328.9	3995.2	480.4	72.2
1995	2278.9	3501.2	1883.3	4179.1	1086.3	1142.1
Average	995.2	1392.6	825.1	3323.3	815.5	416.8

Source: Original data were from Zhang, Q. et al., 1997.

soybean intercropped with maize followed by rape (Treatment B) had the highest soil erosion, while ridged tillage and zero tillage with mulching drastically decreased soil erosion. Meanwhile, the field experiment conducted on purple soil also showed (Table 4.6) that although area of bare land accounted for less than 10% of the total area, its soil erosion accounted for 42.8% of the total, while the area of sloping cultivated lands accounted for 29.7% of area and 41.3% of soil erosion. These data implied that land use had a profound effect on soil erosion; when reasonable land-use patterns were adopted, soil erosion could be effectively controlled.

Effect of Land Use on Soil Fertility

Land use had also a profound effect on soil fertility. As shown in Table 4.7, soil fertility of wasteland and bare land declined, while soil fertility of paddy rice field and pasture improved, and forestland and garden land improved in available P and available K but degraded in organic matter and total N.

As an investigation on soil fertility properties and their evolution under various land-use systems and at different periods of one land-use system had shown (Table 4.8), most red soils on eroded lands and on lands with sparse shrubs and grasses had integrated fertility near the "bottom" value. In general, soil fertility increased as soon as the soils on wastelands were reclaimed no matter whether they were converted to orchards, tea gardens, or croplands. Except for total K, the content of organic matter, total N, total P, and available P in the soils were all improved rather than degraded. Meanwhile, with the increase of cultivated period, nutrient contents were also increased, although the degree of increment depended on input

TABLE 4.7
Changes of Nutrient Contents in Soil under Different Land Uses after 11 Years[a]

Year	Land Uses	o.m. (g/kg)	t.n.[b] (g/kg)	a.n. (g/kg)	t.p. (g/kg)	t.k. (g/kg)	a.p. (mg/kg)	a.k. (mg/kg)
1983	Wasteland	19.30	0.90	12.32	0.60	15.00	4.97	66.57
	Paddy rice	13.89	0.73	9.16	0.45	10.09	4.00	32.76
1994	Wasteland	17.90	0.85	10.26	0.67	14.50	4.00	59.40
	Bare land	12.83	0.70	10.21	0.53	14.28	4.00	71.7
	Pasture	24.77	1.08	15.98	0.56	15.32	23.13	121.25
	Mixed forest	18.00	0.89	12.19	—	—	5.20	64.95
	Coniferous forest	17.48	0.89	12.56	0.67	13.87	10.20	81.53
	Citrus gardens	13.48	0.77	11.13	0.72	14.22	12.75	95.95
	Paddy rice	21.20	1.09	14.28	0.53	9.58	11.77	41.77

[a] Original data were from Wang, X.J, 1995.
[b] o.m. = organic matter; t.n.. = total N; a.n. = hydrolysable N; t.p. = total P; t.k. = total K; a.p. = available P; a.k. = available K.

TABLE 4.8
Fertility Properties of Soils Derived from Quaternary Red Clay under Different Land-Use Systems

Land Use System	Organic Matter		Total N		Available P		Available K	
	Content (g/kg)	DG	Content (g/kg)	DG	Content (mg/kg)	DG	Content (mg/kg)	DG
Eroded land	2.03	D	0.247	D	0.70	D	100.8	B
Waste grass land	12.1	D	0.647	D	0.52	D	64.2	C
Orange orchard	8.87	D	0.631	C	6.24	A	109.2	B
Tea garden	18.37	B	0.973	C	0.52	D	89.2	B
Upland for 30 years	14.04	C	0.783	C	13.2	A	104.8	A
Upland for 100 years	14.54	C	0.692	C	10.2	A	96.7	B
Rotation of paddy and dry crops for 15 years	21.32	A	1.13	B	10.5	A	75.0	C

DG: A = no evincible degradation; B = slight degradation; C = medium degradation; D = severe degradation.

of nutrients and management levels. Moreover, after being rotated with paddy rice, the integrated fertility of soils improved greatly than in uplands. However, it should be noted that available potassium in soils usually decreased after the uplands were converted to paddy fields. This may be explained by the case study of a representative county Yujiang on nutrient balance of the soil systems. As shown in Table 4.9, in

TABLE 4.9
Field Nutrient Balance in Yujiang, Jiangxi (1994)

	N	P	K
Input (ton)	9444.5	1926.7	2981.8
Output (ton)	7100.6	833.0	3910.5
Balance (%)	+33.0	+131.0	−23.7

this county nutrient input was more than output by 33% for N and 131% for P, but less than output by 24% for K.

Effect of Land Use On Soil Acidification

The land-use patterns and their corresponding soil management practices may have a profound effect on soil acidity. An analysis of soils in the provinces of Jiangxi and Zhejiang (Table 4.10) indicated that soil pH in eroded lands and wastelands did not vary much. Tea and *pinus* plantations caused or accelerated soil acidification, while tea intercropped with peanut prevented further acidification. Soil pH in cultivated lands was much higher than that in other land uses. There were no obvious differences in soil-exchangeable hydrogen among different land uses although differences in exchangeable total acidity and exchangeable Al are distinct under different land uses.

TABLE 4.10
Surface Soil Acidity Variation under Various Land Uses

Land-Use Pattern	pH (H_2O)	Exchangeable Acid (cmol/kg)	Exchangeable Hydrogen (cmol/kg)	Exchangeable Al (cmol/kg)
Extremely eroded land	4.98	4.98	0.39	4.59
Wasteland	4.74	5.17	0.46	4.71
Pinus	4.72	5.58	0.41	5.17
Orange orchard	5.09	4.04	0.46	3.58
Tea garden	4.14	6.09	0.46	5.63
Tea + peanut	4.94	3.80	0.39	3.41
Upland crops for 100 years	5.64	1.02	0.46	0.56
Paddy rice for 10 years	5.16	2.75	0.46	2.29

Besides, the same land use with different fertilization management measures may also have an impact on soil acidity to different degrees.

SUSTAINABLE LAND-USE COUNTERMEASURES

As indicated above, the region of southern China has, on the one hand, highly diversified and abundant biophysical resources with enormous biomass production potentials, but on the other hand, it has been confronted simultaneously by various constraints of agricultural production. It is, therefore, of vital importance to develop sound land-use and management systems which prevent land from continuing degradation and restore the degraded ecosystems. To achieve this goal the following management strategies and countermeasures should be involved.

Optimization of Sustainable Land-Use Patterns

According to the field transection investigation and on-site dynamic monitoring of material transportation, the sloping land can usually be subdivided into three sections, i.e., the erosion section, the transition section, and the accumulation section. It is, therefore, particularly important to develop at a landscape level diversified and stereoecological land-use systems, which rationalize and optimize the temporal and spatial disposal of biophysical resources through adjusting the agricultural structure, namely, changing from a mono- and valley-cropping agriculture to an integrated development of agriculture, forestry, animal husbandry, and fishery. One of the most common models of this kind of land-use system can be described as follows: forest in combination with shrubs and grasses on the top of the mountains and/or hills, fruits and/or cash trees on the hillsides, arable crops and grasses on the lower and gentle slopes and tablelands, rice in the valleys and basins, fish-farming in the ponds, and forage crops around the peripheries of the ponds. An experimented implementation of this kind of pattern in the representative hilly red soil region, Yujian County, Jiangxi Province, showed that the stereoecological land use enhanced matter cycling in the system and increased energy using efficiency by 15 to 30%, and economic benefits by 29.1 to 158.2% (Shi et al., 1995).

Restoration of Soil Fertility on the Low-Yield Lands and Wastelands

Since most of the potentially arable lands with fertile soils have been cultivated and the rest are mainly marginal lands and/or lands suffering from various soil and water constraints, a further increase in crop yields will, therefore, be mainly achieved by increasing the cropping intensity (the present average cropping index is only about 199) and improving the productivity of existing cultivated lands, especially those with red soils on hilly lands. To do this, the following strategies and measures aimed at restoring and improving soil fertility need to be adopted, namely,

1. Adopting the cropping systems and tillage methods. The experiment on upland red soils in Jinghua, Zhejiang Province, indicated that compared with the local traditional cropping system of wheat rotated with sweet potatoes, the strip cropping of three rotations with five or four harvests showed a distinctive increase in crop production, the content of organic matter, total N and P in the soil. Meanwhile, the transformation ratio of

the nutrients also increased by 21.0, 88.1, and 15.3% for N, P, and K, respectively.

2. Increasing the nutrient inputs and the utilization efficiency. As the external nutrient input to agricultural production systems in the region was in general very low, and in some cases even under a state of negative balance, an increase in fertilizer application to the soils, especially to those on the sloping uplands, resulted usually in an effective increase in crop yields and soil fertility. A number of long-term field experiments indicated that the increase in application of organic manure, particularly organic manure in combination with chemical fertilizer, was one of most important strategies for fertility management. As shown by a 6-year field experiment conducted on the upland red soil derived from granite, application of organic manure alone showed an increase in soil organic matter by 241.7%, total N, available P, and available K by 120.0, 200.0, and 82.6%, respectively, whereas application of organic manure in combination with chemical fertilizer showed an increase about by 460, 120, 317 and 110%, respectively.

However, application of only chemical fertilizer did not show an apparent increase in soil nutrients content. In addition, a series of other fertility management practices proved also very efficient, for example, conversion of uplands into paddy fields in the areas with good access to irrigation, rotation of paddy rice with upland crops in the paddy fields, adequate techniques for nutrient recycling in agroecosystems, and new techniques for exploiting the untapped wastelands.

Development of Soil Conservation Systems

Since soil erosion has become one of the most serious land degradation forms in southern China, it is, therefore, essential to develop proper land-use systems which prevent soils from erosion and restore the productivity of the degraded ecosystems induced by erosion. Table 4.11 lists some successful cases of management and restorative measures of watersheds in southern China. Among these systems, the stereoconservation system deserves particular attention. The system can generally be described as restoration of moderately to severely eroded lands with a slope of more than 25% by returning the croplands to forestlands and combining reforestation with engineering measures like contour trenches and check dams; hole planting of cash forest and fruits coupled with fire prevention on the middle and upper parts of slopes; and terracing and strip-planting of cash crops and fruits on lower parts of hillsides. Another important aspect is to develop conservation farming systems and agroforestry technologies and the matching soil surface management packages, such as minimum and no-tillage, mulching, and squamose hole techniques. Should these conservation practices be adopted, great benefits will be obtained. As reported by Yang et al. (1992), in severely eroded Quaternary red clay area, with adoption of biological measures in combination with engineering methods, the runoff, the silt content in runoff, and the silt yields decreased by 52, 85.6, and 94%, respectively. Moreover, experiment on sloping upland with red soil in Yingtan, Jiangxi Province

TABLE 4.11
Effect of Management and Restorative Measures of Watershed on Soil Erosion and Productivity

Watershed	Area (km^2)	Measurers	Soil Erosion	Productivity	Reference
Pannonghe, Suining County, Sichuan Province	135.02	Terracing 8.27 km^2, planting trees 9.63 km^2, forest conservation 5.14 km^2, adopting conservation tillage 58.97 km^2	Decreased from 8371 to 1401 $t/km^2 \cdot y$	The yield increased by 32.4%, pure income by 377.1%	Zhong, 1993; Cheng et al., 1993
Wupohe, Wuhua County, Guangdong Province	23.23	Engineering measures like contour trenches and check dams 416.3 *1000 m^3, planting tree and grasses 9.8 km^2, forest conservation 5.4 km^2	Decreased from 10543 to 1700 $t/km^2 \cdot y$	The yield increased by 5.4%, income per capita by 92.8%	OSWC, 1993
Tangbei, Xingguo County, Jiangxi Province	16.39	Engineering measures like contour trenches and check dams 9.33 km^2, planting tree and grasses 9.73 km^2	Decreased by 67.7%	The yield increased by 20.4%, income per capita by 373.0%	Yang, 1990

also indicated that cropping systems and tillage methods, like ridge tillage and/or no-tillage with crop residual mulching, exerted a profound effect on soil and water conservation.

ACKNOWLEDGMENTS

The project was supported by the National 9th 5-year Key Project of China (No. 96-04-03-12) and National Natural Science Foundation of China (No. 49571043).

REFERENCES

Cheng, J., Wang, H., Wang, J., and Wang, J. 1997. Analysis of benefits on integrated management of Pannonghe watershed, *Soil Water Conserv. China* [in Chinese], 4:51–53, 58.

National Soil Erosion-Soil Productivity Research Planning Committee. 1981. Soil erosion on productivity: a research perspectivity, *J. Soil Water Conserv.* 36:82–90.

OSWC (Office of Soil and Water Conservation of Wuhua County Guangdong). 1993. Analysis of benefits on soil and water conservation in Wupohe watershed, *Bull. Soil Water Conserv.* [in Chinese], 13(2):50–54, 60.

Shi, H., Zhao, Q., Wang, M. et al. 1995. Study on comprehensive improvement techniques of red soil ecological system and sustained development of agriculture, in Wang, M. et al., Eds., *Research on Red Soil Ecosystem,* Series 3 [in Chinese], Chinese Agricultural Science and Technology Press, Beijing, 1–27.

Sun, B., Zhang, T., and Zhao, Q. 1995. Integrated Evaluation of Impoverishment of Soil Nutrients in Red Earth Hilly Regions in South China, *Soils* [in Chinese], 27(3), 119–128.

Wang, M. 1995. Present conditions and machanism of ecosystem degradation and the related counter-measures in low-hilly red soil regions, *Acta Agric. Jiangxi* [in Chinese], 7(Suppl.), 1–20.

Wang, X.J. 1995. Monitoring and Evaluation of Soil Changes under Different Landuse Systems in Red Soil Hilly Region of China, Ph.D. thesis, Institute of Soil Science, Academia Siniac, Nanjing, P.R. China.

Xi, C. F., Ed., 1990. *The Way-out the Mountainous Area of Southern China* [in Chinese], Science Press, Beijing, 133 pp.

Xia, H. and Xu, H. 1997. An evaluation on soil and water conservation effect of comprehensive administration practice in Mengba River watershed, *Bull. Soil Water Conserv.* [in Chinese], 17(4), 8–11, 31.

Yang, Y. 1990. Experiences of water and soil conservation in the typical erosion area of southern China, in Associate of Science and Technology, Eds., *Land Degradation and Its Countermeasures* [in Chinese], Press of Science and Technology of China, Beijing, 189–193.

Yang, Y.S., Shi, D.M., Lu, X.X., and Liang, Y. 1992. Characteristics, improvement and utilization of eroded soils in Red Soil Ecological Experiment Station, in Shi H., Ed., *Research on Red Soil Ecosystem* (series 1) [in Chinese], Science Press, Beijing, 251–257.

Zhang, Q., Yang, W., and Lin, C. et al. 1997. Feature and control of soil and water loss in small catchment of hilly area of central Sichuan, *J. Soil Erosion Soil Water Conserv.* [in Chinese], (3)3, 38–45, 57.

Zhang, T., Zhao, Q., Zhai, Y. et al., 1995a. Sustainable land use in hilly red soil region of southeastern China, *Pedosphere,* 5(1), 1–10.

Zhang, T., Zhang, B., Wang, X. et al., 1995b. Potentials and constraints in relation to sustainable farming systems for using red soil resource in hilly region of southern China, in Wang, M. et al., Eds., *Research on Red Soil Ecosystem* (series 3) [in Chinese], Chinese Agricultural Science and Technology Press, Beijing, 28–43.

Zhao, Q.G. and Liu, L. 1990. Human being acitvity and land degradation, in Associate of Science and Technology, Eds., *Land degradation and its countermeasures* [in Chinese], Press of Science and Technology of China, Beijing, 1–6.

Zhong, Z. 1993. Remarkable effects achieved by persisting in a long-term control of Wupohe Valley at Wuhua County in Guangdong Province, *Bull. Soil Water Conserv.* [in Chinese], (3)2, 47–50.

5 Watershed Management for Erosion Control on Sloping Lands in Asia

Eric T. Craswell and Chalinee Niamskul

CONTENTS

INTRODUCTION

In the extensive hilly and mountainous areas of Southeast Asia, population growth and demand for tropical timber have led to widespread deforestation and intensification of agriculture. Consequently, soil erosion appears to be a serious problem, although reliable information on the human-induced impacts on stream flows and sediment loads is scarce. Research institutions in eight Asian countries have formed a consortium with advanced research organizations in Europe, North America, and Australia to tackle the problem of managing soil erosion. The consortium will build the capacity of involved institutions, while developing and refining a research paradigm utilizing a participatory and interdisciplinary approach. The main outputs of the consortium will be acceptable and sustainable community-based land management systems and reliable information and guidelines for decision makers on the on- and off-site agricultural, environmental, and social impacts of managing soil erosion. In 1996–97 the consortium developed comprehensive guidelines for the selection of sites for the model watershed studies, which will be the core of the consortium program. An interdisciplinary team visited ongoing watershed studies in China, Indonesia, Laos, Nepal, the Philippines, Thailand, and Vietnam. The sites visited vary widely in social, economic, and biophysical characteristics, but the generic research methodology can be applied to all. Methodology and paradigm development are linked globally to the work of three other consortia addressing major problems in other regions as part of the Soil, Water, and Nutrient Management Programme of the Consultative Group on International Agricultural Research.

0-8493-0702-3/00/$0.00+$.50

Land degradation is a major concern to the international community. The Earth Summit in Rio de Janeiro in 1992 and a series of subsequent meetings have pointed out the widespread degradation of water and land resources and called for intensified efforts to reverse this trend. Successful programs on soil and water conservation are needed especially in tropical areas, where the many developing countries have not been able to commit the financial and human resources needed. An international meeting in Zschortau, Germany in 1995 highlighted the need for expanded international efforts to promote sustainable land management and the importance of furthering cooperation between people and institutions (IBSRAM/DSE,* 1995). The Zschortau meeting also underlined the emerging consensus that major changes are needed in the approach to research and development activities to promote sustainable land management in developing countries. This chapter reviews these changes and highlights the case of land and water resources in the hilly and mountainous areas of tropical Asia where population pressures and economic growth are significantly affecting land use. The chapter describes a new consortium on managing soil erosion that brings together scientists from eight Asian developing countries and several developed countries with the common goal of combating soil erosion through integrated watershed management.

THE ASIAN TROPICS

Tropical and subtropical Asia can be physiographically and agriculturally classified into the lowlands, where paddy rice has been cultivated for centuries, and the uplands, which are hilly and mountainous. The steep slopes of the young landscapes are fragile and the region's rivers discharge a total of nearly 7500 million tonnes of sediment to the surrounding oceans annually (Milliman and Meade, 1983). The data on sediment yields in Table 5.1 suggest that the watersheds of the smaller rivers are more erosion prone, but since the data are based on net outflows of sediment to the oceans, the lower sediment yields from larger basins probably reflect the extensive deposition in dams and deltas of the larger rivers. The sediment loads of the large rivers such as the Mekong and the Irrawaddy and of the many smaller rivers add up to a major outflow of nutrients, and economic cost, to the countries concerned. Construction of dams in the region for agricultural and urban uses, and for power generation, is continuing, changing the stream flows and silt deposition patterns, and increasing the potential for off-site damage from erosion in upper watersheds. Nevertheless, reliable information on the human-induced impacts, particularly those due to cultivation, on stream flows and sediment loads is scarce (Figure 5.1)

The problem of land degradation through erosion comes sharply into focus when the pressure of population growth on land resources is taken into account. The population of the Asian region is projected to grow from 3.4 thousand million in 1995 to 5.4 thousand million in the year 2050 (Fischer and Heilig, 1997). This presents a major challenge to agriculture to meet the growth in total demand for food and the changing dietary habits, as incomes grow. Since the high-potential

* IBSRAM/DSE = International Board for Soil Research and Management/Deutsche Stiftung für Internationale Entwicklung.

TABLE 5.1
Sample Data on Sediment Outflows to the Ocean from Rivers in Tropical Asia (Milliman and Syvitski 1992)

River	Drainage Basin Area ($\times 10^6$ km²)	Sediment Load (10^6 t/yr)	Sediment Yield (t/km²/yr)
Irrawaddy (Burma)	0.43	260.0	620
Mekong (Vietnam)	0.79	160.0	200
Angat (Philippines)	0.00057	4.6	8,000
Cimatur (Indonesia)	0.00058	1.9	3,000
Cilutung (Indonesia)	0.00060	7.2	12,000
Agno (Philippines)	0.0012	5.0	4,350
Citanday (Indonesia)	0.0025	9.5	3,700
Cimanuk (Indonesia)	0.0032	25.0	7,800
Solo (Indonesia)	0.16	19.0	1,200
Hungho (Vietnam)	0.12	130.0	1,100

Source: Milliman, J. D. and Meade, R. H., *J. Geol.*, 91, 751–762, 1983. With permission.

lowlands are already intensely used for food production, the upland areas of Asia will be used increasingly for agriculture. This trend will be exacerbated by greater urbanization and land use for periurban agriculture. The resultant pressure on fragile sloping lands will seriously exacerbate changes in the hydrology and sediment outflows from upper watersheds unless efforts to promote sustainable land management in the uplands are successful. The current fragmented and limited knowledge of the on- and off-site impacts of soil erosion in countries of the region is a major constraint to the development of appropriate policies and action plans to address the problem.

Upland farmers have been neglected clients for agricultural research in Asia, and face many problems due to remoteness from markets, lack of investment, and unstable security. In some countries, such as Thailand, Laos, and Vietnam, upland farmer communities are minorities, such as hill-tribes, which have been isolated politically and culturally, but are under pressure from governments to settle and cease their shifting cultivation and nomadic traditions. In other countries, such as Indonesia and the Philippines, resettlement programs bring immigrants from overpopulated areas to upland watershed areas. In most countries the pattern has been to intensify agriculture in upper watershed areas with steep slopes (see example from Thailand in Figure 5.2).

A NEW APPROACH TO RESEARCH

Past research on the problems of land degradation in the tropics has had limited success — clients for the research results have been ignored; social and biophysical sciences have operated separately; researchers have been trained in reductionist scientific methods; small plots have been used; links to policy makers have been vague; the scientific capacity is inequitably distributed; and expatriates have domi-

FIGURE 5.1 Erosion rates from bare cultivated soils on steep slopes are high.

nated the research in some countries. IBSRAM and other international research agencies have recognized these problems and now advocate a new research paradigm, key elements of which are listed in Table 5.2 (see Greenland et al., 1994). This paradigm is the cornerstone of a new systemwide program on Soil Water and Nutrient Management (SWNM) of the Consultative Group on International Agricultural Research. The overall *goals* of the SWNM program are to increase long-term agricultural productivity, reduce poverty, and conserve and enhance land and water resources. SWNM brings together four complementary research consortia on nutrient depletion, inefficient water use in dry areas, acid soils, and water erosion of soils. The program is convened by IBSRAM and the International Centre for Tropical Agriculture (CIAT) in Colombia, and each consortium engages national agricultural research and extension systems, nongovernment organizations, advanced research organizations, and international agricultural research centers. The consortia develop and share innovative tools for the assessment of sustainability and impact, using information technology such as geographic information systems and decision support systems. The outputs of the SWNM program go beyond the improved and

FIGURE 5.2 Steep lands in northern Thailand are often cultivated up and down the slope.

TABLE 5.2
Key Elements of the New Paradigm for Research on Sustainable Land Management

Element	Approach
User orientation	Participatory, community based at all stages from planning to implementation
Policy	Focus on policy and institutional issues that influence farmer and community decisions
Equity	Consideration of equity, including gender analysis, in research planning and implementation
Landscape	Integration of people, soil, and water at every scale from plot to watershed
Research intensity	Linking strategic, applied, and adaptive research with technology development and participatory dissemination
Knowledge	Reliance on both indigenous and scientific sources
Orientation/goals	Linking increased productivity with natural resources conservation

appropriate technologies for sustainable land management, by delivering new research tools and indicators, enhanced institutional capacity, and scientifically sound and relevant information to help decision makers tackle some of the most intractable land degradation problems.

The new research paradigm (see Table 5.2) requires an interdisciplinary approach and work that covers the full range of the research continuum covering policy, social, economic, and biophysical factors (Table 5.3). To deal with this complexity a research consortium model was developed (Figure 5.3). The effective orchestration of the SWNM consortia depends on capturing the interest of collabo-

TABLE 5.3
Key Factors Influencing Sloping Land Management in Thailand (IBSRAM, 1997)

Policy/Institutional Factors	Social and Economic Factors	Biophysical Factors
Border security/stability	Expansion of infrastructure to open new areas for development	High population growth rates and pressure on resources
Hill tribe resettlement schemes and land tenure policies	Permanent cultivation replacing shifting cultivation	Heterogenous resource base and diverse land management in the landscape
Land use conflict with Royal Forestry Department policies for protected watersheds	Cash income opportunities through vegetable and fruit production	High erosion potential due to steep slopes and high erosivity of tropical rains
Fragmentation of responsibilities across four government departments	Higher community aspirations for services including education, health, and infrastructure,	General perceptions of high off-site effects of erosion and runoff due to land clearing
	Heavy workload continues to fall on women	Increased demand for water storage and for flood control downstream

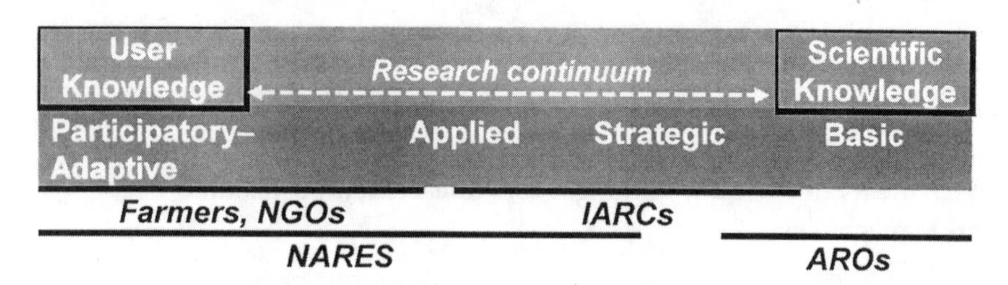

FIGURE 5.3 Primary stakeholder domains along the research continuum.

rating institutions, devising alliances to fill gaps, and securing funds to catalyze the research. As shown in Figure 5.3, advanced research organizations (AROs) have much to contribute at the strategic and basic end of the research spectrum. North American institutions have established reputations for scientific excellence in the fields of soil science and land management, which could significantly bolster the consortia of the SWNM program through collaborative research and training. The SWNM consortia offer a client-centered development context at chosen benchmark sites, where watershed studies focus on social and biophysical aspects of landscape dynamics. The watershed studies are long term and would benefit from supplementary research by AROs on methodologies, technologies, agroecosystem processes, systems analysis, etc. The SWNM global framework also provides opportunities to exchange experience on methodologies and approaches.

THE MANAGEMENT OF SOIL EROSION CONSORTIUM

Surveys by IBSRAM in Asian countries indicated that the capability of national research and extension systems to conduct watershed research varies widely across the region. Institutions in some countries have a strong biophysical research capacity in hydrology and agronomy, but are weak in their research on social and economic issues. Others have well-developed participatory research with communities in all parts of the landscape, but little capacity in the fields of hydrology and soil science. Representatives of research agencies in eight Asian countries expressed interest in joining the consortium, and a series of consortium assemblies was held during 1996–97 to develop a plan for a coordinated research program. The Management of Soil Erosion Consortium (MSEC) will build the capacity of involved institutions, while drawing on the strengths of each. Established model watershed studies are being selected in each country, to provide the focus for the collaborative research. The main outputs of the consortium will be acceptable and sustainable community-based land management systems and reliable information and guidelines for decision makers on the on- and off-site agricultural, environmental, and social impacts of managing soil erosion.

The consortium has developed comprehensive guidelines for the selection of sites for the model watershed studies. An interdisciplinary site-selection team has visited ongoing watershed studies in China, Indonesia, Laos, Nepal, the Philippines, Thailand, and Vietnam. The sites visited vary widely in social, economic, and biophysical characteristics, and in the capacities of national scientists in different fields, but the new research paradigm can be applied to all of them (Figure 5.4).

One example of the consortium sites selected is Khun Sathan, near Nan in northern Thailand (IBSRAM, 1997). The 820 Hmong people living in the watershed are keen to move into the cash economy. They cultivate cabbages on steep slopes, with high inputs of agrochemicals. A watershed of 16 km^2, and several microwatersheds, will

FIGURE 5.4 Modern watershed research involves scientists from many disciplines and countries.

be instrumented and sampled to assess the hydrology and the nutrient and agrochemical fluxes. A review of indigenous knowledge and a biophysical and socioeconomic inventory are being made. After 2 years of calibration, the hydrological balance, dissolved and suspended elements, and bed load of meso- and microwatersheds will be measured. Information on the on- and off-site effects of different land-use practices will be gathered in a participatory study involving the farmers. Land-use practices tested will include soil conservation measures known through research to be effective in controlling soil erosion on steep slopes (Craswell et al., 1998). Institutional and policy arrangements will also be evaluated and an information system based on the biophysical and socioeconomic data developed. The information system will feed into models and decision support systems designed to help land managers and policy makers improve their decision making. The Thai and IBSRAM scientists working at the Khun Sathan site will share experiences and training with institutions from the other seven consortium countries. The ultimate goal is a quantum leap in the capacity of watershed research in the region, which will increase awareness of the scope of the erosion problem and develop effective ways of combating it.

REFERENCES

Craswell, E.T., A. Sajjapongse, D.J.B Howlett, and A.J. Dowling. 1998. Agroforestry systems in the management of sloping lands in Asia and the Pacific, *Agroforestry Syst.* 38:121–137.

Fischer, G. and G.K Heilig. 1997. Population momentum and the demand on land and water resources, in *Philosophical Transactions*, D.J. Greenland, P.J. Gregory, and P.H. Nye, Eds., Biological Series Vol. 352, 1356, Royal Society, London, 862–869.

Greenland, D.J., G. Bowen, H. Eswaran, R. Rhoades, and C. Valentin. 1994. Soil, Water, and Nutrient Management Research — A New Agenda, IBSRAM Position Paper, IBSRAM, Bangkok, Thailand.

IBSRAM/DSE (International Board for Soil Research and Management/Deutsche Stiftung für Internationale Entwicklung). 1995. The Zschortau Plan for the Implementation of Soil, Water, and Nutrient Management Research, Food and Agricultural Development Centre, Feldafing and Zschortau, Germany.

IBSRAM. 1997. Model Catchment Selection for the Management of Soil Erosion Consortium (MSEC) of IBSRAM, IBSRAM, Bangkok, Thailand, 109.

Milliman, J.D. and R.H. Meade. 1983. World-wide delivery of river sediment to the oceans, *J. Geol.* 91: 751–762.

Milliman, J.D. and J.P.M. Syvidski. 1992. Geomorphic/tectonic control of sediment discharge to the ocean: the importance of small mountainous rivers, *J. Geol.* 100: 525–544.

6 Watershed Management for Erosion Control in Steeplands of Tropical Asia and Australia

Cyril A. A. Ciesiolka and
Ghulam Mohd Hashim

CONTENTS

INTRODUCTION

SOME PHILOSOPHICAL CONSIDERATIONS

Deriving an income from land takes on many forms in every political and economic system. As use of land has intensified with technology, the impact has permeated even the masses of air moving across the face of the globe as well as the waters of the oceans.

0-8493-0702-3/00/$0.00+$.50

Management of the Earth's land resources therefore provides humankind of the 21st century with an extremely complex challenge. Literature over the centuries documents examples of exploitation and short-term exhaustion of the land for food production (Vita Finzi, 1977). The major examples of continuous food production from land are the great river valleys where annual flooding provides deposited soil on a regular basis. Thus, erosion of some parts of a landscape as a geological process has supported high population densities in downstream locations. The visible legacies of the exploits of humankind to manage land and water in locations away from river floodplains remind us that previous societies have already grappled with the same question that we face, namely, how to manage landscapes that are net exporters of soil. Although people's appreciation of the concept of a catchment has been greatly enhanced by air and space photography, I believe that the idea is deeply engrained in the human unconscious.

Land ownership evokes the deepest of passions in humankind. Consequently, implementation of land-use policies by any regulatory authority is often frustrated by land users who make day-to-day decisions about deriving a livelihood from land and feel disempowered when directed or cajoled into making changes. The wide variety of land uses is increasing with higher populations and makes consensus more difficult on any issue.

It would appear that the problems of land management are perhaps too difficult to resolve amicably and so governments implement change by legislation. Such actions usually produce a raft of reports and investigations that show such decisions are driven by some interest groups that possess the political, social, and economic skills to manipulate opinions and decision makers. Decisions on the utilization of land and water resources are largely driven by economics of production and distribution, but as science is showing the interrelatedness of physical processes both spatially and temporally, land management at the river catchment scale is vital to the well-being of the human race. This knowledge has generated a profound fear and skepticism in consumers of all agricultural products and raised the cry for "cleaner and greener" foods.

A media-driven society finds it profitable to allocate blame for the state of the environment, and physical and social scientists are busily producing "fix-it" solutions and methodologies for politicians who wish to remain in power and land users who must show a profit to retain ownership of the resource. While there will always be problems to solve, social scientists would advocate that change is best effected by understanding the components of any system and then intervening at one point and trying to make changes. When the term *land use* is narrowed to mean arable and pastoral farming, then choosing the point of intervention necessitates a clear understanding of attitudes, actions, and management decisions at the farm scale and across an industry.

Erosion as a Temporal and Spatial Process in a Catchment

Temporal Processes

Images of soil erosion that have a profound impact upon politicians, landowners, managers, and the public are ones of bare soil on sloping land, rills and gullies, wide areas of deposition in valley floors, and water storages filled with deposits and muddy water. The response of early soil conversation authorities was to change the spatial component of erosion processes by shortening the distance of flow before diverting it across the slope via contour channels and subsequently down grassed waterways into the naturally occurring drainage lines. However, this strategy changed only part of the spatial component of soil erosion. Later studies documented the impact of cleaner water upon existing "natural" channels (Trimble, 1974). In fact, channel hydraulics of naturally occurring drainage networks have been altered by deposition and a new phase of erosion initiated. The erosion problem can be displaced from the hillsides to the valley floors where its social impact is not as visible because people expect to see stream banks being eroded. It is only where eroding meanders undercut productive farmland or threaten urban infrastructure that some interest is aroused.

The temporal aspect of erosion at the catchment scale is not widely communicated by land managers or scientists largely because of the emphasis on highly visible hillside erosion and the longer time frame needed to understand channel erosion.

Channel development passes through distinct phases. Initial incision lowers the bed slope of a channel. The newly formed gullies widen and begin to meander and vertical sides go through a process of declination and revegetation. Sediment that is removed from the upper catchment channels undergoes a sequence of deposition and remobilization progressively down the valleys. In wetter climates or where base flows of streams are important, channel geometry alters to accommodate the sediment fluxes that are generated. In drier monsoon climates, pulsing of sediments down a valley is more observable. Deposits form large alluvial and colluvial fans in valleys and these in turn are reincised and the sediment is pulsed on down the valley to new locations. In time, these deposits go through the same process again.

It is imperative that an understanding of these processes be applied at the landscape scale when considering sustainable management practices of farming in a catchment. It is interesting to note that broadacre zero tillage in sloping vertisols of the tropics eventually undergoes this same process.

Spatial Processes

Most eroded hillslope sediments move to the footslopes and valleys of uplands and are deposited. The distance that suspended loads move from sloping lands down the regional trunk streams is not known.

The morphometry of landscapes that characterize river catchments is largely the product of previous climates and their extremes, geology and its rate of weathering, resistance of soils to erosion, influences of the contemporary vegetation, and changes in the base level of landscapes. Soil erosion processes subsequently vary both spatially and temporally because the physical variables of rainfall, infiltration, vegetative/stone cover, soil consolidation, and surface and subsurface runoff also vary spatially. A drainage density and a set of channel hydraulics evolve in response to the variety of inputs to which the landscape is subjected. Stream order and catchment area are related but sediment movement from each drainage order is not correlated so well. Field observation shows that while some catchments may be gullied, adjacent ones remain stable. If slopes remain about the same length and angle and relief is comparable, in the long term there must be some sort of quasi-equilibrium of processes that give rise to landforms of a similar shape on a particular geology.

The role of subsurface water movement in shaping catchment landscapes and crop production in hard rock geologies has not been a major research topic of tropical lands. Significant contributions to the understanding of the patterns and volume of subsurface flow have been made by Kirkby and Chorley (1967), Pilgrim and Huff (1978), Dietrich et al. (1987), and Loch et al. (1987).

However, soil erosion studies have frequently reported results from experimental plots ranging from 1 to 100 m^2 under both simulated and natural rainfall (Zingg, 1940; Meyer and McCune, 1958; Barnett and Rogers, 1966; Wischmeier, 1966; Soons and Rainer, 1968; Lal, 1976; Edwards, 1983; Loch, 1984; Romkens, 1985; Sheng, 1990). Fewer studies have quantified erosion from farm-sized fields or conservation layouts in Australia (Adamson, 1974; Freebairn and Wockner, 1986) and still fewer authors have reported soil erosion results from "nested" catchment studies (Abrahams, 1972; Douglas, 1973; Ciesiolka and Freebairn, 1982; Loughran, 1984; Oliver and Rieger, 1986; Ciesiolka, 1987). Similarly, there is a scarcity of studies that show conservation at the large scale has increased production and net profit for farmers (Fraser, 1995).

From the volume of publications, it is evident that most effort has been focused on carrying out small-scale studies in order to derive parameter values for models that can be extrapolated to larger scales. Recently, the concept of representative elementary areas has been introduced as a methodology for characterizing the spatial domain that integrates hydraulic and hydrologic process of a landscape (Woods et al., 1995). The fate of sediments upon entering a channel network is not so well documented (Phillips, 1986).

With the realization that erosion processes are both spatial and temporal, this chapter reports on experiments that were set up to identify and measure rates of hydraulic and erosion processes at a variety of landscape scales and then to develop farming systems that took advantage of this knowledge. The approach was to measure the best conservation practices that farmers were using, facilitate their observation of erosion processes at varying scales, and to introduce ideas of farming systems, some of which may have seemed unrealistic.

PROCESS STUDIES

A general summary of information from the sites studied is presented in Table 6.1. In these studies the two processes of soil consolidation and subsurface flow were identified as being far more important than originally envisaged.

SUBSURFACE FLOW

The strongest evidence of shallow subsurface flow came from Malaysia and Australia and the greatest quantities were measured at Imbil, Gympie, Australia on a Lithic Eutropept where slopes steepness ranged between 25 and 35%.

On virgin land growing its first crop of pineapples, 2.5-m-long modified Gerlach troughs with tipping buckets were placed across the footslope of a block of newly planted pineapple tops — Site W3 (Figure 6.1).

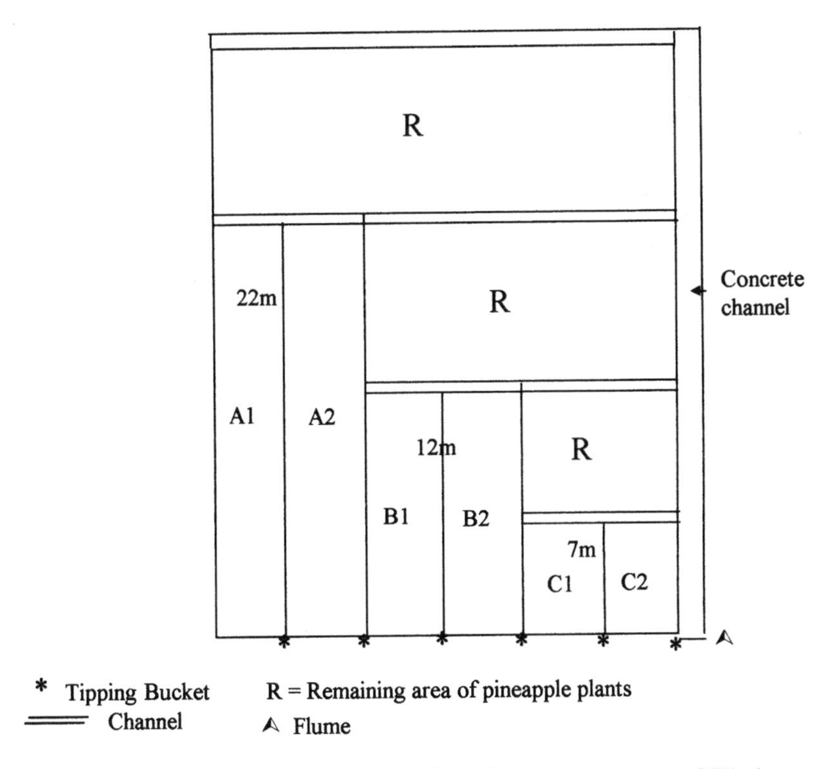

FIGURE 6.1 Site W3 showing layout of row length experiment on 35% slope.

Smaller runoff depths were measured from plot B1 than all others when rainfall was less than 40 to 50 mm. Runoff was also lagged by up to 4 min. For larger falls, especially from tropical cyclones and rain depressions, total depth of runoff from this plot far exceeded rainfall. For example, in the event of April 6, 1989 some 550 mm were measured as runoff from 266 mm of rainfall. The subsurface flow moved in well-defined narrow pathways of less than 2 m width, but in large rainfall events of long duration, an increase in runoff rate was recorded in the adjacent plots indicating that the width of subsurface flow broadened and contracted similar to a surface flood event.

Results from all sites showed that subsurface flows commenced quickly. For some small events when there was no surface runoff, subsurface flow occurred within 15 min. The lag time of maximum subsurface flow rate behind peak rainfall rate varied considerably but was usually less than half an hour. Such a rapid response was unexpected, but often the recession of the subsurface hydrograph continued for 24 h or longer depending upon the size and duration of the rainfall event.

TABLE 6.1
Comparative Soil Erosion and Hydrogolic Data From "Conventional" and "Improved" Soil-management Treatments at Five ACIAR Projects Sites

Site and Treatment	Annual Soil Loss (mm)	Runoff Coefficient (%)	% Suspended Load	% Suspended Load	Suspended Load to Streams	
					Amount (t/ha/yr)	Concentration (kg/m³)
MALAYSIA						
Kemaman, sandy clay loam soil						
Average rainfall = 3000 mm/yr						
Slope angle = 14–17%, length = 20 m						
Conventional: clean weeding and herbicide	80	1110	67	73	58	5.2
Crop: cocoa						
Improved: legume and grass ground cover	8	300	10	74	6	2.0
THAILAND						
Khon Kaen, loamy sand soil						
Rainfall 1991 = 1079 mm						
Slope angle = 3–4%, length = 36 m						
Conventional: clean cultivation up and down slope	3.9	219	20	59	2.3	1.8
Crop: kenaf						
Improved: minimum tillage	0.3	102	9	37	0.1	0.4

TABLE 6.1 ***(continued)***
Comparative Soil Erosion and Hydrogolic Data From "Conventional" and "Improved" Soil-management Treatments at Five ACIAR Projects Sites

Site and Treatment	Annual Soil Loss (mm)	Runoff Coefficient (%)	% Suspended Load	% Suspended Load	Suspended Load to Streams: Amount (t/ha/yr)	Suspended Load to Streams: Concentration (kg/m³)
THE PHILLIPINES						
Los Banos, clay soil						
Rainfall 1993 = 1440 mm						
Slope angle = 14–28%, length = 12 m						
Conventional: clean cultivation up and down slope	105	483	34	0.7	0.7	0.14
Crop: maize						
Improved: hedgerows, clippings, and crop residue used as surface mulch in alleys	5	145	10	2	0.1	0.06
THE PHILLIPINES						
VISCA, clay soil						
Runoff-producing rain = 950 mm/yr						
Slope angle = 50–70%, length =						
Conventional: clean cultivation up and down slope	39	83	9	5	2.0	2.4
Crop: maize						
Improved: hedgerows, clippings used as surface mulch in alleys, minimum weeding, groudnut intercrop	2.7	16	2	5	0.14	0.9

TABLE 6.1 *(continued)*
Comparative Soil Erosion and Hydrogolic Data From "Conventional" and "Improved" Soil-management Treatments at Five ACIAR Projects Sites

Site and Treatment	Annual Soil Loss (mm)	Runoff Coefficient (%)	% Suspended Load	% Suspended Load	Suspended Load to Streams Amount (t/ha/yr)	Suspended Load to Streams Concentration (kg/m³)
QUEENSLAND, AUSTRALIA						
Gympie, loamy sand						
Annual rainfall 1992/93 = 876 mm						
Slope angle = 10–12%, length = 36 m						
Conventional: pineapples planted on beds, and furrows constructed at slope less than maximum hillslope (furrow slope - 5%, length = 36m)	115	141	16	19	22	15.6
Crop: pineapples						
Improved: mulching of furrows	6	89	10	10	1.0	1.1

In spite of the steep slopes, landsliding was not perceived by farmers to be a major problem. The through flow lines provided pathways for subsurface water to move to toeslopes and exfiltrate. Occasional mass movements in large rainfall events shifted whole beds of plants en masse. Simply maintaining mulch on the surface was found to be inadequate. Farmers mulch down the harvested crop but because there are up to 400 t/ha of green, wet manure, two passes of machinery cultivates only the surface 10 cm. Heavy rainfall on this material causes all soil and plant residues to be transported from points of exfiltration into the valley floors. Tines shaped for mole-draining were fitted to deep ripping chisel plows and fields were deep ripped up- and down the hill twice before mulching. This technique was tested under a 2-day rainfall of 494 mm (94 mm/h maximum intensity) and proved that deep ripping and maintaining mulch would be effective. From this experience, deep ripping under each bed of plants to carry water into the lower valley was developed.

In a separate experiment set up to measure surface and subsurface flow down a whole hillside, two complete blocks of the farm layout and the associated roadways were encompassed. Design of the experiment afforded measurement of water from the upper convex slope, the rectilinear midslope, and the concave toeslope (Figure 6.2). While the blocks of pineapple plants were not exactly orthogonal to

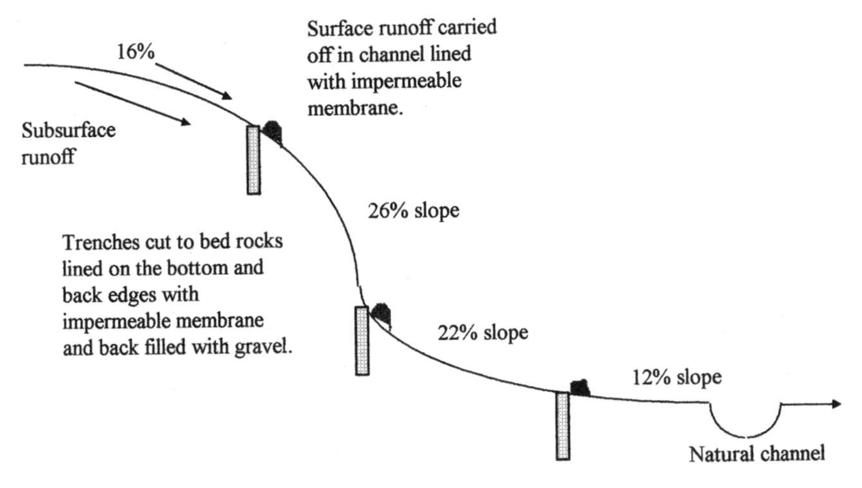

FIGURE 6.2a Hillslope cross section for measuring surface and subsurface flows, Imbil, Gympie, Queensland.

the landslope, results indicated that the preferred pathways of subsurface flow cut across the blocks (Figure 6.3).

Knowledge of this subsurface flow had a number of practical applications. Farmers use a cement slurry for constructing waterways down one side of blocks of pineapple plants. Exfiltration undermines the cement skin and the structures collapse. Second, roads between blocks of plants have a convex shape. These were changed to a concave shape so that water flowed down the middle of the road. The central part of the roadway was lined with green artificial turf and a stoloniferous grass, *Bothriochloia pertusa*, planted on the sideslopes. The central channel inter-

Enlarged Channel Section

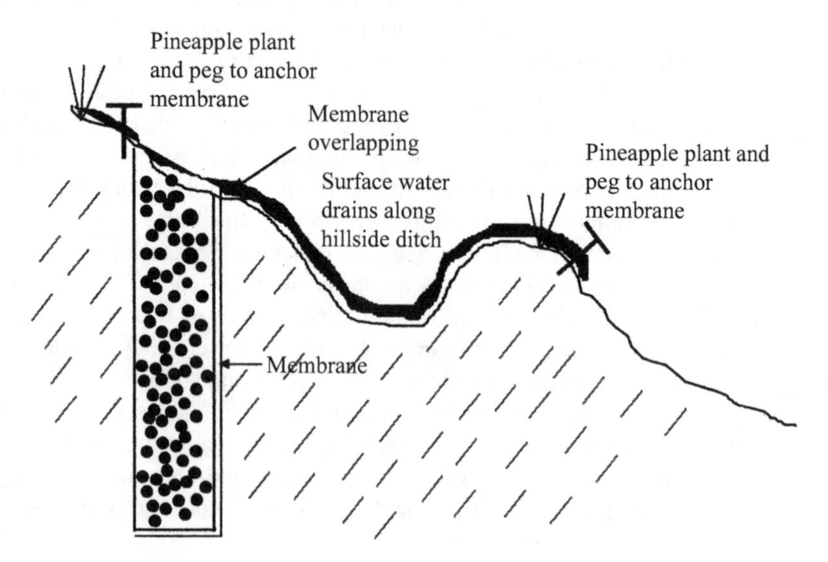

FIGURE 6.2b Cross section plan for measuring surface and subsurface flows.

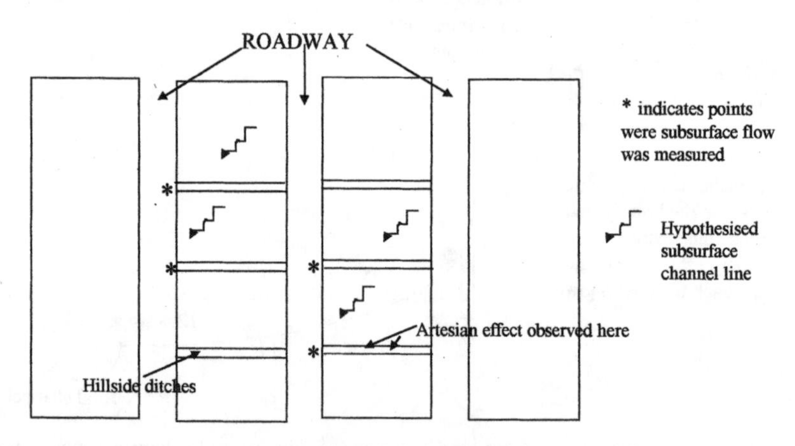

FIGURE 6.3 Plan view of pineapple blocks showing measurement of subsurface flow.

cepted some of the subsurface flow and the water exfiltrated up through the porous material.

There was an unexpected bonus from the concave-shaped roads in the form of tractor safety on steep slopes. On the stony convex-shaped roads, the tractor's front wheels were prone to sudden directional changes requiring the driver to exert considerable diligence. Driving was much easier on the concave-shaped roads because the road shape put most pressure on the outside edges of tires.

Ideally "best-practice" soil conservation layouts could benefit from knowing the likely locations of through-flow lines and points of exfiltration. Subsurface agricultural pipe could then be laid in the lower slopes to drain such points and facilitate trafficability in wet periods.

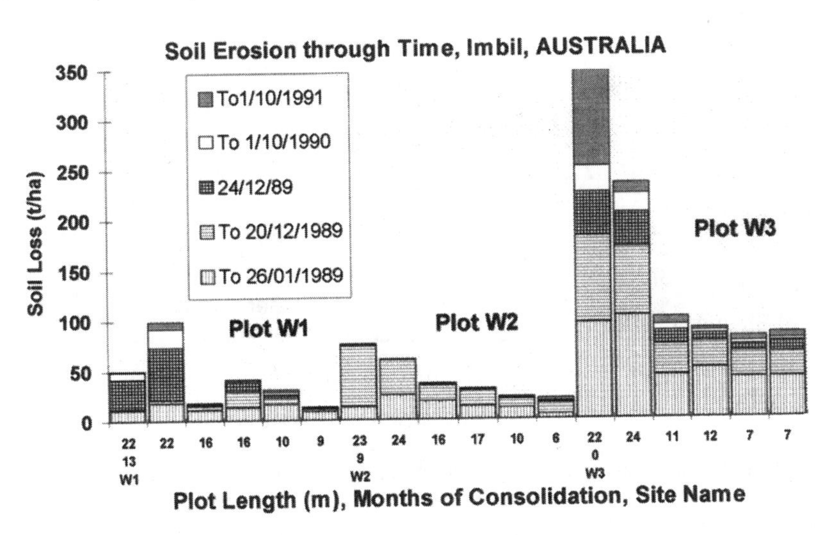

FIGURE 6.4 Soil losses on 35% sloping Lithic Eutropept classified into erosion phases.

Soil Consolidation

In this project two other blocks of pineapples were investigated in addition to the one shown in Figure 6.1 (Site W3). Planting had been 13 months (Site W1) and 10 months (Site W2) earlier on these blocks and results from the earlier planted sites showed less erosion had occurred. From Figure 6.4 it can be concluded that sites W1 and W2 lost less soil than W3, and it would be logical to conclude that vegetative cover was responsible.

A severe thunderstorm on December 24, 1989, however, rilled the longest rows in W1. Once the rills were cut, soil continued to be eroded irrespective of vegetative cover. Thus, it became obvious that soil consolidation by wetting and drying cycles in conjunction with stone armoring created a crusted layer that increased in strength through time and reduced erosion (Figure 6.5).

The temporal dimension of erosion is well illustrated by this study. The Lithic Europept when newly planted has its stone content thoroughly mixed through the finer fraction. The initial erosion events had an abundance of unconsolidated material available for movement and were characterized by high sediment concentrations (Table 6.2). At site W3 rills did not form throughout the first 4 months when there were some typical events. Sediment supply was hypothesized to have drowned rill initiation until the soil attained sufficient strength for the side walls of rills to remain upstanding (Table 6.2, rills formed in event 7). Once initiated, rills grew in size for 4 years. This was largely due to the particular sequence of rainfall events. Progressively throughout the project larger events followed each other until February 20–21, 1992, when 494 mm fell in 2 days. Rill geometry had expanded with each subsequent event and channel hydraulic characteristics and sediment sizes were adjusted to

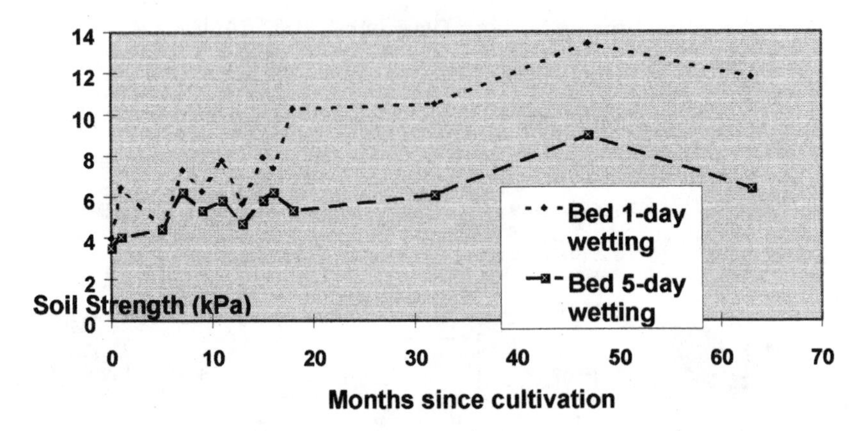

FIGURE 6.5 Soil consolidation by wetting and drying processes on a Lithic Eutropept.

withstand a very large event. Consequently, soil loss decreased to a very small suspended load during the last year of a 5-year evolutionary cycle when the rill sides became highly consolidated and too large to be influenced by mean annual events.

Farmers saw that by creating consolidated conditions in furrows using mechanical means, water could be shed from their fields and so reduce the incidence of root diseases. As a conservation adjunct, pineapple plants were grown at spacings of 8 to 12 m along the furrows after compacting the soil by tractor wheelings.

There is great scope to improve management on such steeplands with machinery. For example, plant residues could be inserted into deep ripped lines using the technology developed by Israeli agricultural engineers for disposing of cotton stubble. Machinery will undergo many new changes as it develops to meet the needs of sustainable farming on broadacres or steeplands. In this project, farmers coined the phrase, "Machinery is the key to conservation."

Mulch in a Conservation Farming System

Malaysia

A second catchphrase that farmers developed during the project was "Mulch is the lifeblood of tropical soils," and results from the trial sites in Malaysia strongly supported this point.

Hillsides were bounded to create minicatchments from crest lines to valley floors. The conventional treatment (T3) was cocoa grown with *Gliricidia*, as a shade tree, and herbicides used to control weed growth while Tl had bananas added as an interculture. The improved treatment, T2, simply included *Indigofera spicata* (a shade-tolerant, nonclimbing legume, but sometimes poisonous to stock) and natural grasses and herbs that were regularly slashed. Annual soil losses from the treatments are shown in Table 6.3.

TABLE 6.2
Summary of Major Erosion Events at the Imbil Experimental Site (1988–1991)

					7m Row				12m Row				22m Row			
Event #	Date	ΣP (mm)	Pmax (mm/h)	Cov (%)	Σ*Q* (mm)	*Q*p (mm/h)	AvC (kg/m³)	Beta	Σ*Q* (mm)	*Q*p (mm/h)	AvC (kg/m³)	Beta	Σ*Q* (kg/m³)	*Q*p (mm/h)	AvC (kg/m³)	Beta
1	26/10/88	25	94	5	6.9	3.9	121.4	1.455	7.8	37	118.1	1.328	6.9	58	265.8	1.238
2	11/12/88	50	110	10	31.6	81	41.1	0.869	34.3	81	31.4	0.906	34.7	93	57.3	0.888
3	19/12/88	205	62	10	142.2	61	0.9	0.092	162.8	40	6.9	0.579	163.9	60	12.2	0.635
4	27/12/88	19	62	10	8	41	18.1	0.616	8.5	50	14.2	0.665	8.7	50	45.6	0.811
5	30/12/88	28	94	12	12.4	60	31.8	0.855	14.5	52	29.9	0.889	14.4	74	89.1	0.961
6	06/01/89	24	77	12	8.6	39	14.4	0.427	11.4	27	11.1	0.641	11.2	35	22.8	0.753
7	24/01/89	28	110	15	12.1	61	61.8	0.977	13.8	81	60.8	1.082	12.6	75	136.9	1.05
8	04/03/89	124	46	20	31	28	2.7	0.095	44.1	26	1.7	0.109	38	28	4.2	0.548
9	01/04/89	185	126	22	120.2	103	6.1	0.389	112.2	96	7.4	0.447	108.2	112	25.9	0.588
10	06/04/89	318	110	22	354.4*	152*	3.3*	0.111*	228.1	97	3.2	0.269	210.4	102	11.1	0.468
11	26/04/89	248	143	30	238.8*	200*	3.8*	0.27*	138	143	8	0.459	116.8	130	34.7	0.731
12	18/08/89	84	46	40	23.7	21	0.3	-0.197	20.4	27	0.7	-0.047	24.9	22	9.1	0.438
13	02/11/89	63	62	50(1)	17.3	38	3.3	0.254	22.4	27	1.9	0.143	22.8	59	19.6	0.584
14	24/12/89	48	336	55(1)	33.8	336	20.5	0.596	33.7	338	24.5	0.591	28.9	338	160.3	0.864
15	16/01/90	77	94	60(2)	51.5*	43*	1.2*	0.039*	36.3*	47*	1*	0.009*	29.6	60	32.7	0.681
16	27/02/90	237	110	65(2)	378.9*	43*	0.6*	-0.17*	87.6*	37*	1*	-0.03*	75.6	45	19.5	0.455
17	27/03/30	176	61	70(3)	104.6*	18*	0.1*	-0.374*	36*	12*	0.6*	-0.104*	22.8	18	1.3	0.049
18	05/11/90	87	158	35(6)	42.6	71	4.6	0.401	36.4	114	1.9	0.16	39.4	111	33.2	0.605
19	07/02/91	54	37	35(7)	22.8	17	2.1	0.113	13.2*	12*	1.9*	0.158*	24.4	25	4.9	0.317
20	09/02/91	56	89	35(7)	33.4	34	1.8	-0.037	28.5	61	1.1	0.26	36.9*	76*	44.1*	0.665*
21	28/03/91	50	130	40(7)	22.8	68	10.9	0.536	25.4	88	6	0.378	39*	138*	130.3*	0.824*

ΣP = event precipitation; Pmax = maximum event precipitation; Σ*Q* = event runoff per unit area; *Q*p = event peak runoff rate per unit area; AvC = event average sediment concentratiion; Cov = areial vegetative cover (contact vegetative cover given in brackets); * = values influenced by throughflow; Beta = an erodibility parameter.

Figures for 1995–96 are incomplete, but bed load from T3 was greater than other sites. Visual assessment of T3 was that the whole catchment has crossed a threshold in 1996 and that erosion was degrading the catchment rapidly because preferred pathways of surface runoff changed into rills. These data support previous conclusions that there is a longer-term temporal effect of erosion which often cannot be found in short-term small plot studies. Tree growth supports the measured changes in erosion (Table 6.4).

Catchment T2 was chosen deliberately for the improved treatment because at the commencement of the project in 1989 it exhibited the poorest surface conditions with some "scalded," bare areas. It was hypothesized that if this catchment could be improved, then the management system would hold credence with farmers. During the period 1989–90, the cocoa trees were planted but growth in T2 was slowest, and when measurements commenced in 1991–92, it still lagged behind the weed-free eroding catchments. When the growth of the cocoa trees began to accel-

TABLE 6.3
Annual Soil Losses (t/ha), 0.1-ha Catchments

Water Year	T1	T2	T3
1989–90	26	12	2
1990–91	91	3	16
1991–92	77	7	34
1992–93	118	16	43
1993–94	51	24	17
1994–Jan 95	68	28	25
1995–96	Data Incomplete as yet		

erate in T2, quite large amounts of earthworms (350 to 450 g/event) were washed into the soil collection trough during prolonged rainfall events and worm castings became highly visible over the T2 catchment. Finally, in January, 1997, cocoa pod production from T2 surpassed the other catchments.

It has taken 8 years for serious degradation and rehabilitation to occur on adjacent areas under the east coast Malaysian environment. A similar time frame for regeneration of degraded cotton soils of Georgia, U.S., has been reported by Langdale et al. (1990).

Australia

At the Australian sites two experiments were carried out on a Typic Eutropept to assess sustainability. The site chosen for the Australian Centre for International Agricultural Research (ACIAR) project had grown two previous pineapple crops

when it was severely eroded at its most vulnerable time before planting. Soil erosion was estimated to be greater than 2000 t/ha from the largest ever recorded 2 days of

TABLE 6.4
Girth Growth Rates of Cocoa Trees (in cm/month)

Year	T1	T2
1991–92	0.28	0.17
1992–93	0.47	0.36
1993–94	0.28	0.48
1994–95	0.46	0.51
1995–96	0.49	0.49
1996–97 (10 months)	0.30	0.38

rainfall (624 mm). More than 5000 t of soil (about 25% of the eroded soil) were carried back onto the field from the deposits in the valley, but it was obvious that the finer fractions of the soil had been lost. A normal application of fertilizer was made to the field and to the crop throughout its growth. The crop yielded 111 t/ha and was regarded as normal for an August plant crop.

The second experiment involved clearing some virgin land and growing a plant crop of pineapples with a normal application of fertilizer and comparing it with adjacent land that had grown three previous crops. The aim was to quantify the effects of organic matter returned to the soil because it had already been concluded that application of the best fertilizer regimes doubled production. Farmers had noted, however, that yields had continued to increase although they were still applying the same amounts of fertilizer.

Chemical analysis of plant leaves showed that they were a large reservoir of nutrients because most nutrients are applied aqueously and therefore taken in through the leaves. As dry plant residues range between 60 and 100 t/ha, the amount of nutrients available to the plant via the organic matter was very significant, e.g., 1.5 t/ha of potash.

These large quantities of organic matter are returned to the soil once every 3.5 to 4 years following a ratoon crop. Farmers had originally raked the residue from fields and burned it believing that it was a source of disease. The common practice by farmers at the commencement of this project was to trash all residues as many times as possible in order to make it decompose quickly. There is now a realization that the plant residues are a major resource for sustainable production especially since EDB (ethelyene dibromide) is being withdrawn from use as a means of controlling harmful nematodes. In collaboration with an industry nematologist, it has been found that leaving thick plant residues on the surface and planting a combination of green manure crops, e.g., *Sorghum sudanence* (forage sorghum), *Brassicas*, *Aracheas*, *Panicum maximum*, and *Bracheria decumbeus* will largely control the problem. Thus, as often occurs, soil conservation is achieved because some other factor of immediate higher economic importance intervenes.

Farmers were especially skeptical of the use of mulch because of the fear of Phytophthora root rot due to increased soil moisture. Consequently, other mechanical measures were tried. Tied ridging proved to be the most promising because machinery is never driven in the rows after planting. It was found that the plants did not suffer root rot when the ponds filled with water because the young plants had a very poorly developed root system. The ponds trapped soil and almost filled with soil during the first 12 months.

The tied ridges were trialed in four blocks over a period of 5 years on the two soil types already mentioned and on slopes of 2.3, 2.5, 4.6, and 28% and soil losses ranged from 0.6 to 6 t/ha.

Results were strongly influenced by the degree of soil consolidation and the sequence in which erosive events occurred. Variations on this practice, such as rolling the tied ridges with an extra pass of the tractor, planting pineapples on every third or fifth ridge, and placing small amounts of pineapple mulch on the ridges, all reduced soil loss by preventing breaking of ridges.

At one site, 520 mm of rainfall occurred over 8 days and tied ridges failed. However, soil loss for the water year was still less than half that of the farmer's conventional layout. Generally, tied ridges were seen by farmers to be a useful tool especially because they trapped the organic matter that was nutrient rich.

At the end of 12 months, canopy cover largely protected the soil and erosion was by entrainment. Table 6.5 shows that in a 16-month-old pineapple crop, sediment concentrations under a 94% canopy cover are approximately 15 to 20% of a bare soil.

These results showed that the supply of sediment and its transport were highly dependent upon rainfall energy. Row lengths of 36 m on 3.5% slopes were too short for runoff to attain a stream power necessary to incise the natural consolidation of the deposited sediments.

As there are large quantities of dry matter at the farmer's disposal, various rates of application and its location were tried. Official trials using coarse and finely chopped mulch in the furrows were laid out on four farms and some farmers began their own trials. Reductions in soil loss using pineapple mulch made soil erosion a nonissue even when row lengths were increased to 67 m on convex-shaped hillslopes that reached a maximum slope of 35%.

A contour layout with 2 to 3% sloping furrows was used on slopes of up to 10% to assess the effects of mulch. Soil eroded from the sides of beds gradually covered the 8 t/ha of mulch in the furrows. At the finish of the crop little surface mulch remained, but soil erosion was still insignificant.

Assistance from the pineapple industry horticulturalists has done much to allay fear of root rot, and other mulch treatments were devised. Mulch was used in conjunction with furrow consolidation, and in the last 8 to 10 m of rows on very steep slopes. All proved to be highly effective.

The real challenge with mulch has been to develop machinery that can handle the large quantities under a range of climatic conditions. New concepts in machinery are being investigated because the advantages of mulch farming are expected to be the cornerstone of sustainable production.

The Scale Problem

Many studies have reported how vegetative cover has reduced runoff in arable farming.

TABLE 6.5
Effects of Conservation Treatments on Sediment Concentration in Pineapples (in kg/m³)

	Treatments					
	BV SITE (planted 4 months) Slope 3.3%			BB SITE (planted 16 months) Slope 5.5%		
Time	Conventional Canopy = 28%, Contact = 1%	Bare Cover = 0%	Plastic Mesh Canopy = 95%, Contact = 0%	Mulch Canopy = 94%, Contact = 60%	Conventional Canopy = 94%, Contact = 2%	Bare Cover = 0%
17:30.0						60.16
17:30.5						69.42
17:31.0						109.78
17:31.5	26.8	40.13	6.81			120.64
17:32	39.54	50.72	5.77		13.11	98.24
17:33	40.43	55.57	4.74		13.63	74.73
17:34	37.17	39.75	3.99			83.09
17:35	37.53	41.95	10.54		12.19	
17:36	40.31	43.07	7.41	7.39	22.84	69.55
17:37	28.76	51.25	6.11			
17:38	29.07	68.36	3.65		21.74	133.62
17:39	42.92	75.57	3.91	4.13	19.16	133.17
17:40	47.55	46.84	4.83	5.59	36.55	107.36
17:41	30.52	49.55	5.54	24.00	43.9	102.33
17:42	46.00	55.08	6.01			126.63
17:43						
17:44				9.51	25.53	106.31
17:45						
17:46				1.07	40.9	106.68
17:47						
17:48				2.38		
17:56				2.36	11.76	160.93
17:57						
17:58	38.81	39.58	6.87	2.43	21.52	127.44
17:59	29.3	27.87	10.9			
18:00	35.87	25.04	8.96		37.15	133.16
18:01	22.96	28.42	13.17			
18:02	37.94	36.37	11.67			
18:03	51.63	55.38	5.4			
18:04	49.53	24.61	3.33			
18:05	45.02	51.51	1.06			
18:06	31.02		0.68			
18:07	7.88		0.53			

In this study infiltration rates of the mulch plots remained high until the end of the crop, 3.5 years later. Mulch reduced runoff from 100 m² plots. Depth of runoff was approximately 30 to 50% of total runoff depth from that of 3000 m² blocks of pineapples for rainfall events of less than 50 mm. In larger events (i.e., >50 mm), more runoff occurred sometimes from the mulched plots but was lagged and depended upon rainfall intensity (Table 6.6). Toward the end of the experiment runoff from all plot treatments exceeded that from the larger blocks. These data highlight

TABLE 6.6
Comparison of Runoff Depths between Plots and Blocks of Pineapples, Gympie, Australia (1992–1995)

Date	Total Rainfall (mm)	Max Intensity (mm/h)	Runoff Plots (100 m²) Mulch (mm)	Conventional (mm)	Bare (mm)	Runoff Blocks 0.35 ha (mm)	0.29 ha (mm)
5/11/92	14.3	37	1.2	2.1	0.6	2.5	2.7
19/11/92	19.0	63	2.7	5.2	7.6	11.4	10.0
21/11/92	56.0	146	24.0	22.0	28.0	32.0	28.0
27/11/92[a]	14.0	116	6.6	5.4	7.7	10.0	8.0
6/12/92	31.0	128	11.0	13.0	19.0	27.0	28.0
7/2/92	56.0	86	6.0	15.0	24.0	31.0	27.0
22/2/93	27.0	106	8.0	11.0	16.0	18.0	16.0
11/3/93	35.0	139	6.0	11.0	19.0	16.0	17.0
2/6/93	33.0	67	4.0	9.0	19.0	9.0	8.0
28/7/93	19.0	99	1.5	5.6	12.0	8.0	9.0
9/11/93	27.0	125	3.4	7.0	16.5	9.5	9.5
24/12/93	52.0	86	12.8	13.7	27.7	13.4	13.5
26/12/93	53.0	179	30.3	27.8	40.0	33.0	35.4
9/1/94	23.0	61	1.0	2.6	7.3	1.9	2.3
20/1/94	25.0	67	2.8	3.5	10.0	3.1	26
3/2/94	29	11	1.1	1.7	3.3	0.75	1.0
27/3/94	32	48	12.6	12.9	14.2	2.6	2.7
5/4/94	48	48	15.0	16.0	19.0	1.5	3.5
6/4/94	28	48	15.0	150	15.0	6.2	6.0
1/5/94	20	46	0.4	0.8	6.0	1.2	0.75
11/7/94	24	67	4.2	9.6	16.3	3.8	3.4
9/9/94	16	73	0.1	1.9	5.3	1.0	4.3
4/11/94	24	132	4.0	7.1	14.2	5.0	4.8
21/11/94	14	79	0.3	2.1	5.2	1.7	1.4
4/12/94	83	67	17.0	24.0	34.0	9.0	10.0
14/2/95	79	79	6.0	12.7	27.0	19.0	22.0
15/2/95	196	36	105	105	128??	44.0	48.0
17/2/95	49.0	105	12.7	16.3	32.0	17.0	10.0

[a]The event on November 27, 1992 invovled a hailstorm.

the problem of extrapolating plot information to catchment scales and that significant temporal changes occur at larger scales.

The cropping enterprises studied in Malaysia and Australia were highly commercialized monocultures that were farmed on whole hillsides. Sediment delivery from whole hillsides was reduced by orders of magnitude (>100 t/ha to <1 t/ha) through improved practices. At one site in Australia, improved conservation practices were laid out on one side of a valley while conventional farmer management was used on the opposite hillside. The valley bottom was planted to *Cynodon nlemfuensis* (African star grass), a plant that is stoloniferous, and *Vetiveria zizaniodes* (vetiver grass), a stiff barrier grass, and sediment concentrations were measured at the top and bottom of the valley. Results concurred with the processes described by Trimble (1974). Within one summer the valley floor began to erode. Previously, the landholder has cemented a channel down the valley but this had been buried by sediment. Flows of cleaner water ($\approx$4 to 6 kg/m^3 compared with 36 to 260 kg/m^3 previously) removed the soil and stoloniferous grass exposing the old layer of cement. The rapidity of the channel adjustment surprised the farmers, but sediment concentrations down the valley were halved to $\approx$2 kg/m^3 through time. The rows of vetiver grass across the valley ponded flows, spread them wider, created deposition on the upstream side, and because of the stoloniferous habit of *C. nlemfuensis* protected soil downstream of the stiff barrier grass. The experiment also demonstrated that a large part of the nutrients was adsorbed onto the suspended sediment because measurements of electrical conductivity were also halved in the downstream direction. This experiment also demonstrated a principle for managing hedgerows on hillslides. When sediment is deposited on the upslope side of a barrier, there will be scouring on the downstream side unless the soil surface is protected.

THE SOCIAL DIMENSION AND CATCHMENT MANAGEMENT

Understanding the spatial and temporal aspects of hydrology and soil erosion processes is only a precursor to catchment management. The literature has reported that similar work has already been done. To make a difference to catchment management, this experimental work was carried out on farms where the owners/managers held credibility throughout the industry. Second, a high-profile field program was carried out so that neighboring farmers became curious about what was happening. Field work involved working the same hours as the farmers. When farmers became so curious that they began to come and look for themselves, field days were held for the wider group of district farmers. Farmers were then encouraged to help in work at the experimental sites. New experiments that farmers suggested were set up, and, finally, field days were held where groups of farmers helped on neighboring properties with some conservation work. Finally, recommendations on management practices were made by all the catchment stakeholders for sustainable pineapple production.

RECOMMENDATIONS FOR SUSTAINABLE PINEAPPLE PRODUCTION

The projects have coined two phrases:

- Mulch is the lifeblood of tropical soils.
- Machinery is the key to conservation.

All of the following recommendations are related to the above two simple propositions.

1. Controlling and managing runoff is the most important practice a farmer can implement. This calls for whole-farm layout plans based upon topography and soil types.
2. Soil surface roughness and depression storage should be maximized to lag any runoff during the early stages of the crop. (Again, pineapple residues are the best option where Phythopthera is not a severe problem. Tied ridges have proved to be an effective answer.)
3. Row lengths should be related to the combination of conservation measures (e.g., pineapple plants in the rows, tied ridges, and mulch. At slope angles of less than 8% row lengths of up to 150 m are permissable where adequate vegetative cover has been used in pineapples.)
4. Row lengths can be related to subsurface flows of water where adequate surface cover is available rather than slope angle.
5. Where an up-and-down slope layout is used, deep ripping under each bed before planting will drain subsurface flows adequately.
6. Roadways should be permanent where possible and concave in shape. Stoloniferous grasses such as *Cynodon nlemfuensis* (African star grass) and *Bothriochloia pertusa* (creeping Indian blue grass) planted on the roadsides will provide protection against erosion.
7. In the valley floors vetiver grass in conjunction with a stoloniferious grass will filter out more than half of the sediment.
8. Farm dams provide sediment sinks and impound runoff from small events, especially thunderstorms which produce the highest sediment concentrations.

The aim was to use the experimental areas for both research and extension so that farmers could understand the processes at work. Consequently, they began to develop their own conservation measures that fell outside of the conventional practices. However, the process still has some way to progress because there remained strong resistance to changing from monocultural practices to more diverse and integrated farming systems. The farmers wish to use all of their land more and more productively for one crop and changing to suitability classifications holds little credence with them because of the low economic returns and uncertainty in agriculture today.

ACKNOWLEDGMENTS

Work in this project was funded from the Australian Centre for International Agricultural Research (ACIAR), the former Queensland Department of Primary Industries, Golden Circle Limited (farmer cooperative for processing and marketing pineapple production in Queensland), Queensland Fruit and Vegetable Growers Association, the National Landcare Programme, and the Mary Valley Integrated Catchment Committee. We wish to thank the project team members, Emeritus Professor Calvin Rose, Dr. Keppel Coughlan, Mr. Banite Fentie, Mr. Tony Judd, and Mr. Ian Drever for advice and dedicated assistance.

REFERENCES

Abrahams, A.D. 1972. Drainage densities and sediment yields in Eastern Australia, *Aust. Geog. Stud.,* 10(1), 19–41.

Adamson, C.M. 1974. Some effects of soil conservation treatment on the hydrology of a small rural catchment at Wagga Wagga, *J. NSW Soil Conserv.,* 32(4):230–249.

Baird, R.W. 1964. Sediment yields from Blacklands watersheds, *Trans Am. Soc. Agric. Eng.* 7:454–456.

Barnett, A.P. and Rogers, J.S. 1966. Soil physical properties related to runoff and erosion from artificial rainfall, *Am. Soc. Agric. Eng. Trans.,* 9:123–125.

Baron, B.C., Pilgrim, D.H., and Cordery, I. 1980. Hydrological Relationships between Small and Large Catchments. AWRC Tech. Paper No. 54, Australia Government Printer, Canberra.

Ciesiolka, C.A.A. 1987. Catchment Management in the Nogoa Watershed, AWRC Research Project 80/128, Australia Government Printer, Canberra, 204.

Ciesiolka, C.A.A. and Freebairn, D.M. 1982. The influence of scale on runoff and erosion, in *Proc. Conf. Agric. Eng.,* Armidale, Institute of Engineers, Australia, Nat. Conf. Publ. 82/8, 203–206.

Dietrich, W.E., Reneau, S.L., and Wilson, C.J. 1987. Overview: "zero-order basins" and problems of drainage density, sediment transport and hillslope morphology, in *Erosion and Sedimentation in the Pacific Rim,* R.L. Beschta, T. Blinn, G.E. Grant, G.G. Ice, and F.J. Swanson, Eds., IAHS Publ. 165, 27–38.

Douglas, I. 1973. Rates of denudation in selected small catchments in Eastern Australia, Occasional Papers in Geography, No. 21, University of Hull, Lowgate Press, Hull, England, 127 pp.

Edwards, K. 1983. Preliminary analysis of runoff and soil loss from selected long term plots in Australia, in *A Perspective in Soil Erosion and Conservation,* El-Swaify, S.A., Moldenhauer, W.C., and Lo, A., Eds., Soil Conservation Society of America, Ankeny, IA, 472–479.

Fraser, K.I. 1995. The profitability of management options for farming dryland hardsetting red soils in New South Wales in Sealing, Crusting and hardsetting soils, in *Productivity and Conservation,* H.B. So, G.D. Smith, S.R. Raine, B.M. Schafer, and R.J. Loch, Eds., Australian Society of Soil Science, Inc. (Queensland Branch), University of Queensland, Brisbane.

Freebairn, D.M. and Wockner, G.H. 1986. A Study of Vertisols of the eastern Darling Downs, Queensland. I. The effect of surface conditions on soil movement within contour bay catchments. *Aust. J. Soil Res.,* 24:135–158.

Gilmour, D.A. and Bonell, M. 1977. Streamflow generation processes in a tropical rainforest catchment — a preliminary study, in Hydrology Symposium, Institute of Engineers, Australia, Brisbane, 178–179.

Kirkby, M.J. and Chorley, R.J. 1967. Throughflow, overland flow and erosion, *Bull. Int. Assoc. Sci. Hydrol.,* 12:5–21.

Lal, R. 1976. Soil erosion problems on an Alfisol in Western Nigeria and their control, *I.I.T.A. Monogr.,* 1:208.

Langdale, G.W., Wilson, R.L., Jr. and Bruce, R.R. 1990. Cropping frequencies to sustain long terrn conservation tillage systems, *Soil Sci. Soc. Am. J.,* 54:193–198.

Loch, R.J. 1984. Field rainfall simulator studies on two clay soils of the Darling Downs, Queensland — an evaluation of current methods for deriving soil erodibilities (K factors), *Aust. J. Soil Res.,* 22:401–412.

Loch, R.J. and Donnollan, T.E. 1983. Field rainfall simulator studies on two clay soils of the Darling Downs, Queensland. II Aggregate breakdown, soil properties and soil erodibility, *Aust. J. Soil Res.,* 21:47–58.

Loch. R.J., Thomas, E.C., and Donnollan, T.E. 1987. Interflow in a tilled, cracking clay soil under simulated rain, *Soil Technol. Res.,* 9:45–63.

Loughran, R.J. 1984. Studies of suspended sediment transport in Australia drainage basins — a review, in *Drainage Basin Erosion and Sedimentation,* R.J. Loughran, Ed., University of Newcastle, New South Wales, 139–146.

Meyer, L.D. and McCune, D.L. 1958. Rainfall simulator for runoff plots, *Agric. Eng.,* 39:644–648.

Morin, J. and Kosovysky, A. 1995. The surface infiltration model, *J. Soil Water Conserv. Soc. Am.,* 50(5):470–476.

Oliver, L.J. and Rieger, W.A. 1986. Low Australian sediment yields — a question of inefficient sediment delivery, in IAHS Publ. 159, R.F. Hadley, Ed., Wallingford, U.K., 355–367.

Phillips, J.D. 1986. Sediment storage, sediment yields and time scales in landscape denudation studies, *Geogr. Anal.,* 18:161–167.

Pilgrim, D.H. and Huff, D.D. 1978. A field evaluation of subsurface and surface runoff. I. Tracer studies, *J. Hydrol.,* 3(8):299–318.

Romkens, M.J.M. 1985. The soil erodibility factor, a perspective, in *Soil Erosion and Conservation,* El-Swaify, S.A., Moldenhauer, W.C., and Lo, A., Eds., Soil Conservation Society of America, Ankeny, IA, 445–461.

Sallaway, M.M., Yule, D.F., Mayer, D. and Burger, P.W. 1990. Effects of surface management on the hydrology of a Vertisol in semi-arid, Australia, *Soil Tillage Res.,* 15:227–245.

Sheng, T.C. 1990. Runoff plots and erosion phenomena on tropical steeplands, in *Research Needs and Applications to Reduce Erosion and Sedimentation in Tropical Steeplands,* R.R. Ziemer, C.L. O'Loughlin, and L.S. Hamilton, Eds., IAHS Publ., Wallingford, U.K., 192, 154–165.

Soons, J.M. and Rainer, J.N. 1968. Micro-climate and erosion processes in the Southern Alps, New Zealand, *Geogr. Ann.,* 50A:1–15.

Trimble, S.W. 1974. Man-Induced Soil Erosion on the Southern Piedmont, 1700–1970, Soil Conservation Society of America, Ankeny, IA.

Trimble, S.W. 1977. The fallacy of stream equilibrium in contemporary denudation studies, *Am. J. Sci.,* 277:876–887.

Vita Finzi, D. 1977. *Chronicling Soil Erosion in Conservation in Practice,* A. Warren and F.B. Goldsmith, Eds., John Wiley & Sons, London, 267–278.

Wischmeier, W.H. 1966. Relation of field plot runoff to management and physical factors, *Soil Sci. Soc. Am. Proc.,* 30:272–277.

Woods, R., Swapalan, M., and Duncom, M. 1995. Investigating the representative elementary area concept: an approach based on field data, in *Scale Issues in Hydrological Modelling* (Advances in Hydrological Processes), J.D. Kalma and M. Swapalan, Eds., John Wiley & Sons, New York.

Zingg, R.W. 1940. Degree and length of landslope as it affects soil loss in runoff, *Agric. Eng.,* 21:59–64.

7 Watershed Characteristics and Management Effects on Dissolved Load in Water Runoff: Case Studies from Western Nigeria

Rattan Lal

CONTENTS

INTRODUCTION

Increase in the use of chemicals for enhancing agricultural productivity is inevitable. Demands of the world population, about 6 billion in 1998 and projected to level off at 11 billion by the year 2100, for food and other basic necessities cannot be met without the judicious use of chemicals. Global food security can be achieved through agricultural intensification on prime lands so that marginal lands can be reverted back to natural ecosystems, and used for wildlife habitat, recreation, and other nonagricultural uses. Agricultural intensification involves use of recommended agricultural practices to produce the optimum crop yields for the ecoregion.

0-8493-0702-3/00/$0.00+$.50

The use of agricultural chemicals, however, is a mixed blessing. While enhancing productivity, it also increases risks of environmental pollution. These risks are especially high for contamination and eutrophication of surface waters and pollution of groundwater. The desirable strategy is to identify management systems that minimize the risks of environmental pollution, and understand the fate of applied chemicals as they move through the soil–plant–water–air components of the ecosystem.

There are few data from long-term experiments in the tropics that have monitored the transport of agricultural chemicals in runoff and seepage water. Methods of deforestation (Lal, 1997), tillage methods and cropping systems (Lal, 1976; 1986), agroforestry techniques (Lal, 1989a,b) affect the amount and rate of transport of plant nutrients in runoff (Lal, 1997a,b). In fact, nutrient loss in runoff is a major cause of fertility depletion of soils of the tropics (Barnett et al., 1972; Lal, 1994).

Although the demand for agricultural chemicals is rapidly increasing in the tropics, neither is the effect of chemical applications on water quality known nor have water quality standards been established for safe use by the human and animal populations. International water quality standards (Van der Leeden et al., 1990) have not been validated for tropical regions, especially in relation to the impact of agricultural practices, yet the quality of surface waters is extremely important in the tropics because a large proportion of the rural population depends on surface water without further treatment.

Dissolved load in surface runoff may be influenced by similar watershed characteristics that also affect runoff rate and amount (Figure 7.1). Important among

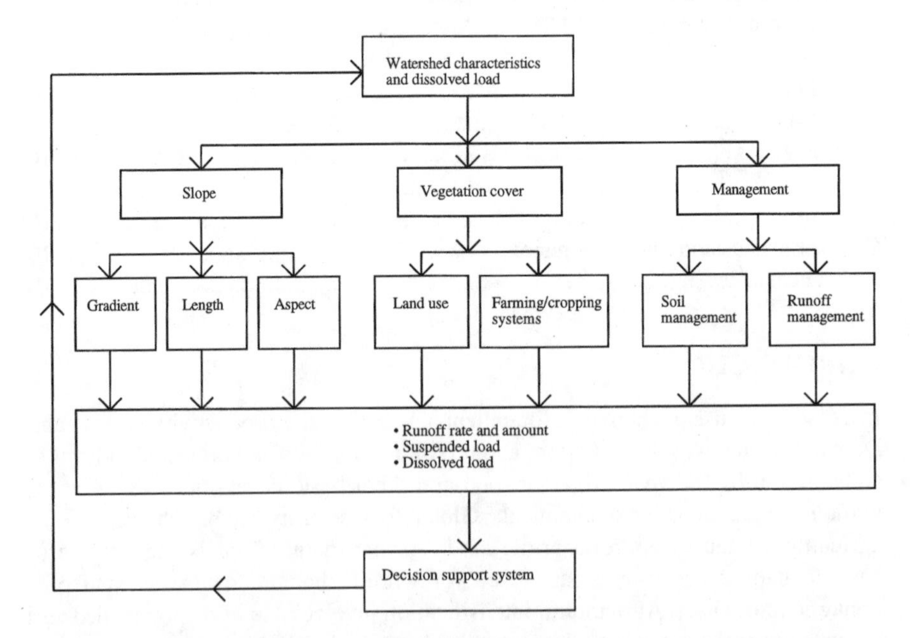

FIGURE 7.1 Watershed and management characteristics that affect the transport of suspended and dissolved loads.

these are slope, vegetal cover, and management. Slope gradient affects runoff and soil erosion through its impact on runoff velocity and its carrying capacity. Increase in slope four times may increase velocity of runoff by about two times, eroding power (detachment of soil) four times, transport capacity 32 times, and size of the material carried 64 times. Slope length may also affect total runoff volume. Ground cover is an important factor, because runoff and soil erosion decrease exponentially with increase in ground cover. The latter is determined by land use and farming/cropping systems. Soil management affects soil structure and infiltration capacity and root system development. Tillage methods and residue management involve the use of engineering techniques, e.g., terraces, waterways, drop structures, gabiens, etc. Site-specific technologies can be identified and locally validated to enhance agricultural productivity while minimizing the risks of soil and environmental degradation.

There is a conspicuous lack of scientific data from well-designed and properly equipped experiments quantifying the impact of watershed characteristics and management systems on the quality of surface water in the tropics. Toward this goal, a series of long-term experiments were conducted in western Nigeria from 1970 to 1988. The objective of this chapter is to collate and synthesize available data on watershed characteristics and management effects on transport of dissolved load.

MATERIALS AND METHODS

This report summarizes the data from three specific experiments. These experiments involved the study of (1) slope gradient on water quality, (2) slope length effects on water quality, and (3) agroforestry and tillage (soil and crop management) effects on water quality. Measurements of nutrient transport in runoff were made from 1970 through 1988 at the experimental farm of the International Institute of Tropical Agriculture (IITA). The IITA is located approximately 30 km south of the northern limit of the lowland rain forest zone roughly between longitudes 3° and 6° E and latitudes 6° and 8° N. This secondary rain forest ecoregion has two distinct growing seasons. The first from mid-March to mid-July ending in a short dry spell of approximately 1 month, and the second from mid-August to early November. The first growing season is characterized by intense tropical rains.

Experimental design of field runoff plots for slope gradient effects on runoff erosion and dissolved and suspend loads were described by Lal (1976). Similarly, experimental setups to study the impact of slope length on fluvial processes have been explained by Lal (1997c), and agroforesty and management effects on dissolved and suspend loads in surface runoff were presented in earlier reports (Lal, 1996, 1997a,b).

Six field runoff plots were established to evaluate the impact of agroforestry and tillage systems on runoff, soil erosion, and transport of nutrients in runoff (Lal, 1989a; b). Six management treatments consisted of vegetative hedges of *Leucaena lencocephala* and *Gliricidia sepium* established on the contour at 4- and 2-m intervals, and no-till, and plow-till systems of seedbed preparation. Test crops grown were maize (*Zea mays*) during the first season and cowpeas (*Vigna unguiculata*) during the second. Hedges of *Leucaena* and *Gliricidia* were pruned periodically to

FIGURE 7.2 The collection system for runoff plots established to assess agroforestry and tillage effects on runoff and soil erosion.

minimize the shading and competition for water and nutrients. These prunings were uniformly distributed on the soil surface as mulch within the rows of maize and cowpeas. Maize received chemical fertilizers at the rate of 100 kg N/ha as urea (⅓ at sowing and ⅔ at 6 weeks after), 30 kg/ha of P as single superphosphate, and 30 kg/ha of K as muriate of potash. Cowpeas grown in the second season received no chemical fertilizers.

All plots were equipped with H-Flume, a water stage recorder, and a runoff collection system (Figure 7.2). Runoff samples were collected after every erosive event for measuring sediment concentration and for analyzing composition of the dissolved load. The samples were stored at 4°C ending chemical analysis. Complete chemical analyses were performed with a Perkin-Elmer atomic absorption spectrophotometer (Page, 1982). Phosphorus was measured colorimetrically using the molybdic acid blue method. The NO_3–N was also measured colorimetrically using brucine (Greweling and Peech, 1965). Mean monthly nutrient concentrations were computed for Ca, Mg, K, NO_3–N and PO_4–P for 1983 through 1985. pH of the runoff samples was also measured. Mean values thus computed were not corrected for the total runoff volume, which varied among treatments. The main emphasis of this report is on concentration of NO_3–N and PO_4–P in surface runoff. High concentrations of these nutrient elements affect human health and eutrophication of natural waters.

RESULTS AND DISCUSSION

SLOPE GRADIENT AND DISSOLVED LOAD IN RUNOFF

The data in Figure 7.3 show the effects of slope gradient (from 1 to 15%) of field runoff plots on concentration of NO_3–N, PO_4–P, K, Ca, and Mg in surface runoff. It is apparent from the data that the concentration of NO_3–N was only slightly

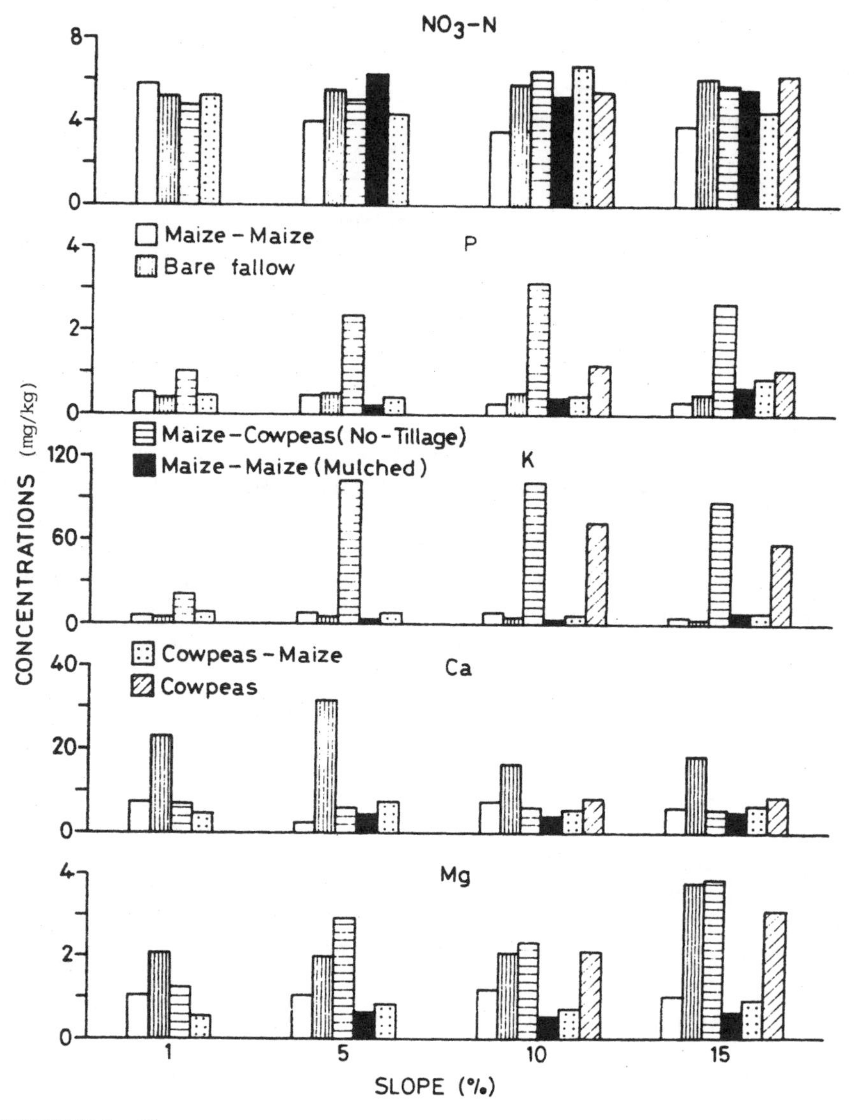

FIGURE 7.3 Slope gradient effects on nutrient concentration in water runoff. (From Lal, R., *IITA Monogr.* 1, 1976.)

affected by slope gradient. Further, NO_3–N concentration ranged between 4 and 6 ppm. In contrast to NO_3, the concentration of P was low in 1% compared with steep-slope gradients. Similar trends with regard to slope gradient were observed in the concentration of other elements. Differences in response of the concentration of

TABLE 7.1
Slope Length Effects on Concentrations of NO_3–P and PO_4–P in Runoff (mg/kg)

Slope Length (m)	PO_4–P						NO_3–N					
	1984	1985	1986	1987	1988	Mean	1984	1985	1986	1987	1988	Mean
60	0.31	0.25	0.40	0.85	0.30	0.42 ± 0.25	2.59	1.80	2.67	2.74	1.02	2.16 ± 0.74
50	0.27	0.22	0.40	1.14	0.30	0.47 ± 0.38	1.75	2.57	1.96	3.87	1.72	2.37 ± 0.90
40	0.37	0.19	0.36	1.39	0.30	0.52 ± 0.49	2.02	2.08	2.05	1.69	0.95	1.76 ± 0.48
30	0.59	0.17	0.40	2.00	0.24	0.68 ± 0.76	2.18	1.58	2.28	1.37	0.96	1.67 ± 0.56
20	0.36	0.16	0.20	1.15	0.18	0.41 ± 0.42	2.18	2.10	3.19	2.02	0.70	2.04 ± 0.89
10	0.28	0.17	0.18	1.37	0.19	0.44 ± 0.52	1.56	1.84	1.67	1.47	1.00	1.51 ± 0.32

Recalculated from Lal, 1997c.

NO_3–N to slope gradient and that of P, K, Ca, and Mg are due to different mechanisms of transport. While NO_3–N is primarily transported as dissolved load, other elements (e.g., P, K, Ca, and Mg) are primarily sediment-borne. Further, the sediment-carrying capacity of runoff increases with increase in slope gradient. It is likely, therefore, that concentrations of P, K, Ca, and Mg in runoff may also increase with increase in slope gradient up to a certain limit. The data in Figure 7.3 support these conclusions. Irrespective of the effect of slope gradient, however, elemental concentrations in runoff were low.

Slope Length Effects on Dissolved Load

Effect of slope length, ranging from 10 to 60 m on about 9% gradient, on concentrations of PO_4–P and NO_3–N are shown in Table 7.1. The data indicate two important points. First, concentrations of both PO_4–P and NO_3–N were extremely low. For 4 out of 5 years, the concentration of PO_4–P in runoff water was <0.59 mg/kg. Only in 1 out of 5 years was the concentration as high as 2 mg/kg. The average concentration for 5 years ranged from 0.41 to 0.68 mg/kg (see Table 7.1).

Effects of Agroforestry and Tillage Methods

The land-use and farming/cropping system may have a strong impact on quality of runoff water. Agroforestry systems and tillage methods are important factors that can strongly influence the concentration and nature of the dissolved load.

TABLE 7.2
Temporal Changes in Mean Monthly NO_3–N Concentration in Surface Runoff, mg/kg (ppm)

Time	*Leucena*		*Gliricidia*			
	4-m	2-m	4-m	2-m	No-till	Plow-till
1983						
May	2.1 ± 1.2	0.5 ± 0.2	1.7 ± 0.7	0.6 ± 0.3	—	0.4 ± 0.1
July	3.7 ± 3.9	1.3 ± 0.01	0.7 ± 0.4	1.2 ± 0.8	3.2 ± 3.6	2.4 ± 1.8
September	6.5 ± 4.9	8.5 ± 3.5	3.2 ± 2.5	3.2 ± 2.5	4.0 ± 1.4	6.0 ± 4.2
1984						
April	7.0	0.6	12.5	0.6	23.5	5.4
May	2.0 ± 1.6	1.8 ± 1.3	12.4 ± 11.6	0.7 ± 0.3	2.5 ± 1.1	0.7 ± 0.7
June	2.5 ± 2.2	2.4 ± 1.9	2.2 ± 1.9	1.3 ± 1.0	2.8 ± 1.3	1.7 ± 0.5
July	2.1 ± 1.6	1.2 ± 0.1	1.9 ± 0.4	1.3 ± 1.0	2.2 ± 0.2	0.3 ± 0
August	4.9 ± 4.6	3.0 ± 3.2	1.7 ± 1.0	0.9 ± 0.6	1.1 ± 0.8	3.2 ± 2.2
September	1.8 ± 1.4	2.3 ± 1.9	2.9 ± 4.2	0.7 ± 0.6	1.8 ± 1.3	7.1 ± 2.7
1985						
May	2.5 ± 1.5	6.0 ± 2.8	2.8 ± 2.6	4.6 ± 4.4	6.3 ± 3.4	—
June	4.7 ± 3.8	3.5 ± 0.5	3.0 ± 1.0	1.7 ± 0.7	4.1 ± 1.1	3.1 ± 3.4
July	23.6 ± 45.6	1.9 ± 0.8	4.4 ± 4.5	0.8 ± 0.4	1.1 ± 1.1	4.1 ± 3.4
August	2.4 ± 2.5	2.5 ± 1.3	4.3 ± 6.1	1.1 ± 0.8	3.0 ± 4.1	1.8 ± 1.1
September	2.1 ± 1.6	3.5 ± 1.2	4.0 ± 3.6	1.4 ± 0.5	0.9 ± 0.6	1.6 ± 1.7

NO_3–N

The data in Table 7.2 show tillage and agroforestry effects in temporal changes in concentrations of NO_3–N in runoff water. The data indicate several points of importance in relation to the water quality.

1. The concentration of NO_3–N was low, and mostly less than 10 mg/kg. The mean monthly concentration exceeded 10 mg/kg for only 3 of the 84 treatment months of analyses. The high concentration of 23.6 mg/kg was observed for *Leucaena* 4-m treatment in July 1985, 12.4 mg/kg for *Gliricidia* 4-m in May 1984, and 23.5 mg/kg for no-till in April 1984.
2. High concentrations were observed in *Leucaena*, *Gliricidia*, and no-till treatments. Although runoff amounts were low in all these treatments (Lal, 1989b), high NO_3–N concentrations were partly due to surface applications of prunings in the case of agroforestry treatments and of the fertilizer in the case of the no-till treatment. Although the total amount of NO_3 lost was low, the NO_3–N concentration in the small runoff volume was seemingly high.
3. The NO_3–N concentrations in runoff were low during the second compared with that in the first season.

4. The concentration of NO_3–N was mostly within the acceptable range for human and animal consumption.

TABLE 7.3
Temporal Changes in Mean Monthly PO_4–P Concentrations in Surface Runoff, mg/kg (ppm)

	Leucaena		*Gliricidia*			
Time	**4-m**	**2-m**	**4-m**	**2-m**	**No-till**	**Plow-till**
1983						
May	0.4 ± 0.3	0.4 ± 0.4	0.4 ± 0.1	0.6 ± 0.3	—	0.4 ± 0.6
July	2.1 ± 0.1	—	2.1 ± 0.1	2.2	2.1 ± 0.1	2.0
September	0.1 ± 0.1	0.2 ± 0.04	0.1 ± 0.1	0.2 ± 0.2	0.01	0.04 ± 0.01
1984						
April	0.15	0.15	0.2	0.15	0.2	0.15
May	0.7 ± 1.0	0.5 ± 0.7	0.5 ± 0.4	1.3 ± 2.3	0.6 ± 0.9	1.2 ± 1.6
June	0.1 ± 0.1	0.1 ± 0.1	0.1 ± 0.03	0.2 ± 0.1	0.1 ± 0.1	0.2 ± 0.1
July	0.2 ± 0.1	0.06 ± 0.06	0.2 ± 0.2	0.3 ± 0.01	0.04 ± 0.02	0.01
August	0.2 ± 0.2	0.07 ± 0.06	0.1 ± 0.1	0.1 ± 0.1	0.1 ± 0.1	0.1 ± 0.03
September	0.1 ± 0.1	0.07 ± 0.07	0.1 ± 0.1	0.1 ± 0.1	0.7 ± 1.1	0.09 ± 0.07
1985						
May	0.1 ± 0	0.09 ± 0.01	0.1 ± 0	0.36 ± 0.46	0.1 ± 0	—
June	0.1 ± 0.09	0.02 ± 0	0.09 ± 0.08	0.1 ± 0.06	0.2 ± 0.1	0.1 ± 0.1
July	0.03 ± 0.04	0.02 ± 0.03	0.02 ± 0.03	0.02 ± 0.03	0.05 ± 0.07	0.03 ± 0.05
August	0.1 ± 0.1	0.07 ± 0.1	0.06 ± 0.09	0.08 ± 0.12	0.06 ± 0.08	0.09 ± 0.14
September	0.09 ± 0.11	0.07 ± 0.07	0.09 ± 0.11	0.05 ± 0.90	0.06 ± 0.06	0.08 ± 0.08

PO_4–P

The data in Table 7.3 show temporal changes in concentrations of PO_4–P. The data indicate the following:

1. The concentration of PO_4–P in runoff was extremely low. With the exception of July 1983 and two analyses in May 1984, the concentration of PO_4–P was <1 mg/kg. Similar results have been reported in other reports for Alfisols in western Nigeria (Lal, 1976; 1997a,b).
2. Because of the low overall concentration, there were no clear trends in concentration of PO_4–P with regard to the treatments imposed.
3. Similar to the lack of the treatment effect, there were also no trends in temporal changes in PO_4–P concentration during the season. While the fertilizer (single superphosphate) was applied in April each year, the concentration of P in runoff was low throughout the season. Expectedly, the concentration was also low during the second growing season.

4. The low level of P concentration in runoff water from plots growing maize and cowpeas is not serious enough to warrant the risks of eutrophication of surface waters.

TABLE 7.4
Temporal Changes in Mean Monthly Ca Concentration in Runoff, mg/kg (ppm)

	Leucaena		*Gliricidia*			
Time	**4-m**	**2-m**	**4-m**	**2-m**	**No-till**	**Plow-till**
1983						
May	2.8 ± 2.7	3.5 ± 3.0	3.7 ± 1.6	4.8 ± 3.7	—	13.0 ± 23.6
July	7.6 ± 6.4	6.9 ± 4.7	4.2 ± 1.8	5.3 ± 3.8	8.8 ± 6.7	4.4 ± 3.2
September	6.8 ± 1.8	7.6 ± 1.5	4.7 ± 2.7	4.8 ± 0.4	8.5 ± 2.1	8.0 ± 0.8
1984						
April	11.8	12.0	12.8	9.4	10.5	10.1
May	5.2 ± 1.9	6.2 ± 2.3	13.3 ± 6.7	4.9 ± 1.8	7.1 ± 1.7	5.0 ± 2.1
June	7.7 ± 4.5	7.3 ± 3.0	9.9 ± 3.6	5.2 ± 1.7	9.7 ± 1.3	8.5 ± 2.7
July	8.7 ± 3.3	9.0 ± 2.1	10.8 ± 3.1	5.5 ± 1.8	10.4 ± 0.6	12.7
August	9.5 ± 4.6	8.7 ± 5.6	8.2 ± 3.5	4.5 ± 1.4	7.1 ± 3.8	9.8 ± 3.8
September	6.4 ± 1.2	7.7 ± 1.9	8.2 ± 3.0	5.2 ± 2.2	7.2 ± 0.9	10.7 ± 3.1
1985						
May	7.5 ± 2.5	12.9 ± 6.2	6.2 ± 2.6	10.9 ± 11.7	14.9 ± 4.0	—
June	9.9 ± 4.9	8.7 ± 5.7	8.8 ± 4.4	9.2 ± 6.1	11.4 ± 9.0	4.4 ± 2.0
July	9.6 ± 4.7	8.0 ± 4.9	12.9 ± 4.7	6.4 ± 2.8	8.8 ± 3.0	7.2 ± 2.9
August	29.1 ± 17.6	22.5 ± 8.0	22.5 ± 7.3	13.0 ± 8.1	31.1 ± 20.0	18.6 ± 7.4
September	12.4 ± 9.3	12.7 ± 4.2	27.6 ± 24.5	8.4 ± 2.8	9.2 ± 1.5	8.7 ± 8.3

Ca

The concentration of soluble Ca in runoff water is shown by the data in Table 7.4. The data support the following observations:

1. In contrast to those of NO_3–N and PO_4–P, the concentration of Ca in runoff was high. It ranged from 3 to 13 mg/kg in 1983, 5 to 13 mg/kg in 1984, and 4 to 31 mg/kg in 1984.
2. The highest concentration of 31.1 mg/kg in August 1984 was observed in the no-till treatment. Since the runoff amount was low (Lal, 1989b), the high concentration did not imply high total loss of Ca from the plot, and the high concentration may be due to the surface application of fertilizers and presence of crop residue mulch.
3. The concentration of Ca was high in August 1988 for all treatments, which may be due to several erosive events experienced during the period. The loss of cations in runoff may depend on the sediment loss.

4. There were no distinct treatment effects or temporal trends in Ca concentration in the runoff.

Mg

Tillage and agroforestry effects on concentration of Mg in runoff are shown by the data in Table 7.5. In contrast to Ca, the concentration of Mg was low and generally

TABLE 7.5
Temporal Changes in Mean Monthly Mg Concentration in Surface Runoff, mg/kg (ppm)

	Leucaena		*Gliricidia*			
Time	4-m	2-m	4-m	2-m	No-till	Plow-till
1983						
May	1.7 ± 1.5	0.9 ± 0.3	1.3 ± 0.4	1.1 ± 0.4	—	1.1 ± 0.9
July	0.7 ± 0.2	0.5 ± 0	0.5 ± 0.1	0.5 ± 0.1	0.6 ± 0.3	0.8 ± 0.0
September	0.9 ± 0.3	1.3 ± 0.2	0.7 ± 0.2	0.8 ± 0.1	0.8 ± 0.1	1.7 ± 0.5
1984						
April	3.5	2.0	2.7	1.1	2.4	2.3
May	1.1 ± 0.5	1.3 ± 0.2	2.7 ± 1.3	0.9 ± 0.3	1.0 ± 0.2	0.9 ± 0.3
June	1.0 ± 0.5	1.1 ± 0.6	1.0 ± 0.6	0.7 ± 0.2	1.1 ± 0.4	1.2 ± 0.4
July	0.9 ± 0.2	1.1 ± 0.1	1.2 ± 0.2	0.7 ± 0.01	0.9 ± 0.1	1.4
August	1.0 ± 0.3	0.9 ± 0.3	0.8 ± 0.2	0.6 ± 0.1	0.6 ± 0.1	1.1 ± 0.1
September	0.9 ± 0.3	1.0 ± 0.4	1.1 ± 0.6	0.8 ± 0.08	0.8 ± 0.2	1.5 ± 0.5
1985						
May	0.9 ± 0.3	1.6 ± 0.1	0.9 ± 0.2	1.2 ± 0.8	1.3 ± 0.3	—
June	1.9 ± 1.4	1.5 ± 0.2	1.6 ± 0.4	1.4 ± 0.8	1.4 ± 0.3	1.3 ± 0.9
July	1.2 ± 0.7	1.1 ± 0.2	1.5 ± 0.8	0.9 ± 0.4	1.0 ± 0.5	1.5 ± 0.5
August	1.5 ± 1.6	1.3 ± 0.7	1.5 ± 1.5	0.7 ± 0.4	1.5 ± 1.1	0.9 ± 0.2
September	1.9 ± 1.7	1.5 ± 0.4	2.8 ± 2.5	1.0 ± 0.2	1.0 ± 0.2	1.1 ± 0.7

<2 mg/kg. An isolated storm in April 1984 caused the highest concentration of Mg in runoff ranging from 1.1 to 3.5 mg/kg. The data support the following observations:

1. The concentration ranged from 0.5 to 1.3 mg/kg in 1983, 0.6 to 3.5 mg/kg in 1984, and 0.9 to 2.8 mg/kg in 1985.
2. Three highest concentrations (3.5, 2.7, and 2.8 mg/kg) were observed in agroforestry treatments. Presence of prunings on the soil surface may have contributed to high concentrations of Mg in runoff.
3. The data for 1983 and 1984 showed declining trends in Mg concentration as the season progressed. The concentration was generally high from April through July and low during August and September. Generally high Mg concentrations in September 1985 were contradictory to this trend.

TABLE 7.6
Temporal Changes in Mean Monthly K Concentration in Surface Runoff, mg/kg (ppm)

	Leucaena		*Gliricidia*			
Time	**4-m**	**2-m**	**4-m**	**2-m**	**No-till**	**Plow-till**
1983						
May	18.9 ± 9.0	9.4 ± 0.9	15.5 ± 11.5	10.8 ± 1.4	—	17.4 ± 16.0
July	13.8 ± 5.3	15.4 ± 13.6	3.3 ± 1.6	6.8 ± 3.2	8.5 ± 2.1	7.2 ± 2.6
September	31.3 ± 23.0	18.8 ± 8.8	15.3 ± 13.8	7.1 ± 2.0	16.3 ± 5.3	8.5 ± 2.2
1984						
April	20	9.6	15	9.2	45	25
May	17.8 ± 14.0	14.0 ± 9.9	13.4 ± 8.5	9.4 ± 9.8	10.2 ± 3.6	5.7 ± 2.4
June	14.2 ± 10.2	9.7 ± 8.7	5.6 ± 2.0	4.3 ± 0.8	11.6 ± 4.7	4.6 ± 0.9
July	20.0 ± 7.1	10.1 ± 3.4	25.0 ± 7.1	7.9 ± 3.0	9.8 ± 0.4	9.9 ± 0
August	24.3 ± 17.0	5.1 ± 2.0	30.8 ± 57.4	4.8 ± 1.0	7.6 ± 1.5	8.8 ± 3.5
September	11.9 ± 5.5	5.5 ± 2.9	15.1 ± 21.7	3.7 ± 0.6	7.2 ± 2.0	8.6 ± 1.9
1985						
May	8.4 ± 2.4	13 ± 0	6.1 ± 2.4	7.5 ± 5.2	10.6 ± 4.1	—
June	14.5 ± 5.6	8.7 ± 1.3	7.4 ± 1.7	7.6 ± 1.8	9.9 ± 4.1	11.0 ± 6.6
July	8.3 ± 3.0	5.1 ± 0.9	7.4 ± 3.9	3.7 ± 1.2	4.8 ± 1.8	9.7 ± 5.6
August	8.4 ± 3.7	5.1 ± 1.7	3.7 ± 2.1	4.2 ± 3.1	5.0 ± 1.8	5.7 ± 3.1
September	7.2 ± 1.2	5.5 ± 1.0	5.0 ± 0.3	5.8 ± 1.6	4.2 ± 1.5	4.3 ± 0.8

K

Similar to Ca, the concentration of K in runoff was also high (Table 7.6). The highest concentration of K was 45 mg/kg. High concentrations reflect high K levels of these soils due to the presence of mica and other K-bearing minerals in the parent material. Similar trends of high concentrations of K in runoff have also been reported from other experiments (Lal, 1976; 1987a,b; 1989b). The data support the following observations:

1. The concentration of K in runoff ranged from 3 to 31 mg/kg in 1983, 5 to 45 mg/kg in 1984, and 4 to 15 mg/kg in 1985.
2. Mean monthly concentration was higher in 1983 and 1984 than in 1985.
3. The highest K concentration of 45 mg/kg was observed in the no-till treatment in April 1984. Surface application of muriate of potash may lead to high concentration of K despite the low runoff volume in no-till treatment.
4. High concentrations of 25 and 31 mg/kg were observed in *Leucaena* and *Gliricidia* 4-m treatments. Surface applications of prunings as mulch may lead to high concentrations of K in runoff.
5. Concentrations of K in runoff were relatively low in the plow-till compared with the no-till treatment, especially so for the events in 1985.
6. There were no distinct temporal trends in K concentrations in runoff within the season.

pH

The reaction of runoff water from all treatments was close to a neutral pH with a range from 6.4 to 7.7. The data presented in Table 7.7 support the following observations:

TABLE 7.7
Temporal Changes in Mean Monthy pH of Surface Runoff, mg/kg (ppm)

	Leucaena		*Gliricidia*			
Time	4-m	2-m	4-m	2-m	No-till	Plow-till
1983						
May	6.7 ± 0.2	6.5 ± 0.3	6.7 ± 0.2	6.6 ± 0.1	—	6.8 ± 0.1
July	6.5 ± 0.5	6.5 ± 0.5	6.4 ± 0.6	6.6 ± 0.4	6.4 ± 0.1	6.4 ± 0.6
September	7.4 ± 0	7.5 ± 0.1	7.4 ± 0.1	7.5 ± 0.1	7.4 ± 0.1	7.4 ± 0.1
1984						
April	6.9	6.8	6.9	6.8	6.8	6.8
May	6.7 ± 0.4	6.8 ± 0.4	6.8 ± 0.2	6.8 ± 0.5	6.9 ± 0.2	6.6 ± 0.6
June	6.8 ± 0.3	6.9 ± 0.2	6.7 ± 0.3	6.7 ± 0.4	6.8 ± 0.3	6.7 ± 0.2
July	7.0 ± 0.3	6.8 ± 0.4	7.1 ± 0.3	6.7 ± 0.6	7.0 ± 0.2	6.8 ± 0
August	7.1 ± 0.4	6.6 ± 0.2	6.9 ± 0.3	6.6 ± 0.2	6.7 ± 0.2	6.7 ± 0.1
September	7.2 ± 0.4	7.2 ± 0.4	7.4 ± 0.7	7.1 ± 0.3	7.2 ± 0.3	7.3 ± 0.5
1985						
May	6.5 ± 0.2	6.6 ± 0.1	6.6 ± 0.2	6.7 ± 0.2	6.5 ± 0.1	—
June	7.6 ± 0.5	7.6 ± 0.2	7.5 ± 0.3	7.4 ± 0.4	7.3 ± 0.3	7.2 ± 0.4
July	7.5 ± 0.4	7.3 ± 0.4	7.7 ± 0.4	7.3 ± 0.2	7.5 ± 0.5	7.1 ± 0.2
August	7.0 ± 0.1	7.0 ± 0.1	6.9 ± 0.3	6.8 ± 0.4	7.2 ± 0.5	6.8 ± 0.3
September	7.5 ± 0.7	7.0 ± 0.2	7.1 ± 0.4	6.9 ± 0.3	7.0 ± 0.3	7.3 ± 0.4

1. The pH of surface runoff ranged from 6.4 to 7.5 in 1983, 6.6 to 7.4 in 1984, and 6.5 to 7.7 in 1985.
2. In general, water pH was higher for 1985 than for 1983 and 1984.
3. There were no consistent trends in pH with regard to the treatment.

GENERAL DISCUSSION AND CONCULSIONS

Important watershed characteristics that affect the magnitude and quality of dissolved load in surface runoff are slope gradient and vegetal cover. Slope gradient affects the concenration of sediment-borne elements (e.g., P, K, Ca, Mg). Increase in slope gradient increases the concentration of such elements in water runoff.

Slope length had little effect on concentration of dissolved load in water runoff. Effects of slope length on soil erosion is also not well defined, and management factors may have a stronger effect on runoff and dissolved load than slope length per se.

Despite applications of fertilizers, residue mulch, and prunings, concentrations of plant nutrients were low in runoff from all treatments. The concentrations of NO_3–N and PO_4–P were low compared with the health standards established for human consumption.

Nutrient concentrations were low in all agroforestry and tillage treatments, despite surface applications of fertilizers in no-till treatments. Agroforestry and no-till systems were effective in controlling runoff and erosion and reducing nutrient concentration in runoff. In addition, actively growing vegetation (maize, cowpeas, *Leucaena*, *Gliricidia*) is an important mechanism for minimizing the leaching losses of plant nutrient elements in runoff.

The data presented support the conclusion that agricultural intensification on appropriate soils with recommended practices of soil management (conservation tillage with crop residue mulch, alley cropping with *Leucaena* or any other appropriate species) that reduce soil erosion risks are also effective in minimizing losses of plant nutrients in runoff.

REFERENCES

Barnett, A.P., Carreker, J.R., Abruna, F., Jackson, W.A., Dooley, A.E., and Holladay, J.H. 1972. Soil and nutrient losses in runoff with selected cropping treatments on tropical soils, *Agron. J.* 64: 391–395.

Greweling, T. and Peech, M. 1965. Chemical soil tests, *Cornell Univ. Agric. Exp. Stn. Bull.* 960.

Lal, R. 1976. Soil erosion problems on Alfisols in western Nigeria and their control, *IITA Monogr.* 1, 208 pp.

Lal, R. 1986. Soil surface management in the tropics for intensive land use and high and sustained production, *Adv. Soil Sci.* 5: 1–109.

Lal, R. 1989a. Agroforestry systems and soil surface management of a tropical Alfisol. I. Soil moisture and crop yield, *Agroforestry Syst.* 8: 7–29.

Lal, R. 1989b. Agroforestry systems and soil surface management of a tropical Alfisol. II. Water runoff, soil erosion and nutrient loss, *Agroforestry Syst.* 8: 97–111.

Lal, R. 1994. Water quality effects of tropical deforestation and farming systems on agricultural watersheds in western Nigeria, in R. Lal and B.A. Stewart, Eds. *Soil Processes and Water Quality,* Lewis Publishers, Boca Raton, FL, 273–301.

Lal, R. 1997a. Deforestation, tillage and cropping systems effects on seepage and runoff water quality from a Nigerian Alfisol, *Soil Tillage Res.* 41: 261–284.

Lal, R. 1997b. Deforestation effects on soil degradation and rehabilitation in western Nigeria. IV. hydrology and water quality. *Land Degradation Dev.* 8: 95–126.

Lal, R. 1997c. Soil degradative effects of slope length with variable area of Alfisols in western Nigeria. II. Soil chemical properties, plant nutrient loss and water quality, *Land Degradation Dev.* 8: 221–244.

Lal, R. 1996. Deforestation and land use effects on soil degredation and rehabilitation in Western Nigeria. III. Runoffs, soil erosion and nutrient loss, *Land Degredation Dev.* 7: 99–119.

Page, A.L. Ed. 1982. *Methods of Soil Analysis,* Part 2, *Chemical and Microbiological Properties,* Monograph 9, 2nd ed., American Society of Agronomy, Madison, WI, 1159 pp.

Van der Leeden, F., Troise, F.L., and Tod, D.K. 1990. *The Water Encyclopedia,* Lewis Publishers, Chelsea, MI, 808 pp.

Technological Options for Watershed Management

8 Sustaining a Rice–Wheat System in the Indo-Gangetic Plains: The Ecoregional Context

Inder P. Abrol, Surinder M. Virmani, and Roy K. Gupta

CONTENTS

INTRODUCTION

The Indo-Gangetic Plains (IGP) of South Asia spread over four countries, Bangladesh, India, Nepal, and Pakistan, and constitute the most important agricultural region of the subcontinent. The region experiences a range of climatic conditions from warm arid in Pakistan in the west to warm humid in Bangladesh in the east (Figure 8.1). Over the past three decades the region has emerged a major food producer — crop yields, particularly of rice and wheat, the staple food grains, have increased dramatically contributing greatly to the food security of one of the most populous regions of the world. Due to a number of factors, both biophysical and socioeconomic, however, productivity increases have not been uniform across the region. Over the past few years productivity growth of food grains even in the region that witnessed high growth in the 1970s and 1980s has slowed. This is cause of

0-8493-0702-3/00/$0.00+$.50

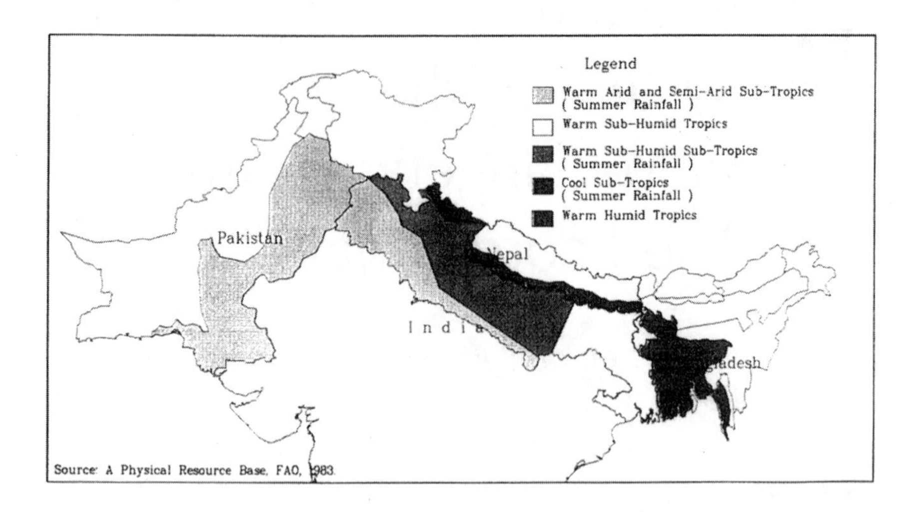

FIGURE 8.1 Regional Ecoregions in Rice-Wheat Belt of South Asia.

serious concern. The reasons for slowdown in productivity growth are varied and not fully understood. Soil-related constraints including nutrient depletion and imbalances appear an important factor. Continuous and intensive cereal cultivation has increased the incidence of pests, diseases, and weeds. Problems arising from changes in water and salt balances induced by the cropping and management strategies across the region are posing a serious threat to sustainability of the production system. This chapter examines rice–wheat sustainability issues of the Indian part of the plains from an ecoregional context.

THE INDO-GANGETIC PLAINS — INDIA

The Indian portion of the IGP extends from its border with Pakistan in the northwest to the state of West Bengal in the east covering the states of Punjab, Haryana, Uttar Pradesh, Bihar, West Bengal, and a small part of Himachal Pradesh. The plains cover an area of about 65 million ha (Table 8.1) which is one fifth the geographical area of the countries and extends over a length of nearly 2000 km along the foothills of the Himalayan mountains with a width ranging from 150 to 300 km, being narrow in the extreme east in West Bengal.

Physiographically, the plains have been distinguished into (1) Punjab Plains, (2) Upper Ganga Plains, (3) Middle Ganga Plains, and (4) the lower Ganga Plains. With an average elevation of about 150 m and ranging from almost sea level in the Bengal delta to nearly 300 m above sea level in the Punjab and Upper Ganga Plains near the foothills of the Shivaliks, the area is characterized by extremely flat plains. Geologically, the region is made up of alluvium brought down by the Himalayan rivers. The alluvial plains constitute one of the richest agricultural resources and a potential resource for groundwater. The aquifer systems are extensive, thick, hydraulically interconnected, and moderate to high yielding. However, quality of ground-

TABLE 8.1
Area of Some River Basins in the IGP by State

State	Area km² (× 1000)	% of State Area Covered by the River Basin of: Indus	Ganges
Punjab	50.4	100	—
Haryana	44.2	22	78
Uttar Pradesh	294.4	—	100
Bihar	173.9	—	83
West Bengal	87.8	—	81

water is an important consideration in the use of groundwater resources for agriculture. An assessment of groundwater quality for such parameters as total salt content, residual alkalinity, sodium absorption ratio, etc. are necessary, as the quality of the groundwater varies considerably. The salinity of the groundwater increases in the southwest direction together with the decrease in the rainfall amount suggesting that groundwater picks up salts as it passes from the recharge to the transition-discharge zones. In semiarid tropical areas in which the annual rainfall is above 600 to 650 mm, the salinity of groundwater tends to be low, with low to medium alkalinity. In the arid–semiarid zone with 400 to 650 mm annual precipitation, the groundwaters have a tendency to have medium to high residual alkalinity, while in the arid drier zones, a high level of salinity, consisting chiefly of neutral salts, is the main content of the groundwaters.

The IGP is served by two river systems, namely, the Indus and the Ganges. The major rivers of the Indus system are *Satluj*, *Ravi*, and *Beas*. These originate in the Himalayas and flow in the southwest direction. The rivers of the Ganges system (also originating from the Himalayas) include *Ganga*, *Yamuna*, *Gomati*, *Ghagar*, *Kosi*, and their tributaries. The rivers that originate in the plateau region and flow in the plain are *Chambal*, *Betwa*, *Son*, and *Ken*. The main Ganges river flows from the west to east direction up to east of Bihar where it takes a turn to the south and bifurcates into two main tributaries which start developing the Bengal basin. The basin extends southward by branching off the tributaries into a number of subtributaries which finally drain in the Bay of Bengal.

Along the northern margin of the plains lie two narrow but distinct strips, the *Bhabar* and the *Tarai*. The Bhabar, a piedmont plain 10 to 15 km wide, is composed of unassorted debris from the Himalayas. The surface streams disappear in this zone of boulders and sands. Immediately below the Bhabar is the 15 to 30 km wide relatively low-lying Tarai region characterized by fine sediments, natural forest cover, emergent and relatively ill-defined water channels, low gradients, and high water table ranging from a few to about 5 m below the ground resulting in swamps and marshes.

The plains, extending east–west through the valleys of the Indus and Ganga system, undergo a gradual transition in the level of land, climatic features, natural vegetation, cropping pattern, etc. Between the arid Rajasthan plains in the west and the perhumid lower Ganga plains in the east lie the semiarid Punjab plains. The subhumid upper Ganga plains and the humid middle Ganga plains each merges into the other imperceptibly. Traditional cropping in the subregion has been millets in the arid Rajasthan, wheat and chickpea in the semiarid subtropical but irrigated areas of the Indian states of Punjab and Haryana, wheat and rice in the subhumid but irrigated upper Ganga plain, rice and wheat or barley in the middle Ganga plain, and rice–jute in the humid lower Ganga plains.

CLIMATIC AND AGROECOREGION

There are large climatic variations across the IGP. The rainfall ranges from 300 mm in the semiarid climate in the west to over 2000 mm in the east. The summer and winter seasons are extreme in the west and relatively moderate in the eastern parts. Based on physiography, growing period, climate, and soils, Sehgal et al. (1990) has divided the IGP of India into five agroecoregions (AERs). Some broad dissipaters of these ecoregions are as follows:

AER No.	Climate	LGP (days)	Areas Covered	Annual Rainfall (mm)
2	Western plains, hot, arid	60–90	Southwestern Haryana and Punjab plains	<300
4	Northern plains, hot, semiarid	90–150	Central Punjab and Haryana and southwestern Uttar Pradesh plains	400–800
9	Northern plains, hot, subhumid	120–180	Northeastern Punjab and Haryana and northeastern, central Uttar Pradesh, southwestern Bihar plains	1000–1200
13	Eastern plains, hot, subhumid moist	180–210	Piedmont plains of central Uttar Pradesh, north Bihar and Avadh plains	1400–1600
15	Bengal plains, hot, subhumid to humid	240–270	Major parts of West Bengal	1600–2000

RICE–WHEAT CROPPING SYSTEM

Rice and wheat are the staple food accounting for some 75% of the total food grains and nearly 60% of the caloric intake by the people of India. In 1995, the area planted to rice and wheat crops was 45 and 25 million ha, respectively, out of India's gross cropped area of 183 million ha (net cultivated area 141 million ha). Nearly 25% of the area under rice and 40% of the area under wheat are currently cropped in rice–wheat sequence largely in the IGP region. In this system rice is grown in the *kharif* (rainy cropping) season followed by wheat in the *rabi* (winter postrainy cropping) season. Although, by definition rice and wheat crops are the system components, in practice the system is highly variable because the farmers commonly include a range of other crops in varying spatial or temporal sequences. This imparts the system a high degree of complexity.

TABLE 8.2
Yield and Production of Rice and Wheat in India

Year	Rice		Wheat		
	Yield (kg/ha)	Production (Mt)	Yield (kg/ha)	Production (Mt)	Relative Production of Wheat as % of Rice
1960–61	1013	35	851	11	32
1970–71	1123	42	1307	24	56
1980–81	1336	54	1630	36	68
1990–91	1740	74	2281	55	74
1995–96	1855	80	2493	63	79

Source: Directorate of Economics and Statistics (1997).
Mt = Million tons.

The area under rice–wheat cropping system has grown rapidly over the past three decades. In the period 1959 to 1962, the area in the rice–wheat cropping was estimated at 4.0 million ha which grew to 9.5 million ha by 1986 to 1989, i.e., at an annual growth rate of 3.2% (Hobbs and Morris, 1996). During this period per hectare yields of rice and wheat also increased significantly, resulting in large overall gains in production (Table 8.2). The states of India in which IGP is located currently account for 44 and 66% of area and 47 and 78% of the production of rice and wheat, respectively.

Increased production and productivity that characterized the Green Revolution of the 1970s and 1980s have come about due to a combination of factors, most important being the expansion of irrigated area by harnessing surface and groundwater resources, introduction and spread of dwarf photoinsensitive high-yielding varieties of rice and wheat, and the increased use of inputs including fertilizers and crop protection agrochemicals. Other supporting elements included expansion and strengthening of research and extension services and an overall agricultural fiscal support policies. Net irrigated area in the countries increased from 25 million to 48 million ha from 1960–61 to 1990–91. During the same period, irrigated wheat and rice areas rose from 4.0 to 20.6 and 13.5 to 19.0 million ha, respectively. From 1960–61 to 1990–91 the annual consumption of chemical fertilizers increased from 0.3 to 12.6 million tonnes, a large fraction of which is consumed in the rice–wheat cropping system. As a result, many rice–wheat farmers use high doses of fertilizers, some up to 500 kg or more nutrients per hectare per year.

PRODUCTIVITY GAINS NOT UNIFORM ACROSS THE INDO-GANGETIC PLAINS

Although the IGP region has been a center of major productivity gains in crop yields as a result of the adoption of "Green Revolution" technologies, the impact has not been uniform across the region. Gains in productivity were maximum in the northwestern states of Punjab, Haryana, and western Uttar Pradesh where the gap between

TABLE 8.3
Area under Rice–Wheat Cropping System and Productivity of Rice and Wheat in the IGP

State	Area (Mha)	Productivity[a] (kg/ha)		
		Rice	Wheat	Rice-Wheat
Punjab	1.8	3370	3956	7326
Haryana	0.6	2730	3639	6369
Uttar Pradesh	5.6	1830	2345	4175
Bihar	1.7	1176	2003	3179
West Bengal	0.3	2011	2109	4120

[a] Average of 3 years ending 1994.
Mha = Million hectare.

Source: Directorate of Economics and Statistics (1997).

the average farm-level productivity and the yield potential of enhanced germ plasm is narrowing. In contrast, in the eastern region represented by eastern Uttar Pradesh, Bihar, and West Bengal the benefits are not evident to same extent as the northwestern IGP region. The productivity levels in the east have remained low, nearly one half to one third of the levels achieved in the northwest (Table 8.3). Yet the region holds enough potential to raise productivity. The reasons for poor performance are varied, the most important being the relatively low level of water management and development of irrigation facilities, poor infrastructure and extension services, etc.

SUSTAINABILITY CONCERNS

Several indicators highlight sustainability concerns. Per hectare yields of rice and wheat rose significantly during the 1970s and 1980s (see Table 8.2). Consequently, both these food grains registered over 3% annual compound rates of growth in production between 1980–81 and 1990–91 which was significantly higher than the annual population growth of 2.1% during the 1980s. However, in the first 7 years of this decade (1990–91 to 1996–97) the annual rate of growth of food grains was only 1.7%, which is lower than the current population growth. Slower growth in production (Table 8.4) is a matter of concern. In the higher-production northwestern region, adoption of high-yielding cultivars is virtually complete. Almost entire wheat and rice crops in the states of Punjab, Haryana, and western Uttar Pradesh are perennially irrigated. Thus, with most traditional sources of productivity growth having being exhausted, future gains will have to come from elsewhere.

At the farmers' level, sustainability concerns are being expressed in several ways. Many farmers believe that the input levels have to be continuously increased in order to maintain yields. In the 1960s and 1970s most farmers used only nitrogenous and phosphate fertilizers to achieve high yields. Due to the occurrence of increasing deficiencies of several secondary and micronutrients most farmers now apply phosphorus, potassium, sulfur, zinc, boron, iron, and manganese to mitigate

TABLE 8.4
Annual Growth in Production (%)

Period	Rice	Wheat	Food Grains
1967–68 to 1980–81	2.90	4.72	2.67
1980–81 to 1990–91	3.35	3.62	2.86
1990–91 to 1996–97	1.52	3.62	1.70

Source: Directorate of Economics and Statistics (1997).

nutrient constraints. Results from many long-term experiments on the rice–wheat cropping system show declining yield trends when input levels were kept at constant. Figure 8.2 shows changes in system productivity and nutrient use in the Ludhiana district of Punjab, one of the most productive districts of the region. It is obvious that the growth rate of system productivity has been slowing, relative to the growth rate of nutrient use. Lowering of the water table due to the intensive rice–wheat system is also forcing many farmers to lower their pumping sets, with consequent increased costs of lifting water. Other emerging problems that threaten sustainability of the rice–wheat cropping system include loss of biodiversity–related issues and groundwater quality. The principal threat to yield stability from modern varieties is their increasing area coverage and their continuous use in a cropping system. Large areas planted to a single variety is a potential cause of concern, no matter how broad the genetic base of the variety. Biodiversity has other intrinsic values and represents a natural balance within an ecosystem. As the diversity is reduced, natural processes that control and affect habitat quality and genetic expression weaken, and for this reason internal and natural controls must be replaced by more externally applied artificial controls in the form of management and other inputs which in due course may lead the system toward unsustainability.

In the intensively cultivated high-production areas, disease and pest problems are also now taking a more serious turn than ever before. These problems are a result of continuous cropping of selected crops, e.g., rice and wheat in the present case. The carryover of some pest and disease complexes between two cereals poses both short- and long-range problems. Some grassy weeds (e.g., *Phalaris minor*) have attained a serious dimension in high-productivity IGP areas. There are reports that this weed has developed resistance to the commonly used herbicide, isoproturon. What it implies is that the farmers are applying increasing amounts of herbicide incurring increasing costs without the benefit of effective control. Similarly, carry-over of stem borer (pink borer) and buildup of soilborne pathogens in continuous cereal-based systems are other examples. Pesticide residues entering the food chain and overall safety in the use of pesticides continue to be a widespread problem.

By far the most serious single threat to the sustainability of the rice–wheat production system is related to issues of water management across the region. An insight into these is important for learning lessons for developing sustainable strategies which can be adopted in the years ahead.

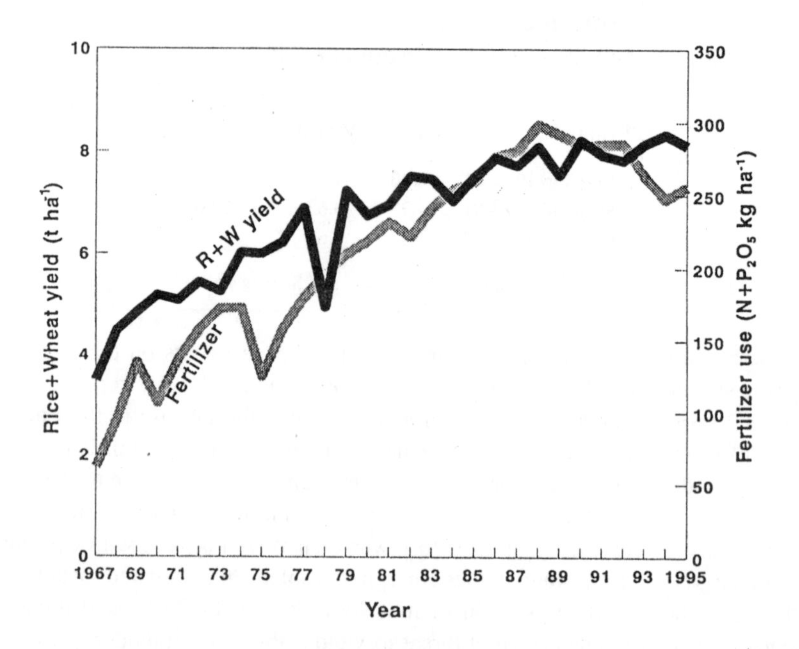

Compound rate of growth (%)		
Period	R+W Yield	Fertilizer $N + P_2O_5$
1967-80	3.9	7.3
1981-90	1.3	3.2
1991-95	1.1	-3.5
1967-95	**2.4**	**4.1**

FIGURE 8.2 Rice and wheat (R&W) yields and fertilizer use in Ludhiana district.

LOW PRODUCTIVITY IN THE EASTERN REGION

Rainfall in the IGP increases from west to east and the winter season in the eastern region is somewhat less extended. Due to higher cloud cover during the later parts of rainy season and due to a shortened winter season, the yield potential of rice and wheat crops is likely to be slightly lower in the eastern region than in the northwest. Thus, Virmani (1996) reported potential rice and wheat yields of 6.4 and 6.8 t/ha for the semiarid Ludhiana district in Punjab in comparison to 5.2 and 4.5 t/ha yields, respectively, for the two crops for the subhumid Rae Bareilly district in the eastern Uttar Pradesh. However, most field observations and experiments reveal that the overall productivity levels in the eastern IGP region have remained far below its agroecological potential. An important factor in continued low productivity of the

TABLE 8.5
The Extent of Irrigation in the IGP (1992–93)

State	Net Irrigated Area (× 1000 ha)	Irrigated Area by Source, %			Net Irrigated to Net Cultivated Area (%)
		Canals	Groundwater	Other	
Punjab	3,861	35	62	2	93
Haryana	2,628	52	47	1	76
Uttar Pradesh	11,322	29	67	4	66
Bihar	3,344	28	51	21	47
West Bengal	1,911	37	37	25	36

Source: Directorate of Economics and Statistics (1997).

rice–wheat system has been the quality of surface irrigation and drainage in the region which has limited growth productivity. For the rice–wheat cropping system the critical need is for an early provision of water before the onset of the monsoon rainy season for raising rice nurseries and their subsequent early transplantation before the main rainy season sets in. This ensures good tillering, higher crop density, and allows for application of fertilizers in time. The monsoon flooding inhibits tillering and restricts the benefits from applied fertilizers. A good rice yield is harvested where such early establishment is practiced. Timely harvesting of rice then permits early establishment of the following wheat crop, which thrives in the cool winter months. Thus good tillering is ensured and yields get a boost. Farmers in the eastern IGP region who are able to ensure timely planting (through tubewell irrigation) are able to obtain yields comparable with those harvested in the northwest IGP. However, groundwater resources in the eastern states have been poorly developed (Table 8.5). Surface irrigation, typically, has been designed as protective irrigation for supplementing watering during the kharif cropping season. Reliance on run-of-the-river diversion schemes with no or minimal storage which are dependent upon the advent of rain means that water availability by itself becomes monsoon dependent. Further, water management difficulties have meant that even during the monsoon cropping season, irrigation is unreliable and poorly distributed between the head and tail ends of commands. Occurrence of waterlogging during monsoon, and flooding in large areas, with little or no provision for drainage infrastructure in place to reduce waterlogging further compounds these problems. Examination of data on water deliveries (Meinzen-Dick, 1994) showed that inadequacy of water deliveries was less of a problem than the timeliness of irrigation water availability. While timely water supplies had a significant positive impact on rice yield, surplus supplies tended to depress yields. Water scarcity had the greatest adverse impact on the production in the middle of the season while water surpluses were most damaging at the beginning and at the end of the season.

In order to sustain rice–wheat yields, the situation calls for overcoming a number of technical, institutional, and social constraints. The two most important measures needed are (1) substantial improvements in the management of existing surface irrigation scheme and (2) increasing drainage to reduce adverse impacts of wide-

TABLE 8.6
Level of Groundwater Development

State	Blocks[a] (No.)	Average Groundwater Development (%)	Range of Development (index)	Blocks Overexploited (No.)
Punjab	118	93.8	43 to 260	53
Haryana	108	92.6	32 to 203	42
Uttar Pradesh	895	37.1	3 to 63	21
Bihar	585	18.6	1 to 41	—
West Bengal	—	24.4	1 to 56	—

[a] Administrative units

Source: Central Groundwater Board (1995).

spread flooding during the rainy season. In most parts of the eastern IGP, lack of drainage is the most critical problem, rather than lack of water or irrigation. In the context of improved water management, efforts are needed to discover ways to provide water at the time of the establishment of rice nurseries, and for subsequent early growth and tillering before the major monsoon rains set in. In order to achieve the improved irrigation–drainage impact, the development of private shallow tube-wells through appropriate support and land consolidation are urgently needed.

NORTHWEST IGP: A CASE FOR SUSTAINING HIGH PRODUCTIVITY

In contrast to poor growth of rice–wheat productivity in the eastern IGP, productivity in the northwest IGP has been highly significant. This has been observed in the states of Punjab and Haryana (see Table 8.3). Average rice and wheat yields in these two states are highest in India and compare well with high cereal yields in other countries. Factors that have contributed to enhanced productivity of the rice–wheat system include almost complete coverage of cropped area with irrigation and large-scale exploitation of groundwater in rice–wheat areas ensuring farmer control of irrigation which has catalyzed the spread of high-yielding varieties and use of high doses of inputs. A major sustainability concern of the northwest IGP is its overexploitation of groundwater resources in the rice–wheat areas (Table 8.6).

It should be pointed out here that before the large-scale development of irrigation, the depth of the water tables in most northwest parts of IGP were, generally, at more than 25 m. Following introduction of irrigation in the late 19th century, water tables typically rose at the rate of 0.2 to 0.3 m annually (see Figure 8.2). When the water table rose to the root zone, it increased evaporation from the soil surface, which invariably led to salinization of the soil or its sodification depending upon the ionic composition of groundwater. The problem became widespread in 1950s and 1960s in most irrigated areas of Punjab and Haryana. Highly sodic "usar lands" widespread in Uttar Pradesh are a result of combined effects of rise in water table and impeded drainage. The reclamation or prevention of further development of "usar lands" depends entirely on improvements in drainage.

At this point, it would be appropriate to recall that large-scale installation of tubewells in Punjab and Haryana in areas where the groundwaters were of low salinity but medium to high alkalinity and extensive spread of alkali soils in the 1970s and 1980s resulted in the evolution of a key strategy of drainage and of providing a source of irrigation. This practice led to the spread of the rice–wheat system. It was also during this period that concerted efforts were made to reclaim large areas of unproductive alkali lands in these two states (Abrol and Gupta, 1990), which further contributed to the spread and enhanced productivity of the rice–wheat cropping system. However, continued expansion of groundwater use in the rice–wheat areas had led to (1) considerable lowering of the water table such that the farmers now need to expend more energy for lifting water; (2) some evident incursion of the influence of saline/alkali waters from adjacent areas or from greater soil depths; and (3) consequences for other components of the hydrologic cycle, e.g., river water flows and, therefore, groundwater changes have to be viewed integratively.

Malik and Faeth (1994) predicted that groundwater depletion was the most costly and seemingly inevitable effect of conventional production practices under rice–wheat rotation in the Ludhiana district in Punjab. The combination of porous sandy soils and semiarid climate regime will inevitably make groundwater use unsuitable. Even at a water use amount of some 20% below the recommended levels, groundwater is likely to decline significantly. Unless production practices are developed which dramatically reduce water use, any rice-based production system may be unsustainable for the region.

As in the case of Punjab, groundwater development in parts of Haryana and Uttar Pradesh where rice–wheat is the dominant cropping system, there has been excessive use of groundwater which has resulted in lowering the water table. In eastern Uttar Pradesh and Bihar, groundwater development has been limited. The main constraint for speedy groundwater development has been inadequate availability of power to energize the wells, the high cost of diesel, and small, fragmented landholdings. However, the experience of Punjab and Haryana should stand out as a valuable lesson for implementing management strategies that sustain natural resources in the eastern IGP.

GROUNDWATER DEVELOPMENT IN WEST BENGAL

Unlike Punjab and Haryana, the level of groundwater use in West Bengal averages only about 25%. Yet, in some areas, a noticeable lowering of groundwater has led to serious environmental problems. According to a recent report (Chakraborti, 1996), arsenic is present in toxic amounts in groundwaters in large areas, spread over some seven districts of West Bengal. It affects nearly 800 villages, and the drinking water of more than 1 million people has been contaminated. In this area, more than 200,000 people are reported to be suffering from varying intensities of arsenic-related diseases. The situation is also serious in the adjoining areas of Bangladesh. While the exact causes of the buildup of arsenic in the groundwater are not clear, most evidence indicates that the source of arsenic is geologic. In Table 8.7 the level of groundwater

TABLE 8.7
Groundwater Development of West Bengal and the Arsenic Problem

Districts	Level of Groundwater Development (%)	
	Range	**Average**
Groundwater Affected by Arsenic (total 7)		
Burdwan, Hooghly, Malda, Murshidabad, Nadia and 24 – Pargana (north and south)	23–56	41
Groundwater Unaffected by Arsenic (total 8)		
Bankura, Birbhum, Cuchbehar, Howra, Jalpaiguri, Medinapur, Purulia, Dinajpur	1–28	12

Source: Chakraborti, D. (1996).

development and the presence of arsenic in the 15 districts of West Bengal is shown. It is obvious that groundwater extraction for irrigation through shallow and deep tubewells *prima facie* is causing conditions conducive for air to enter the aquifer, which in turn leads to oxidation of arsenic-rich pyrite and releases arsenic into groundwater by the action of acids released upon oxidation.

ROLE OF WATERSHED MANAGEMENT IN INCREASING PRODUCTIVITY

It is obvious from the above analysis that a failure to recognize the interdependence among different spatial units, upstream and downstream, in a landscape, in the command area of an irrigation canal in a drainage district, or in a watershed can lead to serious environmental consequences if the management strategies adopted are not carefully chosen to take into consideration both the outside needs and the likely opposite impacts. Such approaches in past have led to a serious unsustainability problem in parts of the watershed. Many of the rice–wheat productivity–related issues, e.g., adverse salt and water balance regimes (declining water tables, drainage congestion, etc.) can be minimized by adopting management strategies that are holistic and based on an understanding of the system response in relation to available options at a point of time.

CONCLUSION

The IGPs developed on alluvium, is agriculturally one of the most important regions in the world. The region is endowed with highly productive soils, rich groundwater resources, and climatic conditions that permit growing two or more crops in a year. The region has contributed very significantly to gains in production of food grains in the Green Revolution era of the 1970s and 1980s. However, these gains have been limited to a relatively small area within the IGP. But even in this region, productivity gains have taken place with attendant huge costs involving natural resource depletion and their serious degradation. This situation poses a threat to the sustainability of the rice–wheat production system. There is urgent need for enhancing productivity from this region to meet the needs of increasing population. A better understanding of the resource endowments, their temporal and spatial variability, and their interconnectedness across the region calls for renewed efforts to devise new farm practices which will, in the future, not only respond to measures aimed at intensification of cropping systems, but also conserve and upgrade natural resources. This is a most urgent agenda for agricultural research in the IGP countries.

REFERENCES

Abrol, I.P. and Gupta, R.K. 1990. Alkali soils and their management, in I.P. Abrol and V.V. Dhruvanarayana, Eds., *Technologies for Wastelands Development,* Indian Council of Agricultural Research, New Delhi.

Central Groundwater Board. 1995. Groundwater Resources of India, Central Groundwater Board, Ministry of Water Resources, Government of India, Faridabad, 147 pp.

Chakraborti, D. 1996. Arsenic in groundwater in several districts of West Bengal, paper presented at the National Meeting on Science and Technology Inputs for Water Resource Management, 8–10 April 1997, 145–153, *Pre-Proceedings Volume,* Department of Science and Technology, New Delhi.

Hobbs, P. and Morris, M. 1996. Meeting South Asia's Future Food Requirements from Rice-Wheat Cropping Systems: Priority Issues Facing Researchers in the Post Green Revolution Era, NRG Paper 96-01, CIMMYT, Mexico, DF.

Malik, R.P.S. and Paul, Faeth, 1994. Rice-wheat production in north-west India, in Paul-Faeth, Eds., *Agricultural Policy and Sustainability: Case Studies from India, Chile, the Philippines and the United States,* World Resources Institute, Washington, D.C.

Meinzen-Dick, R. 1994. Adequacy and timeliness of irrigation supplies under conjunctive use in the Sone Irrigation System, Bihar, in Mark Svendsen and Ashok Gulati, Eds., *Strategic Change in Indian Irrigation,* Indian Council of Agricultural Research, New Delhi, India and International Food Policy Research Institute, Washington, D.C.

Sehgal, J.L., Mandal, D.K., Mandal, C., and Vadivelu, S. 1990. Agroecological Regions of India, National Bureau of Soil Survey and Land Use Planning, Nagpur, India, 76 pp.

Virmani, S.M. 1996. Rice-wheat cropping sequence in the Indo-Gangetic Plains of India: its agroclimatic characterization by GIS approach, paper presented at the Symposium on Sustaining Rice-Wheat Cropping Systems — Emerging Research Agenda, 20 November, New Delhi, 2nd International Crop Science Congress, New Delhi, 17–24 November 1996. [Paper presented with support from Johansen, C., Abrol, I.P., Gangwar, K.S., and Prasad, K.S.]

9 A Landscape that Unites: Community-Led Management of Andean Watershed Resources

E. Bronson Knapp, Jacqueline A. Ashby, Helle M. Ravnborg, and William C. Bell

CONTENTS

INTRODUCTION

The hillside agroecosystem of tropical America covers about 1 million km^2 and sustains an estimated 10 million poor people in marginalized, rural communities. More than half of this area is undergoing rapid environmental degradation as a consequence of deforestation, overgrazing, and destructive agricultural practices. Environmental and economic problems these communities face often transcend the boundaries of individual farms and must be addressed through collective action across entire landscapes. Such action requires that rural communities be able to set clear objectives, quantify the various environmental, economic, and social factors that enter into their decisions, and define with some accuracy the geographic area of interest. A useful unit around which to organize these tasks is the "community watershed." To encourage community-led management of multiple-use watershed resources, two parallel strategies, with examples, are presented. First, there are needs

0-8493-0702-3/00/$0.00+$.50

for increasing the accuracy and precision of data together with convenient decision support tools so that stakeholders can systematically analyze choices for resource use. New methods and tools need to incorporate both strategic principles and local knowledge. The second recommended line of research aims at developing replicable, multi-institutional alliances for establishing consortia with the analytical capacity to plan and support community watershed resource management. In cases of conflicting interests, a process of deal making, in which costs of resource conservation are balanced by concrete incentives, can be catalyzed by a process of successive refinement of information and analysis.

THE CHALLENGE

The hillside agroecosystem of tropical America, which covers about 1 million km^2, is a major contributor to regional food security and is the basis of livelihood for a large proportion of rural poor.

The principal countries (followed by percent area in steepslope agriculture) include Bolivia (40%), Colombia (40%), Ecuador (65%), Peru (50%), Venezuela (70%), Costa Rica (70%), El Salvador (75%), Guatemala (75%), Honduras (80%), and Nicaragua (80%). Tragically, an estimated half of the hillside agroecosystem resource base is degrading as a consequence of decisions that have led to deforestation, overgrazing, and destructive agricultural practices. (UNEP, 1992).

The hillside agroecosystem sustains an estimated 20 million people. Half of this total, 10 million people, are classified as "poor," living in marginalized, rural communities. Moreover, World Bank data show a significant portion indigent, i.e., without means to meet minimal nutritional needs; 23% of Colombian rural population is indigent, 46% in Peru, and 57% in Guatemala (ECLAC, 1990). Female-headed households are a high proportion of the indigent rural population (ECLAC, 1993). Thus, in most countries with significant proportions of area in hillsides, the locus of poverty has yet to shift from rural to urban areas. Furthermore, World Bank figures for the 1990s indicate that rural impoverishment has recently increased in some of these areas.

In the more densely populated and drier areas, fallow periods have been shortened or replaced by organic or chemical fertilizers. When farmers cannot obtain or afford fertilizers, they work off farm, exacerbating the "feminization" of hillside farming in which the real farmers are women managing subsistence or semicommercial small farms. Even in "well-watered" areas, erratic distribution of rainfall can lead to short but critical periods of drought stress. Pest, disease, and weed control are major constraints in annual crops. Degraded fallows, largely synonymous with overgrazed pasture, occupy an estimated 40 to 60% of hillsides. Large farms maintain low stocking rates and sharecrop arable land. This reflects a strategy of investing in land to protect capital. Improved production is frequently not a primary nor an even important objective of large landowners in the hillsides, who make up about 20% of farmers and own 80% of the land. Intensification of production on small farms is an important part of alleviating poverty that drives migrants to colonize, deforest, and degrade increasingly fragile environments.

Environmental degradation of hillsides has serious implications not only for viability of agricultural production in the hillsides themselves, but also for "downstream" lowland agricultural and coastal ecosystems which can be affected by soil erosion and agrochemical contamination. Second, the welfare of rural and urban populations who draw water supplies from watercourses originating in hillside watersheds is also intimately affected by contamination, soil erosion, sedimentation, and major land slippage caused by misguided land-use decisions.

The third and potentially most irreversible damage due to poor planning and management decisions in hillside environments, and that with major social costs, is the loss of biodiversity due to the disappearance of montane forest. This is estimated at between 15% of forest area in Bolivia and 55% in Guatemala. The rate of hillside deforestation is higher than in lowlands, causing an estimated loss of 90% of montane forest by 1990. Montane forest has very high biodiversity, arguably higher than for lowland forest, especially with respect to herbs and shrubs found between 600 and 3000 masl, which are considered important for conserving wild-crop genetic resources *in situ*.

The rapid rate of environmental degradation in hillsides is driven by a number of factors that include the unfavorable structure of incentives for hillside farmers to invest in sustainable management practices. These incentives are shaped by specific agroecological conditions, available technologies, prices of inputs and outputs, opportunities for off-farm employment and migration, as well as cultural and organizational norms of natural resource management. Income-generating activities that permit capital accumulation and agricultural intensification are key to changing rural stakeholders' environmentally destructive management decisions and practices.

In summary, links among agriculture, natural resource management, and economic problems in lesser developed countries were addressed by a recent Consultative Group for International Agricultural Research (CGIAR) task force who concluded that there existed "… an abysmal lack of data to validate … conventional wisdom" (CGIAR, 1997). Clearly, progress toward the goal of more productive, sustainable, and healthy hillside environments is being hindered by a lack of clear objectives, a failure to quantify variables, and a lack of precision in defining physical areas of interest (Lal, 1994), all of which are indispensable for arriving at negotiated agreements for community action as well as reproducing results achieved. Information shortcomings can be largely attributed to a confusing range of temporal and spatial perspectives among stakeholders that so characterize many resource management problems.

A FRAMEWORK FOR RESEARCH

One of the greatest challenges for CGIAR researchers, government and nongovernment organizations (NGOs), and resource-poor farmers is the need to adopt perspectives that transcend field or farm boundaries and accept solutions necessitating some form of collective action among landscape users.

Agroecoregional research requires that the traditional definition of research site and data collection be adjusted to include different spatial and temporal scales targeted by different stakeholders. This is essential for addressing issues through collective action. So far, with notable exceptions (Veldkamp and Fresco, 1996), "across-scale analysis" has resulted in little more than independent characterization of ever-larger geographical areas in less and less detail, from plots through landscapes, as indicated in Figure 9.1.

As suggested in Figure 9.1, research needs to emphasize the organizing principles and functional relationships that structure multiple-scale systems. A key aspect of this strategy, based on hierarchical systems theory, is an increased emphasis on sample surveys and controlled prospective and retrospective studies vis-à-vis laboratory and field experimentation.

Outputs of this research are process-level analytical models that can define and categorize the biophysical and socioeconomic responses upon which agricultural systems depend, responses that may not manifest themselves in the space/time frame of typical crop/soil experimental studies. This information helps identify points of policy and management intervention, and, when combined with local knowledge, provides exante analysis of trade-offs involved in choosing different interventions. In cases of conflicting interests, a process of "deal making," in which costs of resource conservation are balanced by concrete incentives, can be catalyzed by a process of successive refinement of information and analysis.

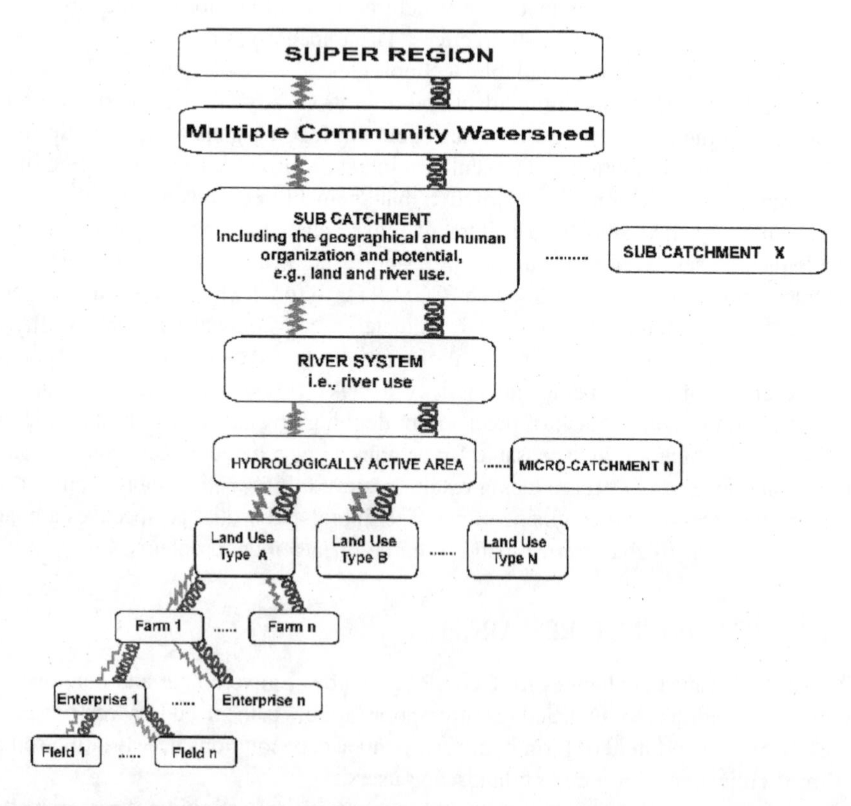

FIGURE 9.1 Schematic diagram illustrating a common physical and social organizational structure found throughout the hillside agroecosystem. Asymmetrical linkages act as control mechanisms affecting the introduction of change.

WATERSHEDS AS AN ORGANIZING UNIT OF STUDY

For the traditional CGIAR center and national agricultural system researchers used to targeting problems, priority areas and beneficiary groups in terms of "single decision makers," the "community watershed" offers a logical statistical population on which to begin addressing multiple stakeholder and common property resource issues.

Technical reasons for targeting community watersheds are well known. Overland and through-flow of water draining through catchments integrates and concentrates the effects of many crop and land management activities. Off-farm effects can be made explicit. Catchment boundaries are specifically definable and reproducible permitting application of systems analysis. Catchments are naturally organized into biophysical and, in some cases, socioeconomic hierarchical systems.

Although watersheds are a useful unit for organizing research, this does not imply that the objective will always result in management plans that optimize water resources. Rather, the objective is to include analysis of water as well as soil and vegetation in the family of indicators which provide a "feedback mechanism" for stabilizing and sustaining hillside production systems (CGIAR, 1996).

The mandate of the CGIAR requires that research centers like CIAT (Centro Internacional de Agricultura Tropical) focus research toward international public goods and services in contrast to developing "site-specific" technologies appropriate for specific community watersheds. In keeping with this mandate, the remainder of this chapter details two parallel research strategies. We begin with examples of procedures that are being established for database use in targeting problems, priority areas, and beneficiary groups and follow with protocols for catalyzing multi-institutional alliances capable of using information technology in planning and supporting community watershed resource management. We conclude with examples using simulation modeling as a tool for stimulating community discussions by characterizing responses to land management interventions.

DATA COLLECTION AND QUALITY CONTROL

As recommended by the CGIAR TAC (1997), "... abysmal lack of data ..." requires major efforts in reviewing, editing, consolidating, and relating/georeferencing biophysical and socioeconomic data into user-friendly databases. These are prerequisites to analyses of relationships, for example, between poverty and environmental degradation. However, requesting proprietary data from national organizations requires extraordinary trust based on long working relationships. This is a comparative advantage of the CGIAR centers like CIAT.

Data collection in tropical America is not institutionalized to the same degree it is in North America. Because of severe resource constraints, national agricultural and population censuses are not carried out regularly nor analyzed to the degree one might wish. Detailed climate and soils data are not usually available for more than the most intensively cultivated commercial areas, and large-scale (1:50,000) topographic maps, when digitized, need to be checked for common errors like rivers overlapping "valley" edges, contour lines at map edges not congruent, incorrect coding of contour elevations, and inconsistent digitizing of river direction. Not as obvious, perhaps, as the aforementioned errors which are passed on to digital spatial coverages, is the issue of identification of original sources used in the creation of digital terrain models (DTMs). In the process of checking DTMs, we have found some models are dominated by the grid of digitized points generated from the original map. This effect can be seen as spikes in the histogram of almost any DTM fitted to digitized or scanned contour data. We have found that some DTMs were quite likely not constructed from the original sources and resolution they were supposed to have been.

Not withstanding potential problems, topographic data are being digitized and edited which allows development of DTMs and topological analysis. Figure 9.2a and b shows a geographic information systems (GIS) analysis of a region of central Honduras for which "community-scale" watersheds of 3000–15,000 ha have been defined. Also shown, as point data, are communities. Since watershed landscapes are hierarchical, there is some latitude when apportioning communities and watersheds. This is an issue that is best resolved through need and "local knowledge."

National censuses do exist and, continuing with Honduras as an example, in collaboration with INEC (Instituto Nacional de Estatistica y Censo) we have resurrected decade-old digital databases of past national population and agricultural censuses. For example, from household-level 1974, 1988, and 1993 censuses, we have reconstructed digital databases allowing statistical characterization and aggregation at the village/aldea scale which is 10 times more desegregated (about 3000 records vs. 300 records) than heretofore available. Until now, population and agricultural census data for countries like Honduras were only available in hardcopy. However, with the availability of reasonably inexpensive hardware and software, and examples of the power of information and interactivity, which mean parameter ranges can routinely be defined, plotted, analyzed, and interpreted, there is demonstrable enthusiasm for making digital databases available.

Enthusiasm, however, for new analytical software, like GIS applications, without training may lead to mishandling of data collection or derivation. Take the example of the large-scale map of watersheds in Figure 9.2. A rather startling revelation is that the areas of two typical watersheds, as determined by a routine analysis of the "flat" map representation, amounted to 11,280 and 5130 ha, respectively (Figure 9.2a). The areas calculated from the three-dimensional representation of the DTM-GIS model are 13,600 and 6115 ha (Figure 9.2b), a difference of 20%!

It can be strongly argued that the most common and successful application for satellite remotely sensed imagery to date has been in mapping land cover and land use (LC/LU). This is particularly so for high-spatial-resolution products where the ground dimensions of a pixel are less than 100×100 m. Many examples can be found in the literature where the reported levels of classification accuracy exceed 80%. It is noticeable, however, that the majority of the applications are based either in regions of large-scale forestry or in intensive agricultural systems of North America and Western Europe. Hillsides are a fundamentally different environment with physical characteristics that present a much greater challenge for deriving LC/LU maps. Current remote sensing research at CIAT using LANDSAT TM (thematic mapper) imagery of mountain hillside systems, characterized by small plots on steep

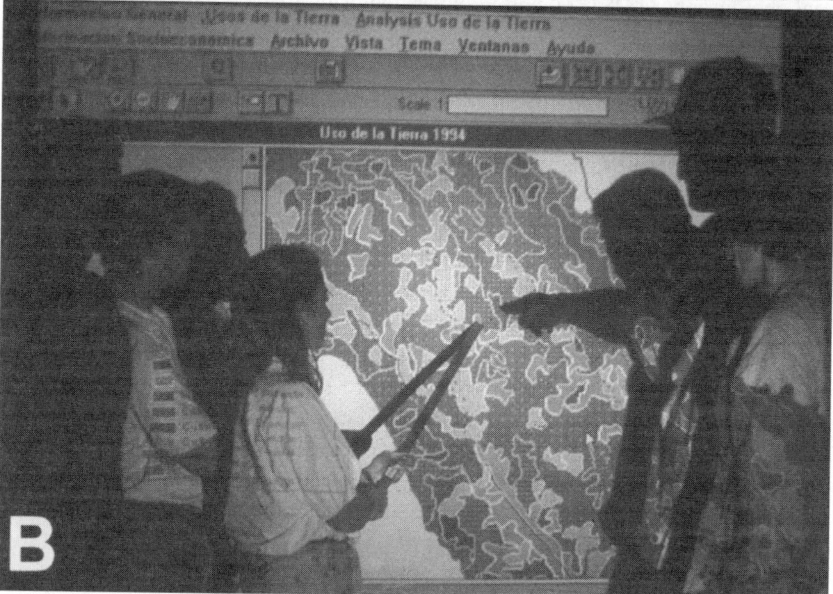

FIGURE 9.2 Examples of how boundaries of community watersheds may be defined. (a) Illustration of first approximations of GIS-derived boundaries that rarely, if ever, correspond to political unit boundaries. (b) Illustration of a DTM for part of the area shown in (a), noting the differing GIS-determinations of area (ha) for two typical watersheds.

slopes, has shown that thematic accuracy of LC/LU mapping as measured by per pixel parameter accuracy can be disappointing, giving an overall percentage agreement in the range of 35%. We have experimental evidence from a Colombian study site with an analytical technique that raises accuracy to acceptable limits using the error matrix itself to adjust or calibrate any bias in LC estimates. More importantly, as the landscape becomes more fragmented with more edge effects and mixed pixels, and as spatial patterns become more intricate, the technique itself becomes more appropriate and more defensible.

Other data collection activities include successive refinement of remotely sensed imagery using Systeme Pour L' Observation de la Terre (SPOT) panchromatic, RADARSAT, and digital air orthophotographs. Work with digital orthophotographs has focused on deriving terrain slope estimates. Terrain slope is one of the most fundamental variables in agriculture as it affects such variables as cost of preparing seedbed, erosion, runoff, mechanization, and access. Despite its importance, slope remains elusive as estimates change continuously depending upon the distance over which it is averaged, grid cell size and scale of the original elevation data (Berry, 1993). A study was carried out for an Andean watershed in southwest Colombia to compare accuracy that can be expected for slope and altitude values derived from low-cost DTMs. Eight gridded DTMs were generated from digitized contour maps at a range of scales (1:10,000, 1:25,000, 1:100,000 and 1:200,000) and a range of contour intervals (25, 50, and 100 m). A control DTM was produced from large-scale aerial photographs (1:28,000) and field verified using 91 deferentially measured GPS ground points. The control DTM showed a vertical root mean square error (RMSE) well within U.S. Geological Survey (USGS) accuracy standards. In addition to cell size and slope relationships, cost of production of DTMs and accuracy of results were determined. Some of the conclusions reached were (1) substantial savings in time and expense in DTM production accrue from digitizing every *n*th contour as long as the new interval is not more than 25 m wider than the original interval when modeling altitude and/or slope, (2) regarding slope determination, contour interval has more influence than map scale, (3) cartographic data sources at scales equal or greater than 1:100,000 and with contour intervals equal or greater than 100 m do not provide sufficient detail to represent slope usefully in community watersheds in hillside agroecosystems.

Decision-support tools are often associated with mechanistic models or goal optimization models. However, many land-use decision requirements may be satisfactorily and economically addressed through a strategy of successive refinement of data on demand. For example, first approximations of potential erosion risk and watercourse degradation were supplied to a local Colombian watershed consortium using a DTM and water-routing analysis. For regions that are contentious and worth the added expense, overlaying land cover on the DTM is an option. Still further refinement is possible by incorporating soil profile characteristics at some additional cost of information. This interactive, heuristic, and descriptive GIS analysis has more short-term operational value for stakeholders managing agriculture-dominated watersheds in the Andean hillsides than USLE-based models using parameters derived for temperate North America.

If a pathological situation is suspected for a given watershed, the next obvious task is diagnosis. Diagnosis might follow a sequence of inquiries such as: (1) assessing the extent, severity, and rate of progress of soil and water degradation; (2) assessment of impact on agronomic productivity and health of water-related processes; and (3) evalt ation of economic impact of the degradation.

We have explored a few diagnostic routines within a peri-urban watershed in southwest Colombia. Analogous to the first request of a medical doctor preparing for a diagnostic examination, we analyzed the "patient's" history. Table 9.1 is an example of the history of land cover/land use (LC/LU) for our watershed recreated from air photo interpretation for three dates, spanning over 50 years. Important conclusions about the "aging" of the watershed followed from the analysis. First, on balance, proportions for major classes of LC/LU aggregated at the three dates, including severely degraded land, seem to have changed little over the past 50 years. Second, analysis of LC/LU of specific, georeferenced mapping units across the three dates proves individual fields have been rotated into and out of different LC/LU over time. The unmistakable conclusion is that there are important, scale-dependent dynamic temporal processes at work throughout the watershed that are not revealed from analysis of time-series snapshots of LC/LU classes aggregated at "large" spatial scales. This led us to the next diagnostic task, assessing the impact of field-scale LC/LU histories on agronomic productivity.

TABLE 9.1
The History of Land Cover/Land Use (LC/LU) for the Río Cabuyal Watershed. LC/LU was quantified in the traditional manner using air photo interpretation at a working scale of 1:30000 resulting in a resolution of interpretation of approximately 3 ha.

Land Cover/Land Use	1991	1970	1946
Natural Forest (Secondary Regrowth)	7	6	9
Reforested in Pine	1	0	0
Mature bush fallow and Improved pasture	43	46	41
Young fallow and poor grazing land	32	38	47
Multiple-cycle crops (e.g., coffee, sisal, sugarcane, monocropped and intercropped cassava)	10	2	1
Intensive cropping (horticultural crops, drybean, maize, land in preparation)	4	0	0
Overgrazed and eroded land, landslides, rock outcrops	3	8	2
TOTALS (%)	100	100	100

Traditionally, agronomic commodity-constraints research has focused on improving cropping productivity by, for example, measuring fertilization efficiency evaluated in terms of economic marginal rates of return for various fertilizer levels. This approach is an undeniably important element in predicting the acceptance and rejection of new technologies by individual decision makers. However, an arguably more relevant task for assessing impact of soil degradation across a wide spectrum of land users, within a watershed context, is identification of land and crop management practices that put the soil resource at risk of exceeding a threshold of irreversible loss of soil productivity. For this study, "irreversibility" was defined as a change of state of the soil which cannot readily be overcome by application of fertilizers, a normal human response to market forces responding to product shortages, and a definition consistent with evaluation frameworks found in the literature (Acton and Padbury, 1994; Conway, 1988; FAO, 1976 and 1993; Keulen et al., 1986; Lal, 1994; Riquier et al., 1970).

In many Andean hillside watersheds, like the Colombian watershed in which these studies were carried out, soil maps at scales of 1:50000 are generally available. At this scale, soil maps differentiate soils by association that generally means the mapping unit aggregates soils formed under conditions of similar climate, parent material, native vegetation and topographic sequence. For this study we used site elevation and topographic classes as proxies for localized weather to impose further stratification. A wide range of human induced land use pressures, either degrading or regenerative, are found within the watershed. Our task was to assess whether land use decisions had irreversibly changed, for better or worse, the productivity potential of defined mapping units. We selected a range of six land use types (LUT), from 50-year old secondary regrowth forest to a long-term cassava-fallow system traditionally associated with soils with greatly depleted fertility. Indicator crops of bean, cassava and maize were seeded for six crop cycles and nonlimiting fertility, weed and pest management applied.

Figure 9.2 shows the directed graph of "path" coefficients (normalized partial regression coefficients) for variables analyzed for their effects on attainable maize yields. Limiting interpretation of the analysis to the sign and general magnitude of the coefficients, it can be seen land use type is "causing" significant variability in potential economic crop yield. Nevertheless, season-to-season variability, interpreted as predominantly 'climate' and represented by the variable 'semester,' is a more important source of variation in yield than LUT. Contrary to popular opinion, the data collected show the trend is for the more intensively cultivated LUTs to have higher attainable yields than the "noncultivated" LUTs given nonlimiting levels of fertility. Diagnostic soil chemical analysis for the two extreme LUTs mentioned above showed significantly lower exchangeable acidity and higher organic carbon in the long-term cassava-fallow system.

Evaluating economic impact from direct measurements of soil and water degradation at the watershed scale is a daunting task and one we have not yet undertaken. In practice, we propose following an analytical strategy of successive refinement of data collection based on value of information as it pertains to negotiating collective action among stakeholders. Following this pragmatic strategy, we recommend an initial screening of LUTs to assess the "financial fitness" of different cropping scenarios found in a specific watershed. For example, in the Colombian example, we asked the question, "how would financial sustainability of a specific, representative farm be affected by different scenarios for soil loss"? Table 9.2 shows the results of a simulation indicating there would be no significant loss of financial sustainability over a fifteen-year period even with assumed soil losses of up to 50 Mg ha-1 yr-1 (Hansen, 1996). The significance of this study is that it highlights potential conflicts that exist between private benefit and social benefit. From the individual's perspective, erosion control strategies may not be economically attractive while, from the perspective of social benefit, there are few alternatives to saving the public water system from degradation. With inferences drawn from this analysis, the population of farms in the watershed can be screened and "at risk" areas in the watershed targeted.

TABLE 9.2
Soil loss as predicted 15-year sustainability ($\surd(15) \pm SE_{\surd}$) of erosion scenarios and G-test statistic ($G_{l,adj}$) for differences from the *Erosion @ 0 Mg ha⁻¹ yr⁻¹* scenario. The McNemar test statistic was always undefined (Hansen, 1996)

Scenario	---- Soil Loss (cm) ---- Annual	Total	$\surd(15)$ $SE_{\surd}$	$G_{l,adj}$
Erosion @ 0 Mg ha⁻¹ yr⁻¹	0.00	0.00	0.68 ± 0.047	H
Erosion @ 25 Mg ha⁻¹ yr⁻¹	0.56	8.33	0.60 ± 0.049	1.0 n.s.
Erosion @ 50 Mg ha⁻¹ yr⁻¹	1.11	16.67	0.54 ± 0.050	3.5 n.s.
Erosion @ 100 Mg ha⁻¹ yr⁻¹	2.22	33.33	0.29 ± 0.045	29.5**
Erosion @ 150 Mg ha⁻¹ yr⁻¹	3.33	50.00	0.16 ± 0.037	56.2**

H Comparison does not apply to the Erosion @ 0 Mg ha^{-1} yr^{-1} scenario.

Finally, it is important to reiterate that CGIAR centers like CIAT receive virtually all their funding from humanitarian aid agencies that demand practical applications of results for improving the well-being of poor farmers. A contentious issue is the ill-defined cause–effect relationship between poverty and resource degradation. Addressing this objective and monitoring development project impact require a multidisciplinary approach to identify who the poor(est) are, a description of their environmental conditions and relationships between the two. As a first step, it was necessary to develop a methodology for making regional poverty profiles based on local indicators of poverty or well-being. By using local indicators identified for different levels of well-being, a parametric model was constructed for the Rio Cabuyal catchment (7000 ha and 1000 families) in the Colombian Andes mountains.

MULTI-INSTITUTIONAL ALLIANCES

Watershed management involves the integrated management of a multitude of resources such as cropland, pastures, forests, and water to each of which a multitude of often conflicting interests relate. These interests arise from shareholders inside as well as outside the watershed. The identification and negotiation of these interests therefore is an important element in the design and development of appropriate technology and its adoption in the context of improved watershed management. Based on experiences with inter-institutional consortia and an experiment organizing local-level management of the Rio Cabuyal watershed in Colombia, six functions have been identified as essential for local-level watershed management organizations. Of these, at least three appear to be specific to watershed management. Besides being important in themselves, these functions provide some principles for the process of organizing for local-level watershed management.

1. Identifying stakeholders and ensuring their representation in management effort

The first of these functions is to identify the distinct local-level interests or stakeholders that relate to the use and management of resources within the watershed and ensure their representation in management efforts.

Local-level organizations can be either community or interest group based. In cases where the individual resource manager's interests are determined by his or her geographical location, community-based organizations are likely to be representative. However, when other factors such as ethnicity or a resource manager's access to resources determine stakeholders' interests, chances of community-based organizations being representative are limited. Our analysis in Rio Cabuyal confirms a caveat expressed elsewhere, that organizational participation in community-based organizations tends to be skewed toward resource-rich households (Pretty and Chambers, 1993; Bebbington et al., 1994).

In watershed management, representation of diverse interests may be vital to institutional effectiveness, due to the interdependency that exists among different users; i.e., one group's use of a resource directly or indirectly affects other groups' possibilities for using the same or other resources within the watershed. This makes the participation of all interest groups or stakeholders relevant to a given resource important to planning and implementation. Thus, a stakeholder-based rather than a community-based organization is essential to effective watershed management.

2. Providing forums for analysis and negotiation of diverse interests

Once diverse stakeholders are identified and have representation, the second function which local-level watershed management organizations should perform is to provide a forum or "platform" (Roling, 1994), where these interests can be analyzed and negotiated. In the first place, this means specifying time and place for such negotiations as well as who should participate.

Because of conflicts of interest relating to watershed management are not easily overcome, such forums cannot assume that stakeholders share a common goal. For example, in the case of the Rio Cabuyal watershed, a conflict arose over slash-and-burn agriculture. Stakeholder analysis showed that very concrete interests, led by the poorest households who are either short of labor or renting land and feeling no incentives to engage in long-term land improvements, prefer burning as a method of cleaning fields despite their awareness of the risks.

During negotiations, participatory techniques are required to draw attention to conflicts and different interests. The principal role of the facilitator organization is that of the devil's advocate to stimulate such analyses. Examples of such techniques are described in Guba and Lincoln, (1989). In most cases, the facilitation skills necessary to lead such negotiations do not exist locally, but will have to be provided from outside, at least in the early stages of organization, underlining the importance for local-level organizations for advice and skill formation with respect to the organizational process as such.

3. *Defining rules and norms for the use of resources within the watershed*

An important function of forums for analysis and negotiation is the definition of norms and rules for use of specific resources within a watershed, as well as sanctions for noncompliance. This is the third function of local-level watershed management organization and is shared with other types of local-level resource management. Thus, Ostrom (1994) asserts that rules regulating resource use need to be carefully tailored to the local conditions by specifying time, place, technology, and quantity of resource units, as well as rules specifying resource input obligations, to support management activities relating to common-pool resources. Ostrom argues that uniform rules established for an entire nation or region cannot take into account such specifics and are therefore bound.

Experiences with creating buffer zones to protect water springs and watercourses in Rio Cabuyal provide a case in point. For many years, the regional water authority attempted to mandate buffer zones in the watershed by applying national laws prescribing buffer zones of 50 m around water springs and 20 to 30 m along watercourses. Their efforts met with little acceptance from the local population. As a result of the involvement of the new watershed "stakeholder" organization, adherence to the general "rules" was relaxed and negotiated on a case-by-case basis, often being determined by the existing boundary between natural vegetation and cultivated area. This has initiated the fencing of several thousand meters of buffer zones by the local population using community labor.

4. *Initiating a process of local-level resource monitoring research*

A fourth function that should be undertaken by local-level watershed management organizations is to initiate local research for monitoring purposes. The watershed consortium was successful in mobilizing local labor for reforestation and creation of protected areas, but did not set up procedures for monitoring results. Conflicts with slash-and-burn land management in the watershed illustrated the need for local problem diagnosis ("locals" interviewing "locals" about their reasons for burning) followed by local monitoring of compliance. Once the need for sanctions based on monitoring was recognized, the additional problem of where to locate enforcement in the organizational structure had to be resolved.

The Rio Cabuyal experience shows that providing information about the "state of resources" is itself an important part of the negotiation of conflicting interests and the definition of compromises and rules for resource use. Local monitoring research has specific importance in watershed management because the interdependence between different resources within watersheds is usually poorly understood by local decision makers. Efficacy of efforts to regulate use or to conserve resources in watersheds is often determined by that factor which is least understood rather than most thoroughly considered (D. Walker, personal communication). For example, in Rio Cabuyal, local people believe that the upper watershed tributaries determine the water flow of Rio Cabuyal while simple mapping shows that tributaries all the way down to the tail end of the watershed are as important as the upper ones. This information helps stakeholders to target better the creation of buffer zones referred to above.

5. *Formulating and exerting demand for services from external institutions in support of local management efforts*

The fifth function that CIAT research finds should be undertaken by local-level watershed management organizations is to articulate local demand for external organizations such as NGOs and government organizations providing services to local communities. As was the case in Rio Cabuyal, local populations are often confronted with an array of organizations each having their own agenda, resulting in a supply-driven rather than demand-driven provision of services, be they technical, social, or organizational.

One of the tasks of local-level watershed management organizations is to attempt to change this situation by formulating agendas, identifying problems and/or defining concrete proposals for action to which external organizations can respond. To be successful, this obviously requires willingness on the part of the external organizations to listen and respond to such demands as well as an institutional mechanism through which such demands can be communicated. The creation of a mechanism through which local organizations could "pull in" services lacking in the upper-watershed communities was critical to the success of their motivational campaign to protect the upper-watershed water sources.

6. *Negotiating internal vs. external watershed interests*

Without the process of organizing for local-level watershed management, attempts to accommodate external interests in watershed management are likely to fail.

The sixth and final function to be carried out by local watershed management organizations is to negotiate internal vs. external interests relating to the use of resources in the watershed. Interests in improving watershed management in Rio Cabuyal originate as much from stakeholders outside the watershed (such as urban populations in need of drinking water or downriver commercial, irrigated farms and industrial users) as from stakeholders within the watershed. Just as in the case of negotiating interests originating within the watershed, the likelihood of reaching a shared sense of a common goal is limited. Instead, based on a process of acknowledging the existence of legitimate but often conflicting interests within as well as outside the watershed, compromises will have to be made that provide incentives for watershed farmers to erode less or for urban and semiurban populations to waste less water.

A strategic result of catalyzing local, multiple stakeholder consortia is to capture the creativity and extraordinary breath of proposals for improving family well-being and environmental health that flow from these groups. The following section briefly lists some of the activities developed and led by a unique community consortium in the Rio Cabuyal watershed in the Andean hillsides of southwest Colombia.

COMMUNITY-LED MANAGEMENT OF WATERSHED RESOURCES

The following are some new activities initiated by a prototype community consortium in the Rio Cabuyal watershed in the southwest Andean hillsides of southwest Colombia. In recognition of its accomplishments, the consortium has received two prestigious awards including the 1997 national "Premio Nacional de Ecologia: Planeta Azul" (National Ecology Award: "Blue Planet").

- Thus far 23 community projects for a value of $7.0 million pesos (about U.S.$7000) have been approved, benefiting 230 families in the Cabuyal River catchment. Of the 23 projects, 65% have had an emphasis on production, 24% on conservation of natural resources, 7% on training, and 4% on postharvest processing.
- An association of Cabuyal River microprocessing units was formed around the processing of dairy products. In three communities, organizations of some 25 people each have been formed around agricultural production projects, managing special Seed Capital Funds.
- Over a 2-year period (1994–95), 30,000 lineal meters (equivalent to 100 ha) have been isolated and another 15,000 m (55 ha) are in the process of being protected. Most of these protected zones (90%) have been in the upper reaches of the catchment while many of the 2800 farmers that have contributed labor have been from the middle and lower zones. This is an example of the degree of agreement and harmony of activities of general interest for the community. The communities' contribution in labor for this activity is equivalent to U.S.$12,000.
- From 1994 to 1996 a total of 172,220 trees were planted in 15 of the 22 villages in the Cabuyal River catchment, with the participation of 1050 farmers, school teachers, and students. The farmers' participation and access to information and technologies have made it possible to increase the number of native trees. Of the trees planted, 60% have been fast-growing species as an energy source, an important need in the zone. The remainder were native species, bamboo, ornamentals, and fruit trees.
- There has been a process of spontaneous adoption, i.e., with no direct intervention by extension agents, of live barriers. Barriers most acceptable to farmers are imperial grass (*Axonopus scopanus*) (91%), sugarcane (62%), citronella (48%), pineapple (30%), lemon grass (26%), and vetiver grass (14%). This selection of materials is related to their double purpose: not only do they retain soil nutrients, but they also serve as feed for animals, food, or for making essential oils, and they do not compete with the main food crops.
- The community has contributed 750 days of labor toward activities restoring two lagoons used in the past for recreation. Activities included clearing aquatic weeds, planting ornamental trees, and improving surrounding ground cover.

- A proposed curriculum in environmental education began a cultural process within the 32 elementary and secondary schools. This will gradually be extended to include community adult education.
- Establishment of rural agroindustries as an alternative for generating employment is a high priority. Milk that was previously sold raw, with no value added, is now processed into seven products including creamy white cheese, two types of farmers' white cheese, smooth cottage cheese, kumiss, provolone, and yogurt

A LANDSCAPE THAT UNITES

Throughout tropical American hillsides, the vast majority of watershed inhabitants scratch a living from small farms on infertile soils, often on steep slopes vulnerable to erosion. Shortages of just about everything — land, water, labor, inputs, cash, credit, schools, clinics, roads, transport, and communications — frustrate their daily efforts to escape from poverty. In this difficult setting, what difference can local community consortia, like the one in the Rio Cabuyal watershed, make? One of the most significant contributions so far has been to give local people a tangible, physical view of the landscape that unites them. Taking up much of the floor space in the consortium's small office is a cumbersome but colorful relief model of their watershed. Built in Styrofoam and papier-mâche, using maps generated by GIS, the model was taken to each of the watershed villages in turn, where local people painted in the streams, roads, fields, houses, and other familiar features (Figure 9.3a). A standard color coding was used to denote land use — brown for coffee, deep green for pasture, dark brown for areas cleared by burning, bright red for eroded land, and so on. The result is a powerful tool for stimulating group discussion and creating a sense of community (CIAT, 1996).

Add to the relief model the power of strategic scientific knowledge and interactive decision-support systems (Figure 9.3b), and the result can help stakeholders ranging from farmers to local government agencies or national policy makers assess more realistically the possible outcomes of different resource management choices. We close with two examples of results of sophisticated simulation studies analyzing the financial sustainability for a 15-year period of an actual hillside farm from the Rio Cabuyal watershed. The sustainability probability index of the base cropping system scenario was set at 0.64 and increases dramatically to 0.99 (scale 0 to 1.00) with the substitution of irrigated tomato for a rainfed crop (Hansen, 1996). An obvious question that arises from this analysis is, "What happens if 200 farmers all demand irrigation?" A follow-up analysis of the trade-offs of introducing irrigation in this particular watershed is that there is predictable risk of shortfalls of water needed for domestic purposes. Furthermore, the risks and sacrifices for stakeholders are not distributed equally across the watershed (Beinroth et al., 1998). As with the physical relief model, stimulation models are another powerful tool for stimulating group discussion and creating a sense of community.

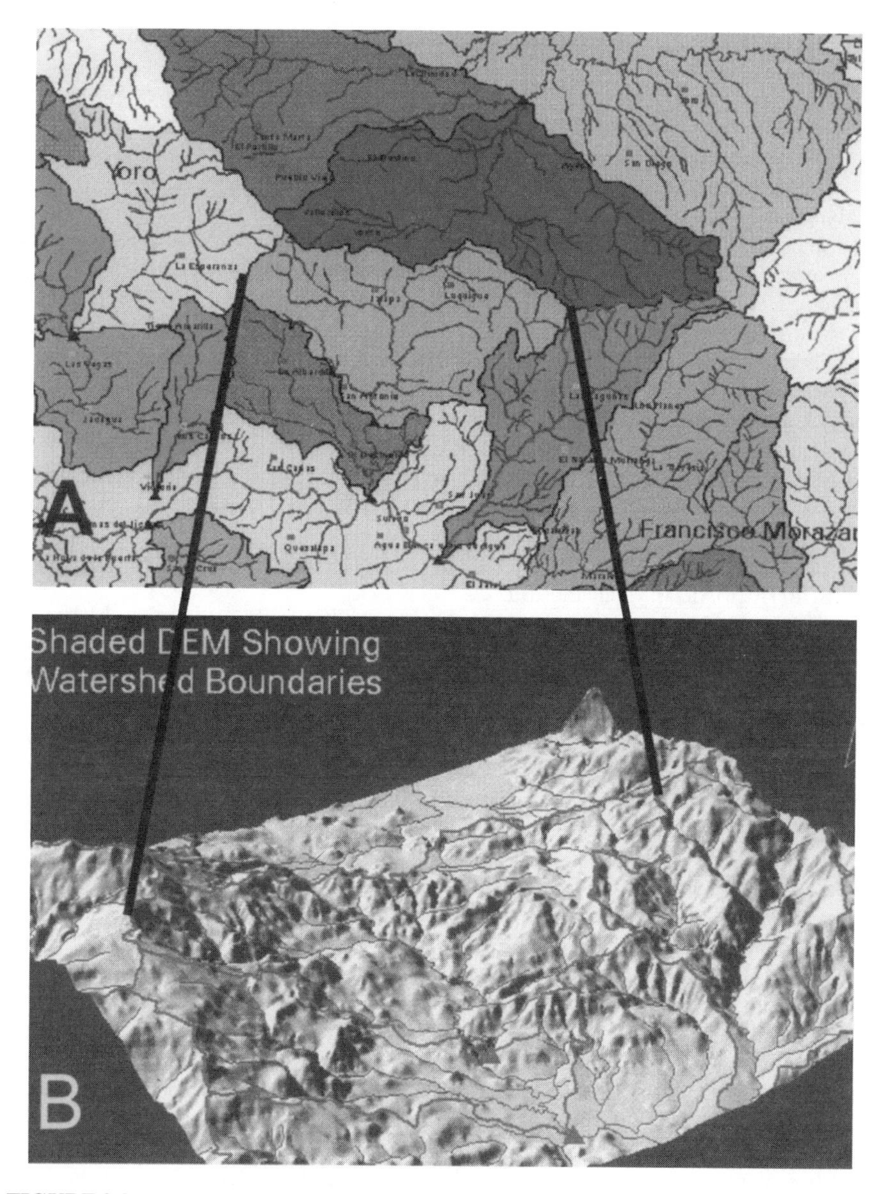

FIGURE 9.3 Two examples of powerful tools for stimulating group discussions and creating a sense of community responsibility for management of watershed resources. (a) Relief model of the watershed inspired by the local consortium and constricted with enormous community involvement. (b) Illustration of a more analytical but complementary approach using interactive GIS and simulation.

REFERENCES

Acton, D.F. and Padbury, G.A. 1994. A conceptual framework for soil quality assessment and monitoring, in Acton, D.F., Ed., *A Program to Assess and Monitor Soil Quality in Canada: Soil Quality Evaluation Program Summary,* Centre for Land and Biological Resources Research Contribution No. 93-49, Ottawa, Ontario, Canada.

Bebbington, A., Merril-Sands, D., and Farrington, J. 1994. Farmer and Community Organisations in Agricultural Research and Extension: Functions, Impacts and Questions, ODI Network Paper 47, Overseas Development Institute, London, 1994.

Beinroth, F.H., Jones, J.W., Knapp, E.B., Papajorgji, P., and Luyten J. 1998.Evaluation of land resources using crop models and a GIS, in Tsuji, G.Y., Hoogenboom, G., and Thornton, P.K., Eds., *Understanding Options for Agricultural Production,* Kluwer Academic Publishers, Dordrecht, The Netherlands, chap. 14.

Berry, J. 1993. There's more than one way to figure slope, in *Beyond Mapping, Concepts, Algorithms and Issues in GIS,* GIS World Books, Fort Collins, CO, 147–151.

CGIAR. 1996, Priorities and Strategies for Soil and Water Aspects of Natural Resources Management Research in the CGIAR, CGIAR TAC Doc. No. SDR/TAC:IAR/96/2.1, April.

CGIAR. 1997, CGIAR Research Priorities for Marginal Lands, CGIAR TAC Draft Doc. SDR/TAC: IAR May, unpublished.

CIAT. 1996. CIAT in Perspective, 1995–96, CIAT, Cali, Colombia.

Conway, G.R. 1987. The properties of agroecosystems, *Agr. Sys.*, 24: 95–117.

ECLAC. 1990. Magnitud de la Pobreza en America Latina en los Affos Ochenta, U.N. Economic Commission for Latin America and the Caribbean, Santiago de Chile, Chile.

ECLAC. 1993. Cambio en el Perfil de las Familias: La Experiencia Regional. U.N. Economic Commission for Latin America and the Caribbean, Santiago de Chile, Chile.

FAO. 1976. A framework for land evaluation. FAO Soils Bulletin #32.

FAO. 1993. An international framework for evaluating sustainable land managment, Dumanski, J. and A.J. Smyth, Eds., FAO Soils Bulletin.

Guba, E. and Lincoln, Y.S. 1989. *Fourth Generation Evaluation,* Sage Publications, London.

Hansen, J.W. 1996. A Systems Approach to Characterizing Farm Sustainability, Ph.D. dissertation, University of Florida, Gainesville.

Keulen, van H., Berkhout, J.A.A., van Diepen, C.A., van Heemst, B.H., H.D.J., Janssen, B.H., Rappoldt, C., and Wolf, J. 1987. Quantitative land evaluation for agro-ecological characterization, in Bunting, A.H., Ed., Agriculture Environments: Characterization, classification and mapping, C.A.B. International, Wallingford, U.K.

Lal, R. 1994. Methods and guidelines for assessing sustainable use of soil and water resources in the tropics, SMSS Technical Monograph No. 21.

Ostrom, E. 1994. *Neither Market nor State: Governance of Common-Pool Resources in the Twenty-First Century,* Lecture Series 2, International Food Policy Research Institute, Washington, D.C.

Pretty, J.N. and Chambers, R. 1993. Toward a Learning Paradigm: New Professionalism and Institutions for Agriculture, IDS Discussion Paper 334, Institute of Development Studies, Sussex.

Riquier, J.D., Bramao, L., and Cornet, J.P. 1970. A new system of soil appraisa in terms of actual and potential productivity (1st approximation: AGL: TESR/70/6) Soil Resources Development and Conservation Service, Land and Water Development Division, FAO, Rome.

Roling, N. 1994. Communication Support for Sustainable Natural Resource Management, *IDS Bull.* 25(2), 125–133.

UNEP. 1992. *World Atlas of Desertification,* United Nations Environment Program, World Health Organization, Nairobi.

Veldkamp, A. and Fresco, L.O. 1996. CLUE: a conceptual model to study the conversion of land use and its effects. *Ecol. Modelling* 85, 253–270.

10 Agroecological Impacts of Five Years of Practical Program for Restoration of a Degraded Sahelian Watershed

Andrew Manu, Thomas Thurow, Anthony S.R. Juo, and Ibrahim Zanguina

CONTENTS

0-8493-0702-3/00/$0.00+$.50

INTRODUCTION

Increased human and livestock populations in recent decades have resulted in serious erosion, rapid decline of soil fertility on croplands, and loss of vegetative cover on grazing lands in the Sahel. This project adopted a watershed as a unit to test a natural resource management approach that integrated the efforts of scientists, administrators, and the farming community. After assessing watershed attributes through intensive surveys, information was integrated into a GIS to identify land management units. Water-harvesting structures were used to capture runoff water and eroded soil for the establishment of keystone species (*Acacia holosericea*) on degraded plateau surfaces and escarpment. Survival rates of plants ranged from 77 to 97%. These agroforestry activities induced autogenic production of herbaceous species which served as alternate source of fodder for livestock. Results from a 3-year protected fallow trial indicated a good potential for soil fertility rejuvenation. Available P levels increased by 65% under protected fallow. There was an 150% increase in organic matter during the fallow period. Farmer-participatory, on-farm research and demonstration trials showed that millet and cowpea yields can be increased from 70 to 185 kg/ha though the use of high densities, inorganic fertilizers, and manures. This project demonstrated a clear case of progress through partnership to improve food crop production while maintaining the integrity of the natural resource base. However, to be effective, the approach and results of the project need to be transferred to other degraded watersheds to generate wide, long-term ecological and socioeconomic impact.

A gradual decline in production potential has occurred in many regions of the Sahel over the last several decades (Le Houérou, 1996). Continued land degradation is further threatening the ecological and economic stability of the region. Environmental degradation is often blamed on recurrent droughts. However, ecosystem change associated with drought cannot be interpreted independently of the concurrent dramatic increase in livestock and human pressure (Cissé, 1981). The human population of the region has been growing at approximately 3% per year (WRI, 1992). This increase coincides with increased intensity and extent of cultivation, grazing, and deforestation. In many regions of the Sahel, these intensified patterns of land use have created an unrelenting stress with little concern for maintenance of the health of the ecosystem. This has led to accelerated erosion, runoff, and decline in fertility. When comparing the impact of droughts from the early part of the century with recent droughts, it appears that intensified land-use pressure is the primary reason for ecosystem degradation (Laya, 1975; Lamprey, 1988), with drought being a stress that combines with the intensified land-use stress to push a system beyond a "threshold" of stability into cycle of degradation (Holling, 1973; May, 1977).

The population of this ecologically sensitive region has been aware of the long-term consequences of the environmental degradation processes. In many cases, the local people have attempted to practice natural resource management within the context of their socioeconomic limitations. There is also a history of research and development projects, funded in large part by international donors, seeking to halt the ongoing degradation. There has been extensive research in the use of mulches for the improvement of the physical and chemical characteristics of the surfaces of degraded Sahelian soils (Chase and Boudouresque, 1987; Valentin and Bresson, 1992; Mando et al., 1996). On a degraded soil in the upland soils in the Casamance, Senegal, a combination of mineral fertilizers, organic amendments in a cereal (millet), and legume (groundnut) rotation produced a significant increase in soil fertility with a consequent sustained yields (Diatta and Siband, 1997). Both the indigenous and outside-funded efforts have generally not approached their expected potential. Most of these efforts have taken a "reductionist" approach to management and restoration, whereby efforts have been designed and implemented for specific sites on the landscape with boundaries defined by ownership patterns or farming systems. These efforts fail to view the restoration of the target sites as integral components of a highly interconnected landscape. Also, the proposed interventions often do not consider the socioeconomic factors that are vital to sustainability of a practice. It has been observed that restoration success can be significantly improved by adopting a holistic approach that considers the social structures and the dynamics of ecological processes across the landscape (Thurow and Juo, 1995; Whisenant and Tongway, 1995).

This chapter reports on the approach and impact of a land management project in Niger, West Africa. The goal of this project was to develop a methodology that would enhance sustainable management of the natural resources in a manner that would better meet the needs of the current residents without destroying the ability of future generations to benefit from the natural resources. This project focused on a watershed toposequence as the management unit because this scale of landscape characterization embodied the water and energy flow patterns which integrate the various component parts of agroecosystem. The objectives of this activity were to (1) determine the surface hydrology attributes through inventories of the natural resources and indigenous knowledge of farmers, (2) delineate functional land management units on the landscape, (3) integrate, test, and demonstrate improved technologies, and (4) disseminate project methodologies and results to researchers, planners, farmers, and private volunteer organizations (PVO) through appropriate technology transfer.

THE SETTING

The watershed is located near Hamdallaye (13°34′N, 2°35′E) which is 45 km NE of Niger's capital city, Niamey (Figure 10.1) and occupies and area of 500 ha. The average annual rainfall (450 mm) is dispersed between May and October. Rainfall is characterized by extreme temporal and spatial variability. Typical of the Sahel, this variability is related to the randomness of the prevailing convective storms (Nicholson, 1983). The watershed is located in a plateau and valley geomorphic zone. The drainage of the area extends from the flat (0 to 1%), laterite-capped, degraded plateau, down a 40-m-high sparsely vegetated escarpment with 2 to 8% slope down and into a broad, gently sloping (0 to 1%) sand valley.

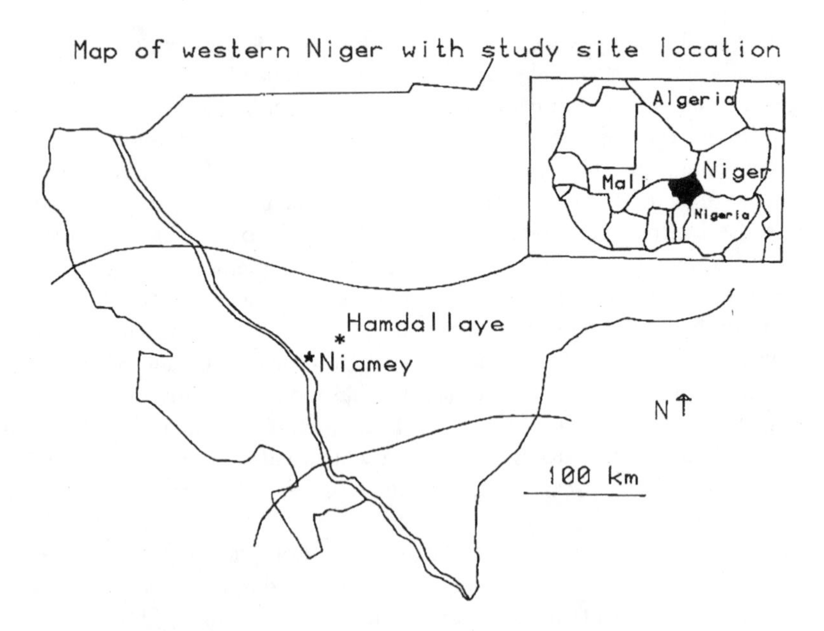

FIGURE 10.1 Map of western Niger showing the study site location.

THE WATERSHED MANAGEMENT APPROACH

MATERIALS AND METHODS

The project began with the collection of data for biophysical and socioeconomic characterization of the watershed. Surveys of farming families, traditional chiefs, and local administrators were also conducted to assess farming systems, traditional methods of soil and crop management, land tenure, gender issues, and other socioeconomic factors operating in the watershed. In particular, farmer views on constraints to sustainable agricultural production were solicited and documented (Taylor-Powell et al., 1991). Soil-mapping units were delineated from a preliminary survey along five transects traversing the watershed along a toposequence from plateau to the valley (Figure 10.2). Profiles were described along the transects and sampled. Soil samples crushed to pass a 2-mm sieve were analyzed for texture by the pipette method of Kilmer and Alexander (1949). Soil pH was determined with a glass electrode in a 1:1 (w/v) ratio of soil and water suspension (Soil Survey Staff, 1972). Available P was measured using the Bray-1 method as outlined in Olsen and Sommers (1982). Organic matter content was determined by the Walkley–Black (1934) wet oxidation method. Exchangeable bases (Ca, Mg, K, and Na) were displaced with NH_4OAc (Thomas, 1982). Concentrations of Ca and Mg were determined by atomic absorption spectrophotometry, while K was determined using flame photometry. Exchangeable Al was measured after extraction in 1 *M* KCl as described by McLean (1982).

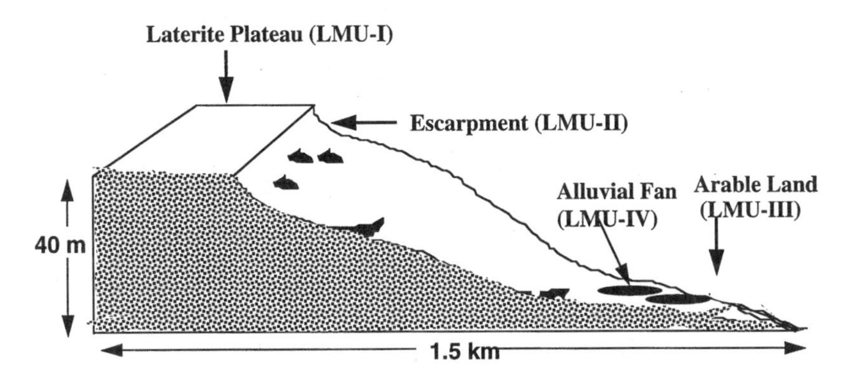

FIGURE 10.2 Schematic representation of the cross section of the study site showing the land management units.

Information obtained from surveys and analyses was integrated into a geographic information system (GIS) (Manu et al., 1994). The data layers integrated into the GIS included topography, soils, drainage network, vegetation, land degradation status, and land use and management. This activity culminated in the identification of land management units (LMUs) which formed the organizational foundation for participatory research and demonstration activities in the watershed (see Figure 10.2). The intervention phase of the project was used to carry out research and demonstration activities that integrated risk-reducing and resource-conserving components of the traditional agropastoral systems with site-compatible elements of modern technology. The field research and demonstration activities were to a large extent determined by the perspective of the farmers and their indigenous technologies. In the crop production trials, millet and cowpea grain yields were determined for the different treatments.

The GLM procedure of SAS Institute (1989) was used to test differences between means by the Duncan multiple range test.

Results and Discussion

Watershed Attributes

Vegetation

The plateau and escarpment (LMU-I and LMU-II) served as source of fuelwood and as a communal grazing land for the local community. Vegetation on the plateau occurred in a "tiger bush" pattern which is typical for the semiarid tropics (White, 1971; Cornet et al., 1992). Various theories that have been advanced to explain the formation and sustainability of the tiger bush (Manu et al., 1994; Ludwig and Tongway, 1995). The stripes of vegetation on the Hamdallaye site which could be 20 to 40 m wide, were separated by 40 to 60 m wide sparsely vegetated corridors. Vegetation in the tiger bush stripes was dominated by 2 to 5 m high deciduous trees, the most common being *Combretum nigricans* and *Acacia macrostachyna*. The following climatic, edaphic, and geomorphic conditions are have been identified by Cornet et al. (1992) as being necessary for the formation of this vegetative pattern:

TABLE 10.1
Summary of Physical and Chemical Properties of Soils Associated with Tiger Bush on Laterite Plateau

Depth, cm	Clay, %	Sand, %	pH, H_2O	Available P, mg/kg	Exchangeable Cations, cmol$_c$/kg			
					Ca	Mg	K	Al
				Vegetated Zone				
0–18	17.4	65.5	5.2	6.50	1.19	0.76	0.22	0.52
18–60	37.7	47.6	5.3	1.77	0.85	0.63	.029	1.60
				Gaps in Thickets				
0–20	20.3	80.7	5.1	5.5	0.35	0.13	0.03	0.82
20–72	37.6	78.3	4.6	1.48	1.29	0.52	0.12	1.87
				Bare Corridor				
0–12	16.3	71.3	4.9	0.56	0.74	0.41	0.25	0.48
12–54	21.2	52.0	4.3	1.52	0.28	0.47	0.11	1.52

(1) an arid or semiarid climate with few but high-intensity rainfall events, (2) gentle but regular slopes (0.25 to 1.0%), and (3) soils with low permeability and a relative abundance of fine particles that produce intense sheet runoff. These conditions exist in the Hamdallaye watershed.

The native vegetation in the valley was mainly confined to fallow fields. This vegetation was dominated by shrubs and grasses that quickly colonize disturbed sites and discourage grazing through various defense mechanisms such as bad taste, sharp seeds, or high silica content. The dominant shrub was *Guierra senegalensis* and the dominant grasses were *Aristida longiflora*, *Aristida pallida*, and *Eragrostis tremula. Acacia albida* and *Acacia senegal* are legally protected tree species and were scattered throughout the valley as a remnant of the original savanna vegetation community.

Soils

Soils on the plateau follow the configuration of the tiger bush. While surfaces of bare areas in the corridor and gaps were shallow and severely crusted, soils associated with gaps within the thicket were significantly deeper to the underlying laterite and had good surface structure. There was little evidence of biological activity in soil profiles of the bare corridor. Soils in the degraded thicket gaps had higher pH values and were relatively rich in bases and available P (Table 10.1).

Infiltration rates determined using a rainfall simulator were low and not significantly different between the dry season (April) and the rainy season (July) (Figure 10.3). This could be attributed to existing surface crusts and the rapid surface sealing when the soils are wetted. Infiltration rates within the thicket were higher than the degraded bare corridors and degraded thickets, and they were higher in the dry season than in the rainy season. The higher infiltration rates in the thickets meant that they were able to accommodate the runoff from the bare corridors. Thus the bare corridors were runoff areas and the thickets were areas that harvested this runoff.

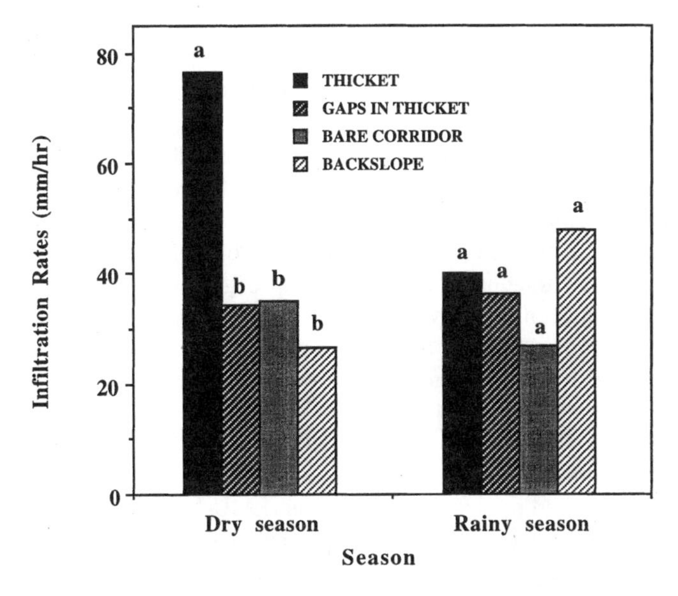

FIGURE 10.3 Mean infiltration rates of different sites on the laterite plateau and backslope. Means with the same letter for the same period are not significantly different ($p < 0.5$) according to the Duncan multiple range test.

TABLE 10.2
Summary of Physical and Chemical Properties of Soils in the Sand Valley of the Hamdallaye Watershed

Depth, cm	Clay, %	Sand, %	pH, H_2O	Organic C, mg/kg	Total N, mg/kg	Avail P, mg/kg	Exchand Cations, $cmol_c$/kg			
							Ca	Mg	K	Al
0–20	5.3	91.4	5.3	0.12	0.01	1.98	0.36	0.08	0.04	0.28
20–100	7.7	85.3	5.3	0.04	0.01	0.55	0.31	0.11	0.02	0.29

The seasonal variation in infiltration rates within the thickets were related to seasonal differences in litter cover and standing herbaceous biomass. Leaf fall from deciduous trees during the rainy season served as surface mulch which facilitated detention storage of surface runoff. The litter thus protected the surface structure by dissipating raindrop energy. Due to enhanced termite and microbiological activities during the rainy season, litter biomass was rapidly decomposed.

The arable soils in the valley have sandy to loamy sand texture, with a low moisture-retention capacity. They were extremely low in organic matter content, slightly to strongly acidic, and had poor structure (Table 10.2). These soils were often subjected to severe wind and water erosion that led to surface sealing (crusting) of the exposed subsoil.

Farming system

The principal crop is millet (*Pennisetum glaucum*) which is often intercropped with cowpea (*Vigna unguiculata)*. Most of the farmers in the watershed owned livestock (goats, sheep, and cattle). The average landholding is about 10 ha/household. About half of the land was fallow (2 to 4 years) at any one time (Taylor-Powell et al., 1991). Farmlands in the watershed were administered under four different villages. Farmers in the valley recognized that their land was impacted by degradation caused by runoff and erosion generated in the upper portions of the watershed. They realized that it was necessary to address some aspects of degradation that were beyond the administrative control of an individual village.

Problems within the Watershed

Overgrazing and overharvesting of fuelwood in the upland resulting in erosion

The land and associated products of the plateau (LMU-I) and escarpment (LMU-II) were considered to be communal property by the local population. This communal ownership is distinguished by individual and independent use of the resources. Individual users had the right to resource exploitation, but had no responsibility for management.

The vegetation on the plateau and backslope had been subjected to heavy cutting, browsing, and grazing; thus most of the larger trees with desirable fuelwood and timber qualities (e.g., *Acacia nilotica*, *Prosopis africana*) as well as most of the perennial grasses and palatable shrubs had been eradicated. This heavy use was also causing the continuity of the vegetation stripes to break apart, allowing runoff and sediment to flow unchecked through the breaks and off the plateau. Torrents of water running off the plateau cut incipient gullies into the adjacent escarpment and through the arable lands (LMU-III) (see Figure 10.2). Sediment eroded from the upslope portion of the landscape was deposited at the end of the gullies creating wide alluvial fans (LMU-IV), which could be 3 to 4 m wide and more than 100 m long. Both the gullies and the alluvial fans disrupted crop production efforts in the valley.

Fertility decline

The sustainable traditional pattern of land use in the watershed was to crop a field for a maximum of 4 years followed by a fallow period of 10 years or more so that the fertility of the field could recover. Due to the dramatic increase in human population and the shortage of arable land, this land-use pattern was no longer possible. Consequently, the land rotation system was faltering or in some instances completely eliminated. This trend, combined with the complete removal of crop residue (used to meet partially the demand for fodder and construction material), disrupted nutrient cycling and hindered the natural regeneration of soil fertility and structure.

All the farmers in the watershed understood the benefits of inorganic fertilizers and 60% of these farmers had occasionally used fertilizers (urea and triple superphosphate). Lack of cash and inability to obtain credit for fertilizer purchase prevented use of fertilizer or resulted in fertilizer applications that were not optimal in either rate or frequency for maximizing crop production.

Supply of manure as an alternate soil amendment was limited by fewer migrating herds passing through the area in recent times. This is the result of decreased availability of native fodder for livestock on the plateau and increased harvest of crop residues for household use.

Inadequate technology transfer

Over the last several decades, several research projects and institutions, both national and international, have made considerable progress in technology development to improve food production in the Sahel. Many of these technologies, however, have only been validated under controlled conditions at experiment stations. While results show the potential of these technologies under the ideal on-station conditions, on-farm adoption has often been undercut by site-specific physical/chemical soil problems and socioeconomic constraints. Conducting on-farm trials of some of the technologies that have the potential to enhance sustainable production in the region was therefore also a part of the efforts in the watershed.

THE WATERSHED RESTORATION APPROACH

Degraded Plateau and Backslope (LMU-I and LMU-II)

The restoration strategy was predicated on testing the hypothesis that reestablishment of the tiger bush vegetation stripes on the plateau would substantially reduce flooding and erosion throughout the watershed. The rationale for this hypothesis was that the degraded plateau was the source area for flash floods which negatively impacted the entire watershed. Restoring vegetation across the entire plateau was economically impractical. We hypothesized that restoring vegetation in the degraded portions of the original tiger bush stripes would improve soil structure and infiltration and thereby prevent runoff from flowing through the vegetation stripes. Therefore, by reestablishing the integrity of the tiger bush stripes, the runoff and erosion would be reduced, both on the plateau and on the downslope sites of the toposequence.

Seedlings were not able to survive if planted into the crusted soils of the degraded thicket gaps. Therefore, it was necessary to create structures that would harvest runoff and the associated suspended nutrients. A variety of microcatchment designs were tested in both the bare and degraded gaps on the plateau. Species to be planted in the microcatchments were selected using the following criteria: (1) easy establishment and rapid growth, (2) perennial growth form that would accelerate amelioration of the environment by trapping windblown soil and litter throughout the year, (3) an extensive root system for effective nutrient cycling, (4) tolerance to both drought and saturated conditions, (5) resistance to browsing, and (6) provision of valued products such as timber or fuelwood.

Species tested in the microcatchments included *Acacia holosericea*, *Prosopis juliflora*, *Ziziphus mauritiana*, *Bauhemia rufuscens*, and *Andropogon guyanus*. We hypothesized that establishment of these species would initiate autogenic successional processes through modification of the harsh environment of the degraded portions of the plateau.

TABLE 10.3
Survival and Growth of Species 16 Months after Being Planted in the Microcatchments on the Laterite Plateau

Species	Survival (%)	Height (cm)	Basal Diameter (cm)
Acacia holosericea	96	209	5.2
Ziziphus mauritania	89	80	1.1
Bauhinia rufescens	97	75	1.6
Prosopis juliflora	89	73	0.6
Andropogon guyanus	77	173	6.3

The most successful design was a 2-m^2 microcatchment (Manu et al., 1994). Survival rate of all species planted in the microcatchments was very good, ranging from 77 to 97%. *Acacia holosericea* showed the highest growth and survival rates. The growth rate of the other tree species was negligible during the first year but had accelerated during the second year (Table 10.3). Several arid land tree species initially put most of their energy into establishing a root system before increasing vegetative growth. To date, the rapid growth of *Acacia holosericae* has promoted the autogenic regeneration of herbaceous and shrub species, especially species that disappeared as a result of environmental degradation. *Andropogon guyanus*, a bunchgrass, although slow in its initial growth, grew at a faster rate during the second year. The dense bunch growth form favorably stabilized the microcatchment by obstructing overland flow.

A preliminary evaluation of the microenvironment around the catchments shows enhanced microbiological activity which could be related the increased organic matter and nutrients availability through nutrient cycling. The previously crusted soil is now friable as a result of the cover, organic matter input, and increased microorganism activity. Natural succession of a variety of native species is evident, as is the reappearance of wildlife within the dense vegetation. The researchers and local residents were both surprised by how quickly trees could become reestablished on a denuded site (Figures 10.4a to 10.4d) and how quickly an autogenic succession sequence could be initiated when microcatchment technologies were used to concentrate water and nutrients. Part of the success of the plant reestablishment effort was that the existing patterns of runoff and nutrient flow were exploited to concentrate scarce resources in the microcatchments, thereby creating fertile islands that enhanced production potential and altered the microenvironment allowing for natural succession to proceed.

Runoff from the portion of the plateau with the restored tiger bush stripes has been reduced to the point that the farmers in the valley no longer suffer from the damage associated with gully formation and creation of alluvial fans in their fields. A testament to the success of this treatment is that farmers have approached repre-

FIGURE 10.4a Microcatchments installed on the degraded plateau. Microcatchments create a favorable environment for tree establishment by harvesting runoff and the associated sediment and nutrients.

FIGURE 10.4b *Acacia holosericea* seedling planted in the microcatchment. This species is resistant to grazing, fixes nitrogen, and grows rapidly. It was the most successful of the tree species tested.

FIGURE 10.4c *Acacia holosericea* 6 months after the seedlings were planted.

FIGURE 10.4d *Acacia holosericea* 15 months after the seedings were planted. At this stage some of the tree branches can begin to be harvested in a controlled manner to provide for the fuelwood needs of the watershed residents while still maintaining runoff and erosion control.

sentatives of the Ministry of Agriculture requesting that, if tools would be lent and seedlings provided, they would volunteer their labor to install additional microcatchments for the purpose of reducing runoff from other areas of the plateau that threaten their fields.

Degraded Arable Land (LMU-III): Natural Fallow

Due to the breakdown of the traditional fallow system, farmers in some areas of Africa have shifted from open grazing of fallow land to fallow systems which are fenced to exclude livestock. By protecting the fallow from grazing pressure, natural succession and reestablishment of soil fertility is accelerated (Holt, 1989). In addition to influencing soil fertility through nutrient cycling, vegetation would trap and hold dust blowing across the site. In the Sahel where dust storms are frequent, dust entrapment constitutes an important nutrient renewal vector. Chase and Boudouresque (1987) reported that autogenic revegetation on barren crusted plateau could be enhanced through trapping of eolian material. This material, which could be local or distant, could provide 6 kg/ha/yr of Ca, 1 kg/ha/yr of Na, 3 kg/ha/yr of K, and 8 kg/ha/yr of Mg (Drees et al., 1993).

To test the efficiency of protected fallow, exclosures were erected on a site in the arable sand valley following continuous cultivation of millet for over 5 years. Surface soil fertility parameters were evaluated after 3 years of protected fallow. These were compared with similar parameters from an adjacent unprotected site which had been in continuous cultivation during the same 3 years as well as for over the previous 5 year period.

Protected fallow for 3 years significantly improved soil fertility. Available phosphorus (P) increased by 65% when compared with the continuously cropped site (Table 10.4). Calcium, the dominant cation on the exchange complex of the soil, also increased significantly. A significant 150% increase in organic carbon was also reflected in a high total N level in the soil. The increases in organic carbon and available P in the surface soil were especially significant because soils in the region have inherently low levels of organic matter and widespread P deficiency. Millet yield responses to soil fertility enhancement resulting from the protected fallow were also highly significant.

TABLE 10.4
Summary of Selected Chemical Parameters of Fallow and Continuously Cropped Soils[a]

Soil Property	Fallow	Continuous Cropping
pH	4.6a	4.5a
Bray-1 P (mg/kg)	3.8a	2.26b
Total N (mg/kg)	130a	60b
Organic C (mg/kg)	0.15a	0.06b
Exchangeable bases		

[a] Values with the same letter are not significantly different ($p > 0.10$).

Improvement of Crop Production on Arable Land (LMU-III): On-Farm Research and Demonstration

The objectives of this activity were (1) to test various crop and soil management technologies as influenced by farmer skills as opposed to environmental factors and (2) to generate public awareness of improved crop production through the integration of traditional and imported management systems.

On-farm research and demonstration activities based on guidelines suggested by Shanner et al. (1981) were established on arable lands. In all, 16 farmers were involved in farmer-managed trials (FMT) which were designed to determine yields on fields using extension advice but no direct input by researchers. Similar technologies tested by farmers were also implemented by a researcher group (RMT). Choice of management packages and treatments was based on a reiterative and protracted process which incorporated results from the earlier diagnostic watershed survey and previous on-station research in the region (Peiri, 1985; Bationo et al., 1989; Geiger et al., 1992). Identification of technological packages was also based on research results of national and international research organizations such as Institut National de Recherches Agronomiques du Niger (INRAN), Soil Management Collaborative Research Support Program (SMCRSP), International Crops Research Institute for the Semiarid Tropics (ICRISAT), International Fertilizer development Corporation (IFDC), and PVOs. Additional considerations for choice of interventions were based on their technical and economic feasibility and sociocultural compatibility with the watershed residents' desires. The five management technologies tested on the millet–cowpea intercrop system are presented in Table 10.5.

TABLE 10.5
Agronomic Treatments Tested in the Watershed

Treatment	Designation	Amendment	Pocket Density[a]
Traditional millet/cowpea Intercrop System	TC	None	FMT = 10,000 RMT = 17,778
High density of millet/cowpea intercrop with fertilizer	HF	22.5 kg P_2O_5/ha	FMT = 10,000 RMT = 17,778
Crop Rotation Cowpea — 1992	CR	22.5 kg P_2O_5/ha	FMT = 35,555 RMT = 35,555
Millet — 1993	None		FMT = 8,889 RMT = 8,889
Manure fertilization with high plant density millet/cowpea intercrop	MAN	5 tons of farmyard manure/ha	FMT = 17,778 RMT = 17,778
Millet residue mulch with high density millet/cowpea intercrop	MR	4 tons of millet residue/ha	FMT = 17,778 RMT = 17,778

[a] Number of planting holes per hectare. FMT = farmer-managed trials; RMT = researcher-managed trials.

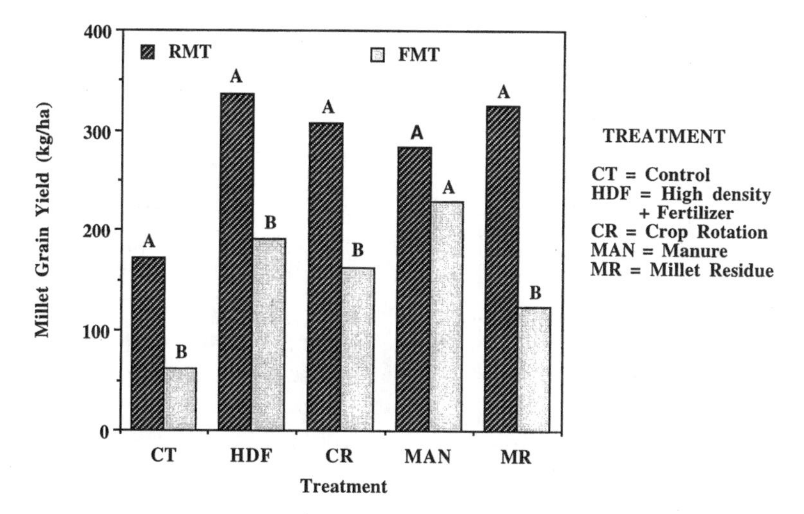

FIGURE 10.5 Millet grain yield of FMT and RMT as a function of treatment in 1993. Bars labeled with the same letter within the same treatment are not significantly different ($p < 0.5$) according to the Duncan multiple range test.

Millet Yields

With the exception of the manure treatment, millet yields associated with all treatments were higher in the RMT than in the FMT in both years of the experiment (Figure 10.5). This demonstrated farmers' familiarity with and mastery of the use of farmyard manure. However, many farmers delayed implementation of recommendations of when to weed, thin, and apply urea. These cultural practices demand much labor and were the bottleneck of the production system.

The traditional millet/cowpea intercrop system (TC) had the lowest millet yields of any of the treatments and the FMT TC had a large range between different field plots (7.0 to 287 kg/ha in 1992; 0.8 to 292 kg/ha in 1993). This wide range in yields on TC could be primarily attributed to the large variation in skill and diligence among farmers and the great spatial variability of soils within fields (Scott-Wendt et al., 1988; Manu et al., 1996). Yields obtained with inorganic fertilizers were lower in the FMT than the RMT. This could be the result of farmers' unfamiliarity with the use of fertilizers.

Cowpea Yields

Because of erratic rain in 1993, cowpea yields were only obtained in 1992 (Figure 10.6). Unlike the millet crop where the RMT consistently had greater yields than FMT, there was no significant difference between managers in cowpea yield, except for the monoculture cowpea fields of the crop rotation treatment. This large disparity in sole cowpea crop yields between the RMT and FMT could be due to the difference in efficiency of insect control between managers. While insecticides

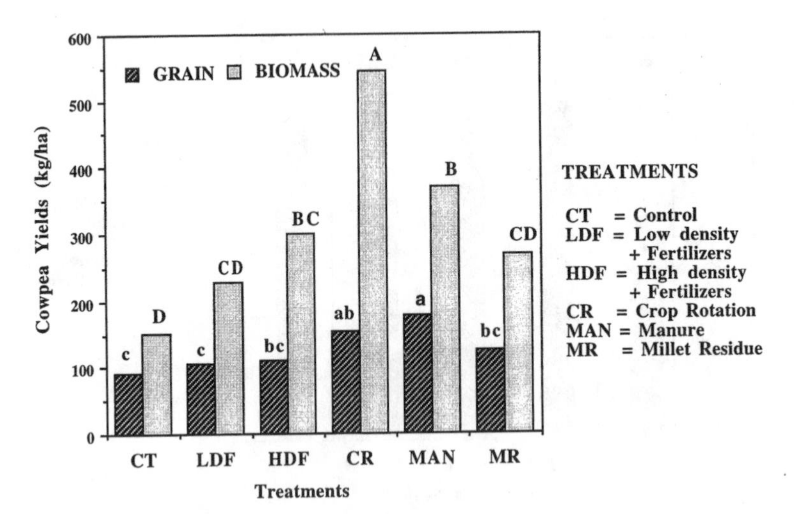

FIGURE 10.6 Cowpea grain yield of FMT and RMT as a function of treatment in 1992. Bars labeled with the same letter within the same treatment are not significantly different ($p < 0.5$) according to the Duncan multiple range test.

were adequately applied in a timely manner by the researcher group, it was generally observed that the timeliness and efficiency of insecticide application varied among farmers. Second, farmers consider cowpea hay as a higher potential source of revenue than the grain and therefore give more emphasis to cowpea fodder than grain harvest.

TECHNOLOGY TRANSFER

As a participatory approach to the efficient management of landscapes, farmers were made an integral component of all facets of the project. Participant farmers played a significant role in the diffusion of project findings and innovations to nonparticipant farmers. Impressions of project activities were relayed at social gatherings and through farmers' show-and-tell approach in the field.

For a wider diffusion of project results to a broader range of clients, yearly field day exhibitions were organized. In addition to farmers in the entire canton of Hamdallaye and beyond, other participants in these field day celebrations included representatives from many national ministries, and a wide variety of international organizations — U.S. Agency for International Development, Food and Agricultural Organization of the United Nations (FAO), the World Bank, CARE International, ICRISAT, PVOs, and nongovernmental agencies. These field days facilitated a dialogue that aided information transfer among the farmers, administrators, and scientists. They served as a forum for soliciting and obtaining feedback from farmers and researchers.

The U.S. Peace Corps used the project site as part of its orientation of incoming volunteers. Consequently, volunteers adopted and extended some of the project findings to farmers in the villages of the other cantons.

CONCLUSIONS

- The objective of this project was to test a model for the restoration and sustainable management of a small agricultural watershed through an integrated effort of scientists, administrators, and the farming community.
- The project identified a watershed as being the logical scale for sustainable management efforts because of the natural interrelationships that exist among component parts of the watershed.
- Watershed attributes were identified and documented through a series of extensive surveys of soils, vegetation, hydrology, and indigenous knowledge. Information obtained from surveys were integrated into a geographic information system to identify land management units which formed the organizational foundation for participatory research and demonstration activities.
- Restoration techniques designed to exploit the natural water, nutrient, and energy flows showed tremendous potential for rejuvenation of the degraded lateritic uplands. Small microcatchments (2 m^2) were very effective at concentrating water and nutrients necessary for establishment of keystone species capable of initiating an autogenic restoration. This restoration process also eliminated the source of flooding and water erosion that degraded downslope sites.
- A livestock exclusion fallow trial was established to test a natural restoration approach of impoverished sandy cropland soils in the lowlands. Compared with a continuously cropped field which was grazed in the dormant season, there was a significant increase in the livestock exclusion fallow trials in the following soil fertility parameters: available P, exchangeable Ca, organic C, and total N. There was a highly significant millet yield increase in response to soil fertility enhancement from protected fallow.
- Results from other on-farm agronomic research and demonstration activities indicated that there is a good potential for improved and sustained production of millet and cowpea. The technologies adopted integrated modern technologies with indigenous soil and crop management practices. The promising integrated technologies included the combined use of farmyard manure and fertilizers, cereal–legume rotations, and appropriate use of millet residue as surface mulch. The pertinent management practices identified were planting with the earliest sufficient rain, use of adequate planting densities, timeiness, and keeping an appropriate weeding calendar.
- This project demonstrated the desirability of taking a watershed approach to management and restoration of degraded lands. Investment in targeted, practical water erosion control techniques not only initiated autogenic restoration of the uplands, but also stopped the erosive runoff that was partially responsible for degrading the downstream croplands. This action, combined with efforts to combine indigenous knowledge with modern technologies, enhanced the sustainable production potential of the croplands.

REFERENCES

Bationo, A., A.U. Mokwunye, and C.B. Christianson. 1989. Gestion de la fertilité des sols sableux au Niger: un aperçu de quelques resultat de recherches de l'IFDC/ICRISAT, in *Les Actes du Séminaire National sur l'Aménagement des Sols, la Conservation de l'Eau, et la Fertilisation,* Tahoua, Niger, 177–194.

Chase, R.G. and E. Boudouresque. 1987. Methods to stimulate plant regrowth on bare Sahelian forest soils in the region of Niamey, Niger. *Agric. Ecosys. Environ.* 18:211–221.

Cissé, S. 1981. Sedentarizations of nomadic pastoralists and "pastorization" of cultivators in Mali, in D. Aronson, Ed., *The Future of Pastoral People,* CDRI, Ottawa, Ontario, Canada.

Cornet, A.F., C. Montana, J.P. Delhoume, and J. Lopez-Portillo. 1992. Water flows and the dynamics of desert vegetation stripes, in A.J. Hansen and F. di Castri, Eds., *Landscape Boundaries: Consequences for Biotic Diversity and Ecological Flows,* Springer-Verlag, New York, 327–345.

Diatta, S. and P. Siband. 1997. Evolution des sols sous culture continue: le cas ses sols rouges fertilitiques du sud du Sénégal, in G. Renard, A. Neef, K. Becker, and J. von Oppen, Eds., Soil *Fertility Management in West African Land Use Sytems: Proceedings of a Regional Workshop,* University of Hohenheim, March 4–8, Niamey, Niger, 221–229.

Drees, L.R., A. Manu, and L.P. Wilding. 1993. Characterization of eolian dusts in Niger, West Africa, *Geoderma* 59:213–233.

Geiger, S.C., A. Manu, and A. Bationo. 1992. Changes in a sandy Sahelian soil following crop residue and fertilizer additions, *Soil Sci. Soc. Am. J.* 56:172–177.

Holling, C.S. 1973. Resilience and stability of ecological systems, *Annu. Rev. Ecol. Syst.* 4:1–23.

Holt, R.M. 1989. Promising Technologies and Approaches for Sustainable Range and Agroforestry Development in Warm Arid Areas of Central Somalia. Ministry of Livestock, Forestry and Range, Somalia, 167 pp.

Kilmer, V.J. and L.T. Alexander. 1949. Methods of making mechanical analysis of soils, *Soil Sci.* 68:15–24.

Lamprey, H.F. 1988. Report on the desert encroachment reconnaissance in northern Sudan, *Desertification Control Bull.* 17:1–7.

Laya, D. 1975. A'écoute des paysans et des éleveurs au Sahel, *Environ. Afr.* 1:53–101.

Le Hourérou, H.N. 1996. Climate change, drought and desertification, *J. Arid Environ.* 34:133–185.

Ludwig, J.A. and D.J. Tongway. 1995. Spatial organization of landscapes and its function in semi-arid woodlands, Australia, *Landscape Ecol.* 10:51–63.

Mando, A., L. Stroosnijder, and L. Brusaard. 1996. Effects of termites on infiltration into crusted soil, *Geoderma* 74:107–113.

Manu, A., T.L. Thurow, A.S.R. Juo, I. Sanguina, M. Gandah, and I. Mahamane. 1994. Sustainable land management in the Sahel: a case study of an agricultural watershed at Hamdallaye, Niger, *TropSoil Bull.* 94-01, Soil and Crop Sciences Department, Texas A&M University, College Station.

Manu, A., M. Salou, L.R. Hossner, L.P. Wilding, and A.S.R. Juo. 1996. Soil-related plant growth variability in the Sahel with special reference to western Niger, *TropSoil Bull.* 96-01. Soil and Crop Sciences Department, Texas A&M University, College Station.

May, R.M. 1977. Thresholds and breaking points in ecosystems with a multiplicity of steady states, *Nature* 269:471–477.

McLean, E.O. 1982. Soil pH and lime requirement, in A.L. Page et al., Eds., *Methods of Soil Analysis*. Part 2, 2nd ed., Agron Monogr. 9, ASA and SSSA, Madison, WI, 199–224.

Nicholson, S.E. 1983. The climatology of Subsaharan Africa, in *Environmental Change in the West African Sahel,* National Academy Press, Washington, D.C., 71–92.
Olsen, S.R., and L.E. Sommers. 1982. Phosphorus, in A.L. Page et al., Eds., *Methods of Soil Analysis*. Part 2, 2nd ed., Agron Monogr. 9. ASA and SSSA, Madison, WI, 403–430.
Peiri, C. 1985. Food Crop fertilization and soil fertility, in *Appropriate technologies for farmers in semi-arid West Africa,* Ohm, H.W. and Nagy, J.G., Eds., Purdue University Press, West Lafayette, IN.
SAS Institute. 1989. SAS/STA user's guide, Version 6, 4th ed., Vol. 2, SAS Inst., Cary, NC.
Scott-Wendt, J., L.R. Hossner, and R.G. Chase. 1988. Variability in millet (*Pennisetum americanum*) fields in semiarid West Africa, *Arid Soil Res. Rehab.,* 2:49–58.
Shanner, W.W., P.F. Phillip, and W.R. Schmehl. 1981. Farming systems research and development: Guidelines for developing countries, Westview Press, Boulder, CO.
Soil Survey Staff. 1972. Soil survey laboratory methods and procedures for collecting soil samples, *USDA-SCS Soil Surv. Invest. Rep. 1,* U.S. Gov. Print. Office, Washington, D.C.
Taylor-Powell, E., A. Manu, S.C. Geiger, M. Ouattara, and A.S.R. Juo. 1991. Integrated management of agricultural watersheds: land tenure and indigenous knowledge of soil and crop management, *TropSoils Bull.* 91-04, Soil and Crop Sciences Department, Texas A&M University, College Station.
Thomas, G.W. 1982. Exchangeable cations, in A.L. Page et al., Ed., *Methods of Soil Analysis,* Part 2, 2nd ed., Agron. Monogr. 9. ASA and SSSA, Madison, WI, 159–166.
Thurow, T.L. and A.S.R Juo. 1995. The rationale for using a watershed as the basis for planning and development, *ASA Spec. Publ.,* 60, 93–116.
Valentin, C. and L.M. Bresson. 1992. Morphology, genesis and classification of surface crusts in loamy and sandy soils, *Geoderma* 55:225–245.
Walkley, A. and I.A. Black, 1934. An examination of the Degtjareff method for determining soil organic matter and a proposed modification of the chromic acid titration method, *Soil Sci.* 37:29–38.
Whisenant S.G. and Tongway, D.J. 1995. Repairing mesoscale processes during restoration, paper presented at Restoration Ecology Session of the 5th International Rangeland Congress, Salt Lake City, UT, July 23–28.
White, L.P. 1971. Vegetation stripes on sheet wash surfaces, *J. Ecol.* 59:615–622.
WRI. 1992. *World Resources 1992–93.* World Resources Institute, Oxford University Press, New York, 385 pp.

11 Natural Resource Management on a Watershed Scale: What Can Agroforestry Contribute?

Dennis P. Garrity and Fahmuddin Agus

CONTENTS

0-8493-0702-3/00/$0.00+$.50

INTRODUCTION

Successful watershed management is built on two pillars: Sound, practical technical innovation and participatory institutional innovation. Agroforestry has a key role to play in both. Although conventionally seen as a set of technical options applied at the field level, agroforestry is increasingly conceived as a framework for whole-landscape management within a community and ecological context. Watershed degradation poses a threat in many countries in Asia, but past watershed management programs have most frequently been ineffectual. Asian watersheds have the highest sediment loads in the world. Nevertheless, within limits, the evidence indicates that it is possible for smallholders to engage in farming and management of natural forest resources in both a productive and conservation-effective manner. Agroforestry research and development is creating a much wider array of practical solutions that reduce the tension in achieving both the environmental service functions of watersheds and the productivity functions essential to the livelihood of the dense rural populations that inhabit them. "Best-bet" agroforestry systems are reviewed for the three major upland ecosystems within Asian watersheds: the forest margins, *Imperata* grasslands, and permanently farmed hillslopes. The environmental impacts of complex agroforests, smallholder timber- and fruit-tree production systems, improved indigenous fallow management systems, and contour vegetative strip systems are discussed in the context of the above issues. Selected watershed management projects in Thailand, the Philippines, and Indonesia are then examined to draw conclusions on the effective pathways toward effective land husbandry and local natural resource management. Application of the concept of community landscape mosaics as a tool is highlighted. Lessons from these cases, and from two global research programs (ASB and SANREM), indicate that if local communities are allowed to capture the direct benefits of improved systems through tenurial security and involvement in decision making, they will be firm partners in reversing the environmental degradation of Asian watersheds.

STRATEGIC ISSUES IN TROPICAL WATERSHED MANAGEMENT

As increasing populations expand into steeper, more fragile areas in the tropical uplands, more catchments are affected by severe soil erosion, declining soil productivity, and environmental degradation. Watershed degradation now poses a threat to the economies of many countries in Asia, and to the livelihoods of the ever-growing populations that depend on these resources. Unfortunately, past watershed management programs to arrest and reverse this trend have been largely ineffectual. But the lessons learned from these failures have been instrumental in promoting a major change in thinking with regard to watershed management (Douglas, 1996). The two key elements underlying this approach are better land husbandry practices and active people's participation.

Better land husbandry represents a shift in emphasis away from a fixation with soil conservation to a more holistic care of the land for sustained production. It follows recognition that, although there will be trade-offs, the farmer's market objectives can be reconciled with society's watershed objectives such that neither loses and both gain. This affirms that the adoption of appropriate management practices that increase yields can likewise combat land degradation.

Emphasis on active people's participation in watershed management (catchment management in the British terminology) is a recent phenomenon in the tropics. It arose from the glaring pattern of failures observed in past "top-down" methods used by the public sector to implement watershed management projects in which the residents were passive recipients of external interventions. These failures have fostered more serious recognition that success depends upon enhancing rural people's inherent abilities to apply and adapt new and indigenous technologies, and to involve local institutions to manage and conserve resources.

Successful watershed management in the tropics is built on two pillars:

- Sound, practical, suitable technical innovation, and
- Participatory institutional innovation.

Agroforestry has a key role in both. Although conventionally seen as merely a set of technical options applied at the field level, the concept and definition of agroforestry have expanded to envision a role at whole-landscape level. This chapter will explore the role of agroforestry in watershed management in the context of this broader, more holistic vision. This first section summarizes key information concerning watershed management in Asia and some of the major issues that have been highlighted by past experience. The second section explores the role of agroforestry in tropical watersheds, particularly in the context of community landscape mosaics. The third section examines agroforestry in upper watersheds and examines projects in the Philippines and Thailand that are instructive case studies. The succeeding sections look at agroforestry in the context of landscapes dominated by grasslands and continuous cropping, with additional case studies from Indonesia. The chapter concludes by summing up the key points learned that point the way to greater success in future watershed management initiatives.

What has been learned about effective ways of promoting local management of natural resources in the Asian context? Early approaches to soil conservation were developed for large landholdings in temperate regions and were based on structural and engineering treatments (for example, bench terracing). Attempts to apply these approaches to developing country agriculture, characterized by smallholdings, diverse farming systems, extremes of climate and topography, wrenching poverty, weak government institutions, and very limited skills, have been disappointing (Magrath and Doolette, 1990).

Fortunately, alternative technical and institutional approaches are emerging. The concept of conservation-oriented farming in the uplands in which farming systems and realistic farming practices combine to conserve soil and improve total production is now recognized. Two complementary strategies for the development of conservation-oriented upland farming are evolving. The first is the adoption of a problem-solving approach aimed at identifying the key constraints on a site-specific basis. The second is the promotion of a suite of agroforestry-based practices that can form the basis of a comprehensive approach to farming system evolution in the uplands.

One example among these are simple vegetative strip systems that provide a foundation for eventual conversion to tree-based systems. Another is recognition of the immense potential for smallholder complex agroforests that provide robust, sustainable incomes while conserving soil and water resources in ways that closely mimic natural forests themselves.

Conventional approaches to watershed management have had little effect because they were dominated by top-down solutions to problems perceived by external stakeholders, not by the people that live there. External stakeholders, whether national governments or international entities, prescribed solutions, usually large-scale reforestation, on lands managed by local smallholders whose economic implications were diametrically opposed to the de facto land managers food and income security objectives. Forced reforestation has been time and again passively resisted by the destruction or neglect of the plantings. Fire control is essential, and that can only be possible with active and self-interested support of local people. Recognition of reasonable and appropriate land-use rights is also fundamental.

After 50 years of disappointment decision makers have been forced to revisit their assumptions, and to wake up to the potential for collaborating with local farmers on solutions which both increase farm productivity and meet watershed protection objectives. This evokes a new era in which the smallholder is beginning to be seen as a critical part of the solution, not simply the scapegoat for the entire problem.

ASIAN WATERSHEDS

A watershed (or catchment) is defined as the land area drained by a common river system. In Asia, the land area located above 8% slope is operationally considered as watershed area. Land above 30% slope is considered upper watershed. Thus, the conventionally accepted watershed area of Asia is 900 million ha or 53% of the landmass (Magrath and Doolette, 1990). About 65% of the region's rural population of 1.6 billion live in these watershed areas. The managers of these lands are smallholder farmers in rural villages. They are severely constrained by poverty and technology. Therefore, as they seek more farm and grazing land to support their families, they have profound effects on the land and water resources of both the uplands and lowlands.

The population occupying the upper watershed areas is roughly 128 million (Magrath and Doolette, 1990). Increasing populations are accelerating pressure on scarce land and forest resources throughout the region. Approximately 19% of the region is under closed forest. Most of this remaining closed forest is tropical rain forest, the reservoir of about 40% of the biodiversity on the planet Earth. Degradation through overcutting and grazing is reducing productivity on much of the remaining stand (Doolette and Smyle, 1990). The forest cover is receding at a rate of about 1% a year. The most recent estimates suggest that the rate of deforestation is not slowing, but is accelerating. In much of the region, forest resources are integral to the agricultural system as sources of fodder and many other products.

The seriousness of soil erosion is not adequately known, but may be deduced from indirect evidence. The most striking picture is that presented by the rate of sediment passing into the oceans from the major river systems of the world. The global data highlights Asia as being in a class by itself: rates of sediment deposition in the oceans are an order of magnitude higher than from comparable-sized areas anywhere else in the world (Milliman and Meade, 1983). Human pressure on the resource base is by no means the only major driving force for these enormous rates of sediment detachment and deposition. Southeast Asian landscapes tend to be geologically young, and exceptionally steep. These factors are also important; but the densest populations in the world are transforming these watersheds at a tremendous rate, and exacerbating their degradation.

The nations of Southeast Asia are progressively opening their economies, and participation in global markets is accelerating. This is causing profound changes on upland livelihood systems, and on the upland environment. The economies of mainland Southeast Asia are interacting more vigorously than ever before, as borders open and roads and railroads facilitate cross-border trade. World market demand for key perennial tree products produced in insular Southeast Asia is spurring smallholder expansion of rubber, oil palm, tree resins, and various fruits, as well as timber production on farm. These forces will continue to impact land-use change in complex ways well into the future.

Watershed degradation does not have to be an inevitable consequence of using land for agriculture or forestry. It is possible for smallholders to engage in farming and management of natural forest resources in both a productive and conservation-effective manner. Despite the availability of a wide range of options, most development projects have relied on a limited and generally high-cost set of interventions. The issue is the development of the technical capital in resource management, but to an even greater extent it is the social capital to facilitate this process. It is now becoming clear that agricultural productivity in upland areas can be intensified in an environmentally sound and sustainable manner. But new approaches will have to be applied to make this a reality.

Service vs. Production in Watersheds

Outside stakeholders such as lowland populations, national government institutions, and the global community (i.e., all others besides the upland residents themselves) tend to be most deeply concerned about the service functions of watersheds. The attention of national policy makers is naturally drawn to the concerns of the more affluent lowland populations and the impact of upstream–downstream linkages on these groups.

The key *service functions* of concern to outside stakeholders follow:

- *Regulate water flow* to the lowlands to reduce flooding and to provide a dependable water supply to the lower watershed for irrigation and power generation;
- *Prevent soil loss* to protect power generation reservoirs and irrigation structures;
- *Conserve biodiversity* and protect natural ecosystems;
- *Sequester carbon* to alleviate the threat of global warming.

Although these concerns may also be shared to some extent by the resident populations of the watersheds, they are most urgently concerned about the *productivity functions* of watershed resources, that is, to

- *Sustain agricultural production*, and
- *Retain forest resources for local uses: timber, fuel, grazing, nontimber products.*

Can there be practical solutions that can meet both needs? In many circumstances, it is possible to improve the environment and increase the output of goods and services at the same time. One of the major goals of agroforestry research and development in Southeast Asia is to reduce the tension between these two goals by developing a range of choices that are both "service" and "market" oriented (Thomas, 1996). Economic losses from watershed degradation may be divided into on-site and off-site costs. On-site costs derive from the direct effects of degradation on the quality of the natural resources, expressed in terms of declining yields, reduced livestock-carrying capacity, and decreased supply of forest products. Off-site costs result from the indirect effects of degradation on the service functions of the watershed.

The primary justification for watershed management is usually a reduction in off-site costs, particularly when the watershed is upstream from dams or flood-prone valleys or plains. However, it is generally unappreciated that the off-site costs may be of a much lower magnitude than the on-site costs. For example, in Java, Indonesia, annual estimated off-site costs (U.S.$25.6 to 91.2 million) were only a fraction of the on-site costs of U.S.$335 million because of productivity losses. In practice, it is the on-site costs that are the primary economic justification for undertaking a watershed management program. Any reduction in off-site costs should be seen as a secondary justification (Douglas, 1996).

Watershed management involves a range of activities. Each activity would be expected to contribute to the aims of improving the sustained productivity of the natural resources, protecting designated natural ecosystems, and improving rainwater management to provide the quantity and quality of water to meet the different needs of water users within and downstream of the watershed.

THE ROLE OF AGROFORESTRY IN TROPICAL WATERSHED MANAGEMENT

The conventional view of agroforestry is that it is "the deliberate cultivation of woody perennials with agricultural crops on the same unit of land in some form of spatial mixture or sequence." This has led many people to see it merely as a set of distinct prescriptions for land use. This limits its ultimate potential. We now see agroforestry as the increasing integration of trees in land-use systems and conceive it as the evolution of a more mature agroecosystem of increasing ecological integrity. Leakey (1996) proposed that agroforestry be considered a "dynamic, ecologically based, natural resource management system that, through the integration of trees in farm and range land, diversifies and sustains smallholder production for increased social, economic, and environmental benefits." This definition is currently being refined by the International Centre for Research in Agroforestry (ICRAF) as a more holistic concept of agroforestry. It evokes the process of integrating the variety of current agroforestry practices into productive and sustainable land-use systems. Land use becomes progressively more complex, biodiverse, and ecologically and economically resilient. This new vision of agroforestry is transforming the ICRAF approach.

Sanchez (1995) noted that although agroforestry systems have been classified in a number of different ways, ultimately there are two functionally different types, simultaneous systems and sequential systems. Thomas (1996) showed that these may be further classified according to two subcategories based on the land management unit: field-based systems at the household level and landscape-based systems at the village or watershed level. Field-based sequential and simultaneous systems have received dominant attention. These are closely associated with the conventional perception of agroforestry as a suite of farming practices in which trees and crops interact in a field over space and time. Sequential field-based systems are exemplified by fallow rotation or shifting cultivation: crops and secondary (or managed) tree fallows occupy the field in a rotation sequence. Simultaneous systems are typified by alley cropping or complex associations of trees and crops managed in the same field at the same time, such as home gardens or agroforests.

The concept of landscape-based agroforestry systems is much less appreciated, but is most relevant to a discussion of the role of agroforestry at the watershed scale. In these systems, the boundary of the management unit is drawn around a larger landscape unit than an individual field. The determination of an appropriate landscape unit will depend on local conditions, but it would generally extend to the lands in a subwatershed, and directly influenced by whole villages or a group of villages (Thomas, 1996). Landscape-based agroforestry systems incorporate individual fields as components of a broader landscape management system. It moves beyond individual households to include management functions at a community level.

AGROFORESTRY IN UPPER WATERSHEDS: THE FOREST MARGINS

Successful action to mitigate tropical deforestation depends upon a more comprehensive understanding of the forces driving this process. The driving forces vary locally within regions of individual countries. Mitigation efforts must address the interacting forces within each nation and locality. Here we examine two basic issues:

Where forest conversion is inevitable, or has already occurred, what are the preferred land uses that best protect watersheds and provide the environmental services that natural forests would have provided?

Where forest ecosystems have been designated for full conservation, how can the boundaries best be protected?

We examine the agenda for each of these cases in turn.

Retaining Environmental Benefits When Forest Conversion Is Inevitable

In the Indonesian context we observe that a substantial amount of large-scale conversion of natural forests to other land uses is inevitable in the future. This conversion is driven dominantly by pressures to convert forest land to large-scale agricultural and forest plantation enterprises and partially by the unplanned objectives of swelling populations of local smallholders who seek to make a living through the conversion of forests to smallholder farms. We note, however, that the new land uses vary greatly in their ability to substitute for the environmental services provided by natural forests.

Agroforests on Production Forestland. In Indonesia, much of the land designated as "production forest" has been so degraded by inappropriate logging practices that regeneration of forest has not occurred. Devastating fires have intensified the degradative processes. The Ministry of Forestry has sought to keep land designated as production forest from being settled by smallholders. However, hundreds of villages already existed traditionally on lands that were classified as production forest only in recent decades. These villagers have often evolved complex agroforest land-use systems by cultivating mixed perennials with their food crops after slash-and-burn as a logical part of their livelihood strategy. These agroforests are predominantly based on rubber, dipterocarp resin, or fruit species, with timber species husbanded as a component.

Farmer-evolved agroforests often resemble natural secondary forest systems in structure and ecology. The trees provide food, fuel, and cash income. The agroforest accumulates a carbon stock that in some systems may be maintained indefinitely. There are many examples of agroforests in the humid tropics. An outstanding case is the "damar" agroforest system in Lampung, Sumatra, Indonesia (Michon et al., 1995). Over the past century, local populations have extended the cultivation of the dipterocarp tree *Shorea javanica*, which is tapped to yield a resin that is sold for industrial products on the national and international markets. This man-made forest now extends over some 10,000 ha on production forest land in Lampung. It harbors a major proportion of the natural rain forest flora and fauna species (Michon et al., 1995). There are many indirect sources of evidence that soil fertility levels are maintained over long periods, and that soil organic matter levels increase in these mixed systems (Torquebiau, 1992).

These mixed tree-crop systems, particularly complex agroforestry systems or "agroforests," provide an alternative to other land uses to protect watersheds from soil erosion and flooding risk, conserve a greater amount of biodiversity, and provide a greater sustained source of income generation for local communities, than most other forms of crop or tree monocultures. Research on these systems is accumulating empirical evidence that supports this (e.g., Mary and Michon, 1987; Salafsky, 1993; Momberg, 1993; de Foresta and Michon, 1997). Thus, we find that the objectives of smallholder communities practicing such systems are more compatible with those of national governments in protecting watersheds and biodiversity than has previously been assumed.

The most widespread agroforest system in Indonesia is rubber agroforestry (or "jungle rubber") which occupies over 2.5 million ha (Gouyon et al., 1993). In this system rubber trees are the main component, but many other species of fruit and timber trees are combined with rubber, either intentionally or through natural regeneration. The rubber seedlings are established as intercrops planted along with food crops in a slash-and-burn cultivation system. After the 1 to 2 year cropping phase the plot is left alone and the rubber trees mature along with the secondary forest regrowth. Biodiversity levels often approach those of natural secondary forest (Thiollay, 1995; de Foresta and Michon, 1997). Can smallholder jungle rubber systems be made more productive while retaining substantial biodiversity value? We hypothesize that smallholders can substantially increase their rubber yields by incorporating new clonal germ plasm into their present jungle rubber or mixed agroforestry systems, and that this would enable a much broader advance in productivity than investing in intensive monoculture systems. ICRAF and GAPKINDO (the Rubber Association of Indonesia) are investigating this issue in a collaborative project implemented with smallholder rubber farmers in several parts of Sumatra and Kalimantan (Penot, 1996).

Tenure reform. Official recognition of local land and tree tenure systems would underpin their security and enhance the development and expansion of agroforests. Smallholder communities could then contribute substantively to the national production objectives for which the production forests exist. Research and case study experiences are needed as a foundation for policy reform to enable communities to obtain more secure tenure and to make aggregate contributions to the economy. Detailed protocols for developing management agreements between local populations and the national government are essential. ICRAF and its partners in the Alternatives to Slash-and-Burn (ASB) Program are examining how viable instruments may be designed at three levels: the household, the community, and the region. We are focusing on management mechanisms at the community level that provide an appropriate degree of tenurial security while ensuring that the environmental and production objectives of the national government are met.

Environment impacts. Data on the comparability of the environmental services of agroforests vis-à-vis natural or plantation forests will be required. The ASB Program is estimating the implications of alternative land uses, including various types of agroforests, on carbon sequestration, greenhouse gas emissions, and soil erosion (van Noordwijk et al., 1995a). It has been shown, for example, that agroforests are significant sinks for methane, a greenhouse gas generated by wetland rice and other annual crop systems. Tomich et al. (1998) have developed a framework for comparing the economic and environmental impacts of different land-use systems on the forest margins and the trade-offs inherent in land-use choices.

de Foresta and Michon (1997) have emphasized that agroforests are successful only when they meet smallholders' income needs. They note that such a system is usually composed of two sets of commercial tree species suited to local conditions, one set providing regular cash income (e.g., rubber, resin) and the other providing seasonal or irregular cash income. Such composition ensures economic and ecological viability of the forest in the long run, provided that clear tenurial rights on the basic units are recognized. Figure 11.1 is a cross section of an Indonesian smallholder agroforest that illustrates these objectives.

The issue of partially conserving biodiversity outside protected areas through complex agroforests needs to be critically assessed. Conflicts between "nature" and

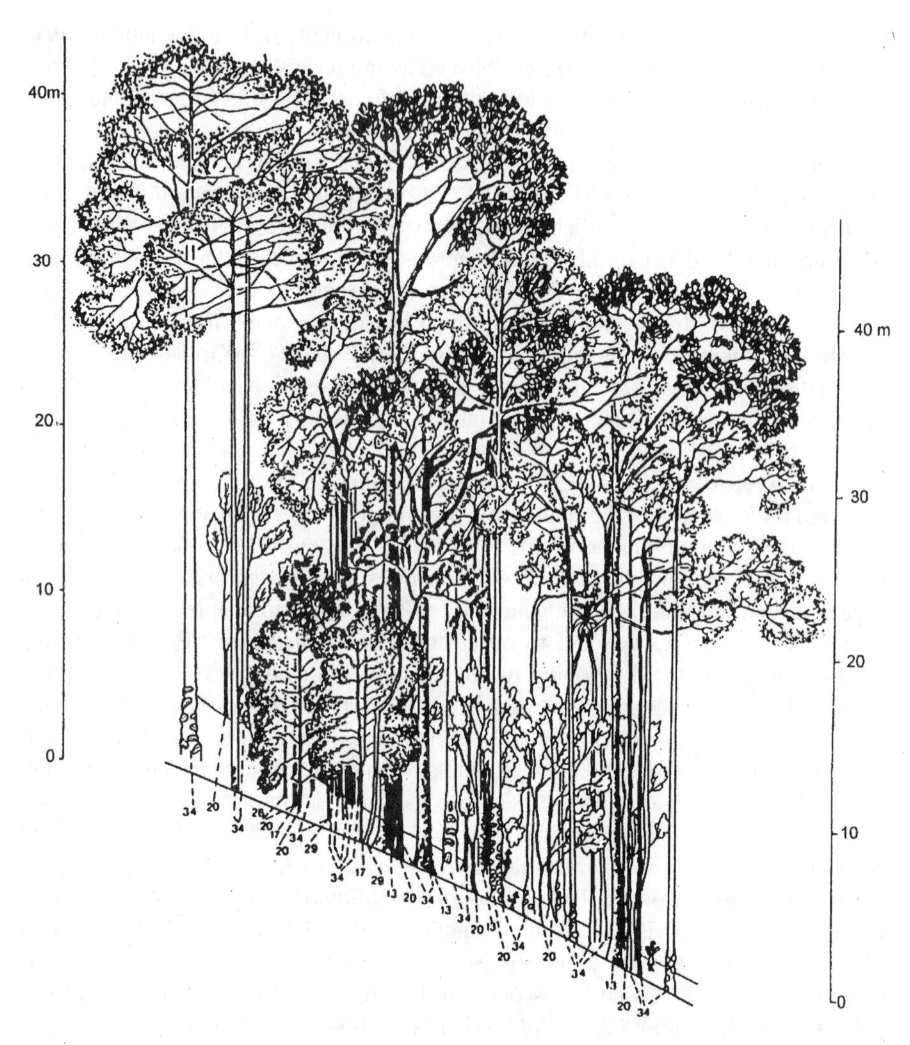

Legend:

34: resin producing tree species, tapped every month and forming the upper canopy as well as the main frame of the forest; valuable timber,
20: understorey fruit tree species, with high commercial value, but seasonal production,
26: understorey fruit tree species, with low commercial value and seasonal production,
17: understorey fruit tree species, with high commercial value, but seasonal production,
29: medium to upper canopy tree species, with valuable timber,
13: upper canopy fruit tree species, with high commercial value and more or less continuous production; valuable timber.

FIGURE 11.1 Example of a mature agroforest designed for both profitability and sustainability in smallholders' conditions.

"agriculture" can be resolved by segregating nature from agricultural land by designating areas of full protection or by integrating nature into the agricultural landscape through production systems that ensure conservation of a major part of the biodiversity of the natural forest (van Noordwijk et al., 1995b). Multifunctional forests and agroforests are examples of the "integrate" option. Mixed strategies may be envisioned where nature reserves coexist with agriculture. Biodiversity will then vary among the facets of the landscape.

Our research is putting strong emphasis on assessing the viability of the integrate option. van Noordwijk et al. (1995b) have proposed a simple model to examine the decision framework. It presumes that relative biodiversity tends to decline more drastically with productivity increases in food crop systems than in jungle rubber agroforestry. This is because the rubber-based system tends to recreate a forest structure that nurtures a range of niches that can be filled by other natural species without detrimental effects on productivity. If this is the case, then agricultural intensification in such types of agroforest systems may be more compatible with a major degree of biodiversity conservation. We are currently examining the implications of this model by surveying the level and nature of the biodiversity maintained in a wide range of agroforest types in Sumatra.

In the highland watersheds of northern Thailand, fruit tree gardens are spreading rapidly (Turkelboom and van Keer, 1996). They are popular because farmers expect higher, more stable income with less labor than with their annual cropping systems. They are also seen as a way to bolster the household's claim to avoid land expropriation by the state. Do fruit tree gardens have similar watershed functions as a forest? Moisture relations were compared over a 2-year period for a litchi (*Litchi chinensis*) garden without undercover, a mixed deciduous forest with limited disturbance, and an annual cultivated field with a short fallow cycle (Figure 11.2). The soil profile was drier in the litchi orchard than in the forest or annual crop field throughout both the wet and dry seasons. The difference was related to human activity in the orchard which caused greater soil compaction that inhibited infiltration in the wet season. In the dry season water consumption was highest in the orchard because the evergreen fruit trees continued transpiration whereas the deciduous forest shed its leaves. These observations suggest that the orchard will increase stream flow in the wet season, and reduce the base flow in the dry season. The data highlight the importance of compaction by human traffic in tree-based systems as a determinant of the water conservation value of an agroforestry practice.

Commercial forest plantations of single, even-aged species have some of the same characteristics of fruit tree gardens (Bannerjee, 1990). Neither may be as good as natural forest, but if forestry departments are content to pursue commercial timber plantations as a means of protecting watersheds, there is little justification for disallowing fruit tree gardens as an alternative for these lands. At this point we have examined the issue of land uses that may best provide favorable environmental services when forest conversion inevitably occurs. What can agroforestry research contribute to protecting the boundaries of natural areas designated for full protection? In such situations we must view the research approach from a very different perspective.

Agroforestry for the Buffer Zones of Protected Ecosystems

National parks and nature reserves are the last-ditch bulwarks of protection for the priceless biodiversity resources of the humid tropical forests. They are under enormous

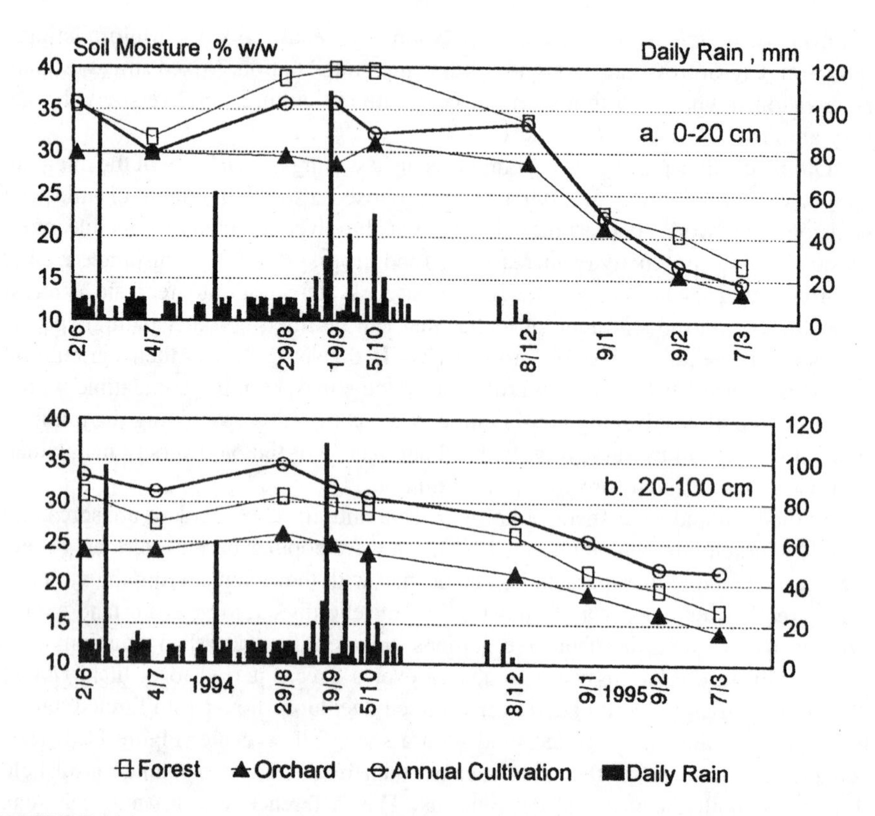

FIGURE 11.2 Average soil moisture content at Mae Sa Mai, northern Thailand, under three contrasting land-use systems. (From Turkelboom, F. and van Keer, K., Eds., *Land Management Research for Highland Agriculture in Transition,* Mae Jo University, Thailand, 1996. With permission.)

threat of encroachment virtually everywhere. The classical method of preserving them has been to declare them off-limits and to enforce the exclusion of local people. Boundaries were delineated and guards patrolled. Unsurprisingly, this is not working. It often results in serious conflicts between the enforcement agency and the local communities. Enforcement alone does not work in most countries because population pressure is too great, the gains captured by local elites by encroachment are too lucrative, or the costs of enforcement are too high. How might agroforestry contribute to alleviating the degradation or ultimate conversion of natural areas? Part of the solution in many cases lies in identifying conditions that reduce or eliminate the economic "necessity" for smallholders (or large-scale operations) to encroach across the protected-forest boundary.

There are now many projects in the tropics called integrated conservation-development projects (ICDPs) that are attempting to save particular natural areas using this approach. Unfortunately, there is a widespread assumption among practitioners of the ICDP approach that people made better off by a development project will refrain from illegal exploitation of a reserve area, even if no enforcement is practiced. Wells and Brandon's (1992) global review of ICDPs found absolutely no evidence to support this. A social contract between communities and outside stakeholders must include enforcement mechanisms in tandem with the development benefits received.

Compensation to communities in terms of development activities may take many forms. Most projects attempt to encourage improved natural resource management practices in the areas outside the reserve. The objectives are to increase people's incomes and to intensify their production systems away from the more extensive, environmentally degrading systems they may currently practice. There is growing interest in the development of more intensive land-use systems on the margins of protected forests and the identification of policy and technology directions to underpin these efforts.

In addition to enforcement and increased land-use intensity there are two other important factors: migration and off-farm employment. If in-migration is occurring, the accelerated population pressure will destabilize the balance between intensification and enforcement. Migration must be controlled in the communities on the boundary. In some areas this has been successfully achieved in mature communities through local land tenure systems (see Cairns, 1994, for an example in mature Minangkabau communities in Indonesia). But in most pioneer communities, local control of migration is problematic.

Conditions in the wider economy play a major role in affecting migration. Off-farm employment for village residents in the buffer zone may be increased or decreased. Figure 11.3 illustrates the park protection problem as a function of four factors: intensity (I) of land use, enforcement (E), migration (M), and off-farm employment (OFE). ICDP programs must consider the implications of all of these factors and their interactions. With the above caveats as a backdrop, improved agroforestry systems have frequently been cited as a path toward appropriate intensification in the buffer zones of protected areas (Wells and Brandon, 1992; Garrity, 1995b; Cairns et al., 1997). Tree planting is often a highly desired intervention by recipient communities near protected areas. Provision of tree germ plasm through nursery programs has therefore been one of the most popular ICDP development interventions. Farm families can increase their nutrition and economic welfare through a greater quantity and diversity of fruit and timber trees in the home garden area and on their farms. In many areas there is an encouraging trend toward tree crop farming as an alternative to practicing shifting cultivation (Garrity and Mercado, 1994).

Where there has been a history of tree crop cultivation in the vicinity of a protected area, the environment of the farming zone outside the boundary develops ecologically favorable characteristics for protection, and even extension, of the biological diversity of the park itself. The "damar" agroforest systems on the boundaries of the Barisan National Park in Lampung, Indonesia, harbor a major proportion of the natural rain forest flora and fauna species (Michon et al., 1995) and effectively act as an extension of the biodiversity of the park itself. Rubber agroforests on the boundary of Kerinci-Seblat National Park in Jambi Province play a similar role (van Noordwijk et al., 1995a). Even in areas where smallholder agroforestry systems do not yield such striking levels of protection or extension for natural biodiversity, the benefits of increased tree cover on the landscape may be important.

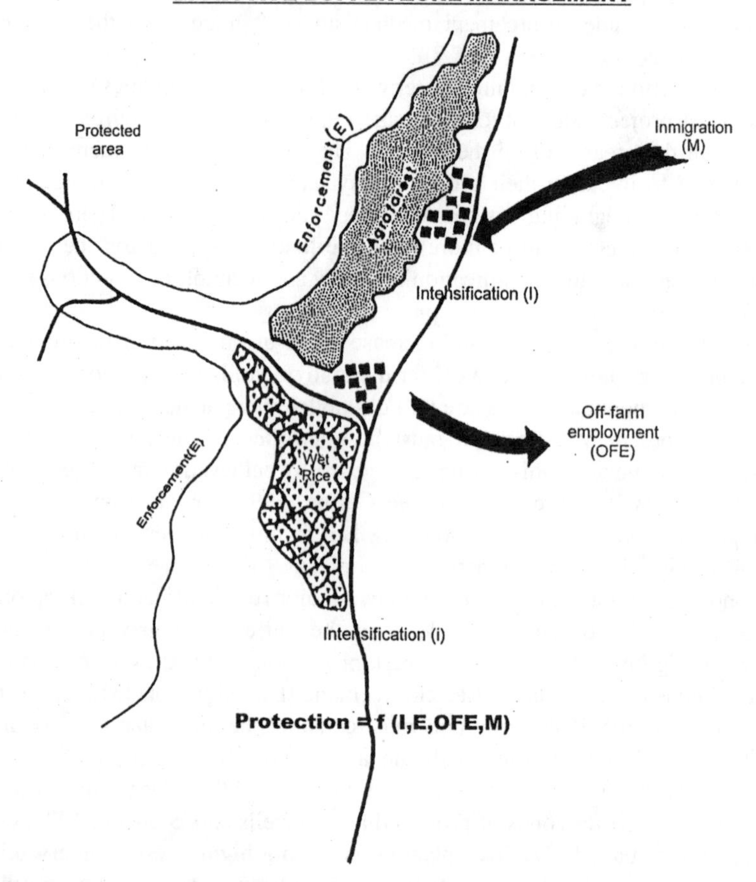

FIGURE 11.3 The protection of a natural area through effective buffer zone management is a function of land-use intensification (I), boundary enforcement (E), off-farm employment (OFE), and migration (M).

Case Study of the Manupali Watershed, Mindanao, the Philippines

Research will play an increasingly important role in providing options and insights for ICDP development. The Sustainable Agriculture and Natural Resources Management (SANREM) Collaborative Research Support Program is a global program that takes a landscape approach with a strong participatory bias. At the SANREM research location in the Manupali Watershed in Mindanao, Philippines, ICRAF is collaborating in a consortium that is developing the elements of a practical social contract for buffer zone management, developing improved agroforestry systems for the buffer zone, and assembling a natural resource management system for the Katanglad National Park. The research team is composed of scientists and practitioners from institutions including ICRAF, non-governmental organizations (NGOs), universities, the tribal community, and local and national government institutions.

We found that the natural resource management strategies of the indigenous Talaandig communities living on the boundary provide a strong foundation for park protection (Cairns, 1995). However, population increase and commercialized vegetable production are causing serious encroachment pressure. The buffer zone area surrounding the park (classified as national production forestland) has high agricultural settlement pressure and is now dominantly grassland and shifting cultivation. The emerging path for household farming systems intensification is small-scale vegetable production combined with timber and fruit tree production. We have surveyed and mapped (Glynn, 1996) the perceptions of local farmers on the performance of current tree species by elevation in the watershed (200 to 1800 m). On the basis of these results we have initiated trials with farmers across this entire transect of elevations to evaluate the most-promising agroforestry species for the range of ecologies and farmer circumstances. We are also working with scores of farmers on conservation farming practices and tree nurseries to elucidate more effective methods of diffusing new practices that will sustain crop yields and increase tree cover. The policy group is tackling the challenge of combining these technical innovations with stronger community-level resource management systems that will support measures to build a "safety net" of active enforcement of the park's integrity. This entails assisting in the development and implementation of a municipal-level natural resource management plan, as well as a management plan for the national park and its buffer zone. The lessons of this approach will be scaled up through partnership with the Integrated Protected Areas Network in the Philippines. Only with democratization and decentralization of power can natural resource management at the local level succeed. Fortunately, this process is well under way in the Philippines. Local governments have begun to have the resources and authority to respond to local needs. In other parts of Southeast Asia such devolution is farther down the road.

In Vietnam some remarkable experiments in participatory resource management are in progress. State lands, including nearly all the forestlands in the country, are being allocated in small chunks to individual households under management contracts. These contracts give the rural family a real stake and responsibility for sustainable management of these lands. The family receives an annual honorarium from the government along with clear and mutually determined management conditions relating to the protection of the ecosystem, the sustainable extraction of forest products, grazing regimes, and others. This is a case of the privatization of natural resource management that is groundbreaking. There will be some remarkable lessons from this experience.

Applying the Landscape-Based Agroforestry Concept: The Case of the Sam Muen Project, Northern Thailand

In Thailand, forest destruction and watershed degradation are of particular concern in the northern highlands, which are the headwaters of all major tributaries of the country's major river artery, the Chao Phraya River. Hundreds of farming villages exist in the upper watersheds, which has spurred the forest department to attempt to reforest lands with timber plantations, to remove populations from protected areas, and to enforce regulations against farming there, resulting in conflict with the resident villagers. These efforts have had limited effect. A framework was necessary that recognized the legitimate rights of communities to reside in upper watersheds and that explored ways in which the service functions of the watershed could be maintained or enhanced while enabling the communities to pursue farming activities that were in reasonable harmony with these objectives.

ICRAF is working with numerous partners to develop landscape management systems in key watersheds. The concept is to move beyond individual households to include management functions at a community level (Thomas, 1996). The agroforestry system is a community watershed land-use mosaic that includes forest, tree, and crop components which interact in numerous ways. The utility of the landscape-based agroforestry concept is illustrated by the experience of the Sam Muen Highland Development Project (Limchoowong and Oberhauser, 1996). This was a pioneering example of the development of a community watershed mosaic system that is having major impact in spurring a rethinking of the whole approach of the Thai government in managing upland watersheds.

The realization gradually advanced that a framework was necessary that recognized the legitimate rights of communities to reside in upper watersheds and that explored ways in which the service functions of the watershed could be maintained or enhanced while enabling the communities to pursue farming activities that were in reasonable harmony with these objectives. A project to experiment with the concept was initiated in 1987.

The boundary was drawn around the perimeter of a small highland subcatchment. A participatory land-use planning approach provide a mechanism for villagers and the forestry department to negotiate and implement a mutually suitable solution. Three-dimensional models of the portion of the watershed occupied by the village proved to be conducive tools by which land-use zoning was done. Watershed committees were established that identified the problems and developed community-enforced land-use rules in place of rigid government regulations. The landscape was categorized into a mosaic of areas for various types of land use, which may include appropriate simultaneous combinations of protected natural forest, managed natural forest, field-based agroforestry, boundary plantings, annual crops, rice paddies, and others (Thomas, 1996). Zones for field agroforestry and annual crops are managed by individual households, subject to necessary conditions imposed by the community. After realistic boundaries were established for protected forests, and the security of land-use rights was confirmed in areas designated for agriculture, the communities became active agents in forest protection. The result has been dramatic improvement in the watershed environment (Figure 11.4). Forest cover has increased substantially and the area in annual cropping has decreased. The establishment of fruit tree gardens has diversified income sources while enhancing soil conservation. Intervillage relations are managed through a watershed management network, which is authorized by the local subdistrict government.

Such a community watershed mosaic system is an agroforestry system at a larger scale. The landscape unit includes forest, tree, and crop components which interact through on-site watershed functions, fire and grazing management, allocation of investments and benefits at household and community levels, as well as through nutrient concentration and cycling, weed and pest dynamics, and other biophysical

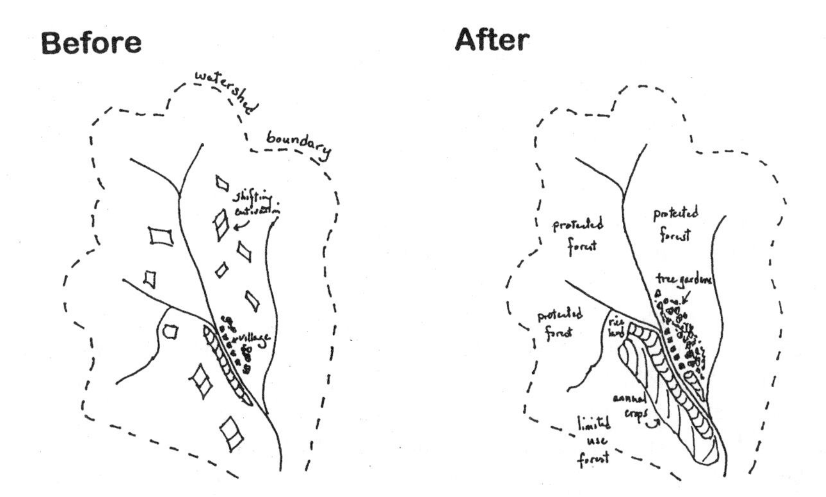

FIGURE 11.4 Land-use change with participatory land-use planning.

factors that interact across field boundaries (as well as within). Such a framework is conducive to the management of land-use rights at the community level that are conditional upon the maintenance of the landscape management system. The experience demonstrated clearly that local communities can become enthusiastic partners with government to solve watershed management problems. This may be particularly true on land claimed by the state on which villagers have tenuous land rights and seek to gain recognition of their *de facto* occupation. However, a major challenge remains in sensitizing the bulk of personnel in the responsible government agencies if the lessons are to be applied on a wide scale in the upper watersheds throughout Thailand.

AGROFORESTRY AND *IMPERATA* GRASSLAND REHABILITATION

When forests are opened for food crop production throughout Southeast Asia the land becomes infested with *Imperata cylindrica* grass within a few years. Farming the grasslands becomes increasingly laborious and crop yields generally decline. Smallholders then leave the grasslands behind and move to the forest margin where their returns from labor are higher. Currently there are about 35 million ha of *Imperata* grasslands distributed among the tropical Asian countries from India through mainland Southeast Asia and Indonesia and the Philippines (Garrity et al., 1997a). Can the initial degradation into *Imperata* be avoided? Can the grasslands be reclaimed and used more intensively?

Tenure insecurity is a key factor depressing effective conversion of *Imperata* grasslands to more intensive uses (Tomich et al., 1997). The ideal response may be to establish secure property rights, but this may not be practical because of the political sensitivity of national governments over their control of state land. A recent international workshop (Garrity, 1997) on improved agroforestry systems for *Imperata* grassland rehabilitation offered a major policy recommendation on this issue. It recommended that for large blocks of state forestland covered by *Imperata* grassland, farmers who convert that grassland by planting and managing trees in small plots should receive property rights over all their products, including the timber. This policy recommendation would apply only to existing grasslands, not to forests cleared in the future, avoiding perverse incentive to accelerate conversion of forest to grassland. ICRAF is implementing a major Southeast Asian regional policy project that will test this recommendation on a pilot scale and provide further analysis to assist the Indonesian Ministry of Forestry and other national governments in implementing it on the ground.

Turning to the question of suitable smallholder technologies for intensification in the grasslands, we find that purely annual crop–based production systems have only limited scope to be sustainable under upland conditions prone to infestation by *Imperata* if animal or mechanical tillage is not available (van Noordwijk et al., 1997). However, where farmers have access to animal or tractor draft power, continuous crop production has been sustained in many *Imperata* areas. A wide range of agroforestry systems that evolve from shifting cultivation have shown particular potential (Garrity, 1997a). We examine three types of promising options.

Building on Indigenous Strategies of Intensification

Natural fallows work eminently well in regenerating soil fertility, if the local fallow species diversity and soil quality have not been degraded (e.g., Szott et al., 1991). But where natural fallows in shifting cultivation systems have been degraded to the point where grasses, particularly *I. cylindrica,* dominate the abandoned fields, natural regeneration is usually infeasible. Annual nutrient accumulation in *Imperata* fallows levels off after 1 to 2 years and is far inferior to that of woody vegetation. The result is a fallow that is incapable of regenerating adequate nutrient accumulation, yet that is very laborious to reopen for cultivation.

Farmers prefer to locate their prospective plots in woody vegetation and avoid grassland whenever possible (Cairns, 1994). *Chromalaena odorata* is an important pioneer fallow species that naturally suppresses *Imperata* in the absence of frequent fires, accumulates many times more biomass, and regenerates crop productivity much more efficiently. Shifting cultivators throughout Southeast Asia find it a highly desirable fallow species (Dove, 1986; de Foresta and Schwartz, 1991), even compared with secondary or primary forest. *Austroeupatorium inulifolium* is a similar nonnative invasive species common at midelevations above 600 m, that has proved very beneficial to farmers practicing shifting cultivation in West Sumatra (Cairns, 1994). It spread widely after its introduction in the late 19th century. Farmers found that it halved the necessary fallow period to regenerate soil fertility. This was a major contribution since land pressure was intense (Stoutjesdijk, 1935). Cairns (1994) showed that *A. inulifolium* fallows accumulated several times more biomass and nutrients beneficial to crop growth in a 2-year period (over 150 kg N/ha and 20 kg P/ha) compared with nearby *Imperata* fallows (25 kg N/ha and 6 kg P/ha) of the same age.

We believe that one of the most-promising approaches to identifying biophysically workable and socially acceptable technologies and to intensifying shifting cultivation is to document, verify, and disseminate cases of indigenous adaptations. If successful indigenous strategies for managing fallow land can be refined and diffused to other upland areas with degrading shifting cultivation, this will enable intensified land use and improved living standards for some of the most marginalized communities in the region. Unfortunately, there is little documentation of such innovations to feed into the national and international research agenda or to inform policy makers. Indigenous strategies are either unobserved or misinterpreted.

ICRAF is collaborating with local partner institutions in a regional research initiative on "Indigenous Strategies for Intensification of Shifting Cultivation." Teams in several countries are investigating a variety of improved fallow systems that have evolved in different ecozones. These vary from herbacious to shrub-based to tree-based fallows. The approach is to identify a range of pragmatic and adoptable solutions for wider extrapolation among communities facing similar swidden degradation problems. Some of the most important indigenous systems combine elements of local practices with new exogenous technology. One example is that of a slash-and-burn community in Indonesia that is using glyphosate as part of the management system (ICRAF, 1997). Farmers independently found that glyphosate enabled them to reopen fields with a substantial amount of *Imperata.* The glyphosate not only controlled *Imperata*, enabling zero-tillage annual cropping, but also shifted the postcropping fallow vegetation from Imperata to *Chromalaena*, a desirable fallow species. They now observe that the area dominated by *Imperata* in the village is declining.

Smallholder Timber Production Systems

In countries such as the Philippines, Vietnam, and Thailand, which are experiencing extreme forest encroachment pressures on the remaining natural areas, there is a concurrent trend toward major increases in the value of farm-grown timber. Smallholders, even shifting cultivators on the frontier, are now engaging in farm forestry for the first time in great numbers, in response to recent strong price incentives (Garrity and Mercado, 1994). This situation dramatically increases the prospect of stimulating smallholder timber production systems as a major vehicle for rapidly increasing overall tree cover in the landscape. In the Philippines, ICRAF is evaluating smallholder timber production systems and more effective methods to disseminate improved tree germplasm, through private and village nursery systems.

Pruned-Tree Hedgerow/Fallow Rotation Systems

Alley cropping was originally conceived as a sustainable replacement for fallow rotation on degrading lands. But pruned-tree hedgerow systems may extend the cropping cycle beyond that possible with natural fallows, but cannot eliminate fallowing unless fertilizer or manure is used. And the labor to maintain the pruning regime is too high. We are conducting studies to characterize the hedgerow/fallow system and its prospective sustainability in both the Philippines and Indonesia. Crop productivity and overall profitability from fields with fallowed hedgerows of *Senna spectabilis* (syn. *Cassia spectabilis*) that were reopened for maize cropping increased compared to cropping in adjacent plots with natural fallow vegetation (Suson and Garrity, unpublished data). We observed a major shift in fallow vegetation in the alleyways as a consequence of hedgerow/fallowing. After 3 years the alleys were intensely shaded and dominated by broad-leaved species. *Imperata cylindrica* was effectively suppressed. Fire is a frequent threat on grassland farms because it is very difficult for the individual cultivators to protect their fields from fires that spread from elsewhere. Experiments in Lampung, Sumatra have shown that *Peltophorum dasyrachis* shows greater promise than *Gliricidia sepium* as the hedgerow component because *Peltophorum* is more resistant to fire (van Noordwijk et al., 1997). We hypothesize that the extrapolation domain for pruned-tree hedgerow/fallowing is most likely to be areas where farmers practice slash-and-burn cultivation in *Imperata* grasslands with manual cultivation and no access to fertilizer inputs.

Applications in Indonesia

Soekardi et al. (1993) estimated the area of *alang-alang* (*Imperata*) in Indonesia to be about 9 million ha (5% of the total land area). Most alang-alang areas are located on land registered as state forestland. Utilization of alang-alang areas for expanding agricultural land is less environmentally destructive than clearing of forests. Adiningsih and Mulyadi (1993) found that for alang-alang land, phosphorus deficiency, aluminum toxicity, and low organic matter content were the main constraints limiting productivity. They demonstrated that soil fertility improvement in the mostly acid alang-alang soils could be attained with the application of phosphate rock and the use of leguminous cover crops. Sukmana (1993) recommended for alang-alang land integrated farming systems involving annual and perennial crops (agroforestry systems) and livestock production. Minimum-tillage practices in combination with cover crop management are superior to conventional tillage in terms of lower labor input and higher crop production. Table 11.1 shows the effectiveness of agroforestry

TABLE 11.1
Soil Loss (t/ha/yr) as Affected by Hedgerow Treatments on a Typic Eutropept with Slopes Ranging from 10 to 15%

	Year of Observation			
Treatment	**1989/90**	**1990/91**	**1991/92**	**1992/93**
Control	66	107	133	68
Caliandra	8	22	19	20
Flemingia	0	0	1	0
Vetiver	12	14	0	2

After Rachman et al., 1995.

systems, especially those using *Flemingia* as hedgerow species, in controlling erosion in Indonesia. Besides its effectiveness in controlling erosion, *Flemingia* produces about four times as large a biomass as vetiver (Ai et al., 1995).

AGROFORESTRY FOR HILLSLOPE CONSERVATION FARMING

Slash-and-burn farmers were the initial adherents of conservation tillage. But as population density increases, fallows are shortened and the biomass and richness of the fallow vegetation declines. Clean tillage for weed control is frequently practiced even on steep slopes, and accelerated soil loss is common. Smallholders who have farmed sloping lands with clean-tillage for some years are well aware of the threat of soil erosion, and are keen to learn about and apply conservation measures (Fujisaka, 1993), as long as such methods are practical, within their very limited resources and labor. Unfortunately, most proposed methods are not practical, in the farmer's eyes. But low-labor, low-investment practices that do the job of saving soil are eagerly awaited by small farmers. This section reviews some of the most-promising directions toward providing conservation tillage practices that may make sense to Southeast Asian upland farmers.

THE CASE FOR NATURAL VEGETATIVE STRIPS

The main conservation farming practice prescribed for open-field intensive cultivation systems in Southeast Asia has been contour hedgerow systems (Garrity, 1995a). Contour hedgerow farming with leguminous trees has thus become a common feature of extension programs for sustainable agriculture on the sloping uplands in Southeast Asia. These systems control soil erosion effectively, even on steep slopes (Kiepe, 1995; Garrity, 1995). Data from the IBSRAM Sloping Lands Network trials in six countries have confirmed that annual soil loss with hedgerow systems is typically reduced 70 to 99% (Sajjapongse and Syers, 1995). There are also numerous reports of increased yield levels of annual crops when grown between hedgerows of leguminous trees. However, farmer adoption of these systems is very low. Constraints include the tendency for the perennials to compete for growth resources and hence reduce yields of associated annual crops, and the inadequate amounts of phosphorus cycled to the crop in the prunings. But the major problem is the extra labor needed to prune and maintain them (ICRAF, 1996). The extra labor did not pay off.

In Claveria, Philippines, some farmers independently developed the practice of laying out contour strips that were left unplanted, and were revegetated by native grasses and forbs. Researchers found that these natural vegetative strips (NVS) had many desirable qualities (Garrity, 1993). They needed much less pruning maintenance compared with fodder grasses or tree hedgerows and offered little competition to the adjacent annual crops compared with the introduced species. They were very efficient in minimizing soil loss (Agus, 1993). And they did not show a tendency to cause greater weed problems for the associated annual crops. NVS were also found to be an indigenous practice on a few farms in other localities, including Batangas and Leyte Provinces.

Installing NVS is quite simple. Once contour lines are laid out there is no further investment in planting materials or labor. The vegetative strips do not need to conform closely to the contour; they act as filter strips rather than bunds. Their biomass production, and economic value as fodder, is lower than many other hedgerow options, but labor is minimized. Vetiver grass fills a similar niche as a low value-added but effective hedgerow species. But for vetiver or any other introduced hedgerow species the planting materials must be obtained and planted out, requiring extra labor. One limitation of low-maintenance hedgerows is that they do not enhance the nutrient supply to the crops. In this respect they do not differ from many other hedgerow enterprises, including fodder grasses or perennial cash crops like coffee. With continuous cropping, NVS or other low-management hedgerow options can only be sustainable with fertilization. They have proved to be popular in northern Mindanao and have been adopted by hundreds of farmers in recent years. A land care movement has evolved around the technology which has resulted in the development of numerous farmer self-help organizations that are spreading the method among their members (Mercado et al., 1997).

A practical ridge-tillage system has recently been developed for smallholder animal-draft farming (Garrity, 1997b). It eliminates the need for primary tillage operations and thus reduces production costs substantially. The system may be used along with contour hedgerows to reduce the rate of soil redistribution from the upper alleyways, which tends to degrade soil fertility in these zones. Ridge-till may also be employed in open fields as well. Soil loss is reduced substantially. Grain yields are sustained, and may even be increased after several years. Thus, the system warrants much wider testing as a promising conservation farming method.

Programs for Intensively Farmed Watersheds in Indonesia

As land pressure increases in the upland areas of Indonesia, farmers have started continuous cropping on steep slopes. The government of Indonesia (GOI) introduced a program focusing on structural and vegetative soil and water conservation activities using the watershed as the basic unit for planning and management. Watershed management and conservation are executed through several projects. The most prominent ones were conducted in the Upper Solo Watershed in the early 1970s, followed by soil conservation projects in the Citanduy River Basin and in Yogyakarta. These assisted farmers in building bench terraces, provided agrotechnology packages of seeds and fertilizers, reforested government land, and constructed check dams and gully plugs.

The main government program in watershed management and soil conservation is the Regreening and Reforestation Program (the R&R Program). It began in 1976 with the following objectives: (1) controlling erosion and flooding, (2) improving land productivity and farmer's income, and (3) increasing people's participation in preserving natural resources. Initially, it supported mainly the supply of seedlings for planting on farmer's land (regreening) and on public land (reforestation). The present approach includes demonstrations of soil conservation and agronomical packages through Soil Conservation Demonstration Units (*Usaha Pelestarian Sumberdaya Alam* = UPSA) and through Sedentary Farming Demonstration Units (*Usaha Pertanian Menetap* = UPM).

The main approach of UPSA and UPM is to increase land cover through promoting agroforestry. Government recommendations are to increase ground cover by perennial crops to 25% on land with slopes between 15 and 25%, to 50% on slopes between 25 and 40%, and to 100% on slopes steeper than 40% (Sekretariat Tim Pengendali Penghijauan dan Reboisasi Pusat, 1996). Implementation is limited by many constraints, particularly the prevalent subsistence mode of farming, insecure land tenure that forces farmers to invest in activities with a fast return, and inaccessibility or uncertainty of markets. Thus, it is now recognized that a combination of annual crops with perennials is the only practical way forward. We briefly examine the appropriateness of the technology options being introduced under the R&R program and other farming systems development programs.

Soil Conservation Demonstration Units

Table 11.2 summarizes the opportunities and constraints of a number of soil conservation measures for Indonesian smallholders. Although bench terracing is implemented at almost every project site, the technique is often unsuitable due to a shallow soil depth, an unstable soil structure, or to very acid subsoil (Sukmana, 1995; Agus et al., 1996). Examples of its unsuitability can be found at a few demonstration plots in shallow and unstable soils in central and south Sulawesi and in East Nusa Tenggara. When bench terraces are physically unsuitable for the site, they deteriorate soil conditions for farming by increasing the rate of mass wasting and by exposing the subsoil. Terracing costs about U.S.$900/ha, while the initial income generated from most rain-fed uplands before as well as after treatment was in the order of $600/ha/year. Such investment costs are clearly unsuitable for smallholders. High costs and labor requirements to maintain these structural conservation measures also often run counter to the reality of smallholder agriculture. The success of bench terracing has therefore been limited to parts of Java (Sembiring et al., 1989) and Bali where there is extreme land pressure and a history of making terraces for paddy fields.

The use of contour hedgerows of grasses or legume shrubs as an erosion control measure and as a source of feed is the most widely adopted packages in Southeast Asia. In general, farmers seek additional benefits apart from erosion control. Vetiver grass as a hedgerow was less preferred, although it shows almost no signs of competition with food crops and is effective in reducing soil erosion. Napier grass (*Penisetum purpureum*) which suppresses the growth of a few rows of food crops adjacent to the hedgerows was preferred because it can be used as fodder (Agus et al., 1997).

Village Nurseries

Farmer-managed nurseries in the R&R program are intended to produce high-quality timber and fruit tree seedlings and seeds of fodder and grass. They receive technical support and limited incentives during the first year. In the second and the following years, the nurseries are expected to become self-supporting. A few nurseries have survived, but most disappear when outside support is withdrawn. Often this is due to an unpopular choice of tree species, which is often in conflict with the current

TABLE 11.2
Opportunities (+) and Constraints (–) of Implementation and Sustainability of Selected Conservation Measures

Conservation Measure	Opportunities and Constraints
Bench terrace	+ Could be recommended if labor availability is high – High costs and labor for establishment and maintenance – Does not increase crop yields in the short run, rather it decreases yields in the first few years after establishment – Being exotic to farmers outside Java
Contour hedgerow system	+ Provide animal feed, firewood, organic matter + If legume species is used, it alleviates the need to apply nitrogen + Effective in controlling erosion if arranged properly – Hedgerows take out part of food crop planting area – Could be very competitive to the alley crops for water, nutrients, and light – Extra costs and labor for establishment and maintenance – Could harbor pests and diseases
Planting of fruit tree crops	+ Familiar to farmers + Won't be cut unless the production or the price drops + Long-term protection to soil compared to planting annual crops + Long-term source of income – Seedling might not be easily available, expensive, or of low quality – Long waiting period – Cannot be adopted by farmers with insecure tenure.
Planting of timber or pulp tree species	+ Long-term protection to soil compared to planting annual crops – Farmers might not have certainty of market availability; long waiting period – Upon harvesting/cutting it creates recurrent problem of land denudation – Cannot be adopted by farmers with insecure tenure – Unavailability of quality planting materials
Minimum tillage and mulching (including green mulching)	+ Reduce requirement for labor + Give protection to soil surface in terms of erosion reduction, moisture conservation, and, to a lesser extent, nutrients contribution – Many farmers do not see direct benefits of mulching – Lack of mulching materials – Extra works for obtaining materials – Mulch might harbor pests and diseases
(Grass) strip cropping	+ Supports livestock production + Effective conservation measure for slopes between 15 to 45% – Requires fertilization and replanting of the grass – Only applicable to areas potential for livestock production

annual crop farming systems or to emphasis on timber species when farmers prefer fruit trees. Where the species fulfill a local need, however, the centralized village nursery may transform into individual nurseries usually managed in home gardens. These individual nurseries appear to be more sustainable because they are easier to manage and can be operated in a self-reliant manner.

Program Adjustments to Meet Farmer Needs

Despite success, there are weaknesses in the R&R program. Among the major weaknesses is the use of less appropriate technology due to lack of involvement of the farmers and a weak linkage between research and extension. There has been too much emphasis on a few conservation measures due to a poor understanding of farmers' circumstances among the program implementors. Rather than matching technology options with the biophysical conditions and the socioeconomic situation of the farmers, a top-down approach has dominated the selection of soil conservation measures. Starting in 1994, the lessons learned have induced an improved integration of the R&R program into the regional plans, and program guidelines have evolved toward addressing local conditions and community needs. But much more needs to be done along these lines. Extension workers should not limit themselves to a very few options, but should take advantage of the wide range of options that are available (Agus et al., 1997).

Paradigm shifts are taking place in hillslope conservation management as indicated by Garrity and van Noordwijk (1995): (1) the engineering approach has shifted to a biological approach, (2) the top-down approach is yielding to a bottom-up approach, and (3) the classical alley farming concept is diversifying to a wider array of agroforestry. In most cases, low-cost vegetative soil conservation measures can be applied. Sanchez (1995) and Lal (1991) warned not to oversell agroforestry technologies as many of their benefits have still to be scientifically proved. Future research should vigorously address these issues.

CONCLUSIONS

What will it take to turn things around for Asia's fast-degrading watersheds? Watershed management requires an integrated and multisectoral approach to sustainable development, but government departments are compartmentalized and geared for top-down operations. They will need to change. Participatory approaches transfer principles rather than standard solutions, and make available a basket of choices rather than a set package of practices. Problem analysis must not simply be done by outsiders for the community, but must be done by the community itself with backstopping by the outsiders. The solution is not to transfer some known technology, but to assist farmers to adapt technologies to their own circumstances. This is predicated on the recognition that rural people, educated or not, have a much greater ability to analyze, plan, and implement their own development activities than was previously assumed by outsiders.

What can agroforestry contribute? As a highly integrative field on the interface between the agricultural, forestry, social, and environmental sciences, agroforestry will play a critical central role in helping to provide key technical and institutional innovations at the landscape scale. As a natural resource management system that involves the increasing integration of trees into the agricultural landscape, it will play a major role, holistically and comprehensively, in the process of providing options that increase rural livelihoods, and yet are conducive to the conservation of fragile watershed resources.

REFERENCES

Adiningsih, J.S. and Mulyadi, D.1993. Alternatif teknik rehabilitasi dan pemanfaatan lahan alang-alang [Alternatives of rehabilitation and utilization of *Imperata* grassland], in S. Sukmana, Ed., *Pemanfaatan Lahan Alang-alang Untuk Usahatani Berkelanjutan,* Prosiding Seminar Lahan Alang-alang, Bogor, 1 Des. 1992. Pusat Penelitian Tanah dan Agroklimat, Bogor.

Agus, F. 1993. Soil Processes and Crop Production under Contour Hedgerow Systems on Sloping Oxisols, Ph.d. dissertation, North Carolina State University, Raleigh, 141 pp.

Agus, F., Gintings, A.N., and Ai, D. 1996. Sumberdaya alam daerah aliran sungai Cimanuk Hulu dan teknologi konservasi [Land resources of the Upper Cimanuk Watershed and conservation technology], in B.R. Prawiradiputra et al., Eds., *Prosiding Lokakarya Pembahasan Hasil Penelitian dan Analisis Pengelolaan Daerah Aliran Sungai,* Pusat Penelitian Tanah dan Agroklimat, Bogor, 113–128.

Agus, F., Abdurachman, A., and van der Poel, P. 1997. Daerah aliran sungai sebagai unit pengelolaan pelestarian lingkungan dan peningkatan produksi pertanian [Watershed as a unit of environmental management and conservation and agricultural production], Prosiding Pertemuan Teknis Pusat Penelitian Tanah dan Agroklimat, Bogor.

Ai, D., Suganda, H., Sujitno, E., Tala'ohu, S.H., and Sutrisno, N. 1995. Rehabilitasi lahan alang-alang dengan sistem budidaya lorong di Pakenjeng, Kabupaten Garut [Rehabilitation of *Imperata* grassland using alley cropping systems at Pakenjeng, Garut District], in D. Santoso, *Prosiding Pertemuan Teknis Penelitian Tanah dan Agroklimat,* Pusat Penelitian Tanah dan Agroklimat, Bogor, 31–41.

Bannerjee, A.K. 1990. Revegetation techniques, in Doolette, J. and Magrath, W.B., Eds., *Watershed Development in Asia: Strategies and Technologies,* World Bank Technical Paper No. 127. World Bank, Washington, D.C., 109–130.

Cairns, M. 1994. Shifting Cultivation on the Perimeter of a National Park, MSc thesis, York University, Toronto, Canada.

Cairns, M. 1995. Ancestral Domain and National Park Protection: A Mutually Supportive Paradigm? International Center for Research in Agroforestry, Southeast Asian Regional Research Program, Bogor, Indonesia.

Cairns, M., Murniati, Otsuka, M., and Garrity, D.P. 1997. Characterization of the Air Dingin-Muara Labuh Area of the Kerinci Seblat National Park: farm and national park interactions in *Proceedings of the Workshop on Alternatives to Slash and Burn,* Central Research Institute for Food Crops, Bogor, Indonesia, June 1995, 135–172 (in press).

Carson, B. 1989. Soil Conservation Strategies for Upland Areas of Indonesia, Occasional Paper No. 9, Environment and Policy Institute, East-West Center, University of Hawaii, Honolulu.

de Foresta, H. and Michon, G. 1997. The agroforest alternative to *Imperata* grasslands, *Agroforestry Syst.* 36:105–120.

de Foresta, H. and Schwartz, D. 1991. *Chromalaena oderata* and disturbance of natural succession after shifting cultivation: an example from Mayonbe, Congo, Central Africa, in R. Nuniappau and P. Ferrar, Eds., *Ecology and Management of* Chromalaena odorata, BIOTROP Spec Publ. No. 44: 23–41.

Doolette, J.B. and Smyle, J.W. 1990. Soil and moisture conservation strategies: review of the literature, in Doolette, J.B. and Magrath, W.B., Eds., Watershed Development in Asia: Strategies and Technologies. World Bank Technical Paper No. 127. World Bank, Washington, D.C., 35–70.

Douglas, M. 1996. *Participatory Catchment Management,* 2 Vols., HR Wallingford, Oxon, U.K.

Dove MR. 1986. The practical reason of weeds in Indonesia: peasant vs. state views of *Imperata* and *Chromalaena*, *Human Ecol.* 14:163–190.

Fujisaka, S. 1993. A case of farmer adaptation and adoption of contour hedgerows for soil conservation, *Exp. Agric.* 29:97–105.

Garrity, D.P. 1993. Sustainable land-use systems for sloping uplands in Southeast Asia, in *Technologies for Sustainable Agriculture in the Tropics,* American Society of Agronomy Special Publication 56, Madison, WI, 41–66.

Garrity, D.P. 1995a. Improved agroforestry technologies for conservation farming: pathways toward sustainability, in *Proc International Workshop on Conservation Farming for Sloping Uplands in Southeast Asia: Challenges, Opportunities and Prospects,* IBSRAM, Bangkok, Thailand, Proc. No. 14, 145–168.

Garrity, D.P. 1995b. Buffer Zone Management and Agroforestry: Some Lessons from a Global Perspective, International Centre for Research in Agroforestry, Southeast Asian Regional Research Program, Bogor, Indonesia.

Garrity, D.P., Ed. 1997a. Agroforestry innovations for *Imperata* grassland rehabilitation, *Agroforestry Syst.* Special Issue 36, 276 pp.

Garrity, D.P. 1997b. Conservation Tillage: A Southeast Asian Perspective, International Center for Research in Agroforestry, Southeast Asian Regional Research Program, Bogor, Indonesia, 26 pp.

Garrity, D.P. 1997c. Addressing Key Natural Resource Management Challenges in the Humid Tropics through Agroforestry Research, International Centre for Research in Agroforestry, Bogor, Indonesia, 29 pp.

Garrity, D.P. and Mercado, A. 1994. Reforestation through agroforestry: market-driven smallholder timber production on the frontier, in Raintree, J.B. and Francisco, H.A., Eds., *Marketing of Multipurpose Tree Products in Asia,* Winrock International, Morrilton, AR, 265–268.

Garrity, D.P. and Van Noordwijk, M. 1995. Research Imperative in Conservation Farming and Environmental Management on Sloping Lands: An ICRAF Perspective, ICRAF, Bogor, Indonesia, 5 pp.

Garrity, D.P., Soekardi, M., Van Noordwijk, M., de La Cruz, R., Pathak, P.S., Gunasena, H.P.M., So, N., van Huijin, G., and Majid, N.M. 1997. The *Imperata* grasslands of tropical Asia: area, distribution, and typology, *Agroforestry Syst.* 36:3–29.

Glynn, C. 1996. Overcoming Constraints to Agroforestry Adoption in Tropical Highlands: Part I: An Investigation of Performance by Elevation Patterns for Some Commonly Grown Timber Species in the Manupaly Watershed, Bukidnon, Philippines, M.Sc thesis, Tropical and Subtropical Horticulture and Crop Science, Wye College, University of London, London.

Gouyon, A., de Foresta, H., and Levang, P. 1993. Does "Jungle Rubber" deserve its name? An analysis of rubber agroforestry systems in southeast Sumatra, *Agroforestry Syst.* 22:181–206.

ICRAF. 1996. Annual Report For 1995, International Centre for Research in Agroforestry, Nairobi, Kenya.

ICRAF. 1997. Annual Report For 1996, International Centre for Research in Agroforestry, Nairobi, Kenya.

Kiepe, P. 1995. No Runoff, No Soil Loss: Soil and Water Conservation in Hedgerow Barrier Systems, Wageningen Agricultural University, Wageningen, the Netherlands.

Lal, R. 1991. Myths and scientific realities of agroforestry as a strategy for sustainable management for soils in the tropics, *Adv. Soil. Sci.* 15:91–137.

Leakey, R. 1996. Definition of agroforestry revisited, *Agroforestry Today* 8(1):5–6.

Limchoowong, S. and Oberhauser, U. 1996. In *Proceedings of the Discussion Forum ""igh-land Farming: Soil and the Future?"* Can villagers manage highland resources well?"Maejo University, Chiangmai, Thailand. pp. 13–22.

Magrath, W.B. and Doolette, J.W. 1990. Strategic issues in watershed development, in Doolette, J.W. and Magrath, W.B., Eds., *Watershed Development in Asia: Strategies and Technologies,* World Bank Technical Paper No. 127, World Bank, Washington, D.C., 1–34.

Mary, F. and Michon, G. 1987. When agroforests drive back natural forests: a socio-economic analysis of a rice/agroforest system in South Sumatra, *Agroforestry Syst.* 5:27–55.

Mercado, A. Jr., Stark, M., and Garrity, D.P. 1997. Enhancing sloping land management technology adoption and dissemination, paper presented at the IBSRAM Sloping Land Management Workshop, Bogor, Indonesia, 15–21 September, 1997, 24 pp.

Michon, G., de Foresta, H., and Aliadi, A. 1995. An Agroforestry Strategy for the Re-appropriation of Forest Resources by Local Communities: The Case Study of Damar Agroforests in West Lampung, Sumatra, International Center for Research in Agrofor-estry, Southeast Asian Regional Research Program, Bogor, Indonesia, 54 pp.

Milliman, J.D. and Meade, 1983. World-wide delivery of river sediment to the oceans, *J. Geol.* 91:1–21.

Momberg, F. 1993. Indigenous Knowledge Systems. Potentials for Social Forestry Develop-ment: Resource Management of Land-Dayaks in West Kalimantan, M.Sc. thesis, Tech-nische Universitat Berlin, Germany.

Penot, E. 1996. Sustainability Through Productivity Improvement of Indonesian Rubber-Based Agroforestry Systems, International Center for Research in Agroforestry, South-east Asian Regional Research Program, Bogor, Indonesia, 13 pp.

Rachman, A., Abdurachman, A., and Haryono, 1995. Erosi dan perubahan sifat tanah dalam sistem pertanaman lorong pada tanah Eutropepts, Ungaran [Erosion and changes in soil properties in alley cropping system on Eutropepts, Ungaran], in *Proceedings Pertemuan Teknis Penelitian Tanah dan Agroklimat Bidang Konservasi Tanah dan Air, dan Agrokli-mat,* Pusat Penelitian Tanah dan Agroklimat, Bogor, 17–30.

Sajjapongse, A. and Syers, K. 1995. Tangible outcomes and impacts from the ASIALAND management of sloping lands network, in *Proc. International Workshop on Conservation Farming for Sloping Uplands in Southeast Asia: Challenges, Opportunities and Pros-pects,* IBSRAM, Bangkok, Thailand, Proc. No. 14, 3–14.

Salafsky, N. 1993. The Forest Garden Project: An Ecological and Economic Study of a Locally Developed Land-Use System in West Kalimantan, Indonesia, Ph.D. thesis, Duke Uni-versity, Durham, NC, 327 pp.

Sanchez, P.A. 1995. Science in agroforestry. *Agroforestry Syst.* 30:5–55.

Sekretariat Tim Pengendali Penghijauan dan Reboisasi Pusat. 1996. Petunjuk Pelaksanaan dan Petunjuk Teknis Bantuan Penghijauan dan Reboisasi, Jakarta.

Sembiring, H., Syam, A., Hardianto, R., Kartono, G., and Sukmana, S. 1989. Evaluasi adopsi teknologi usahatani konservasi lahan kering di DAS Brantas: Studi kasus Desa Srimulyo, Malang [Evaluation and adoption of conservation farming technologies in the upland of Brantas Watershed: a case study at Srimulyo village, Malang], in H. Suhardjo, Ed., *Prosiding Pertemuan Teknis Penelitian Tanah Bidang Konservasi Tanah dan Air,* Pusat Penelitian Tanah dan Agroklimat, Bogor.

Soekardi, M., Retno, M.W., and Hikmatullah, 1993. Inventariasasi dan karakterisasi lahan alang-alang [Inventarization and characterization of *Imperata* grassland], in S. Sukmana, Ed., *Pemenfaatan Lahan Alang-alang untuk Usahatani Berkelanjutan, Prosiding Semi-nar Lahan Alang-alang,* Bogor, 1 Dec 1992, Center for Soils ansd Agrlclimate Research, Bogor, Indonesia, 1–17.

Stoutjesdijk, J.A.J.H. 1935. *Eupatorium pallescens* DC op Sumatra's westkust [*Eupatorium pallescens* DC on the west coast of Sumatra], *Tectona* 28:919–926.

Sukmana, S., Ed. 1993. Pemanfaatan lahan alang-alang untuk usahatani berkelanjutan [The use of *Imperata* land for sustainable land management]. *Prosiding Seminar Lahan Alang-alang,* Bogor, 1 Dec. 1992, Pusat Penelitian Tanah dan Agroklimat, Bogor.

Sukmana, S. 1995. Perkembangan penelitian konservasi tanah dan air di Indonesia [The development of soil conservation research in Indonesia], Dalam D. Santoso et al., Ed., *Prosiding Pertemuan Teknis Penelitian Tanah dan Agroklimat,* Bogor, 10–12 Jan. 1995, Pusat Penelitian Tanah dan Agroklimat, Bogor, 153–169.

Szott, L.T., Palm, C.A., and Sanchez, P.A. 1991. Agroforestry in acid soils of the humid tropics, *Adv. Agron.* 45:275–300.

Thiollay, J.M. 1995. The role of traditional agroforests in the conservation of rain forest bird diversity in Sumatra, *Conserv. Biol.* 9(2):335–353.

Thomas, D. 1996. Opportunities and limitations for agroforestry systems in the highlands of north Thailand, in *Proceedings of the Discussion Forum "Highland Farming: Soil and the Future?"* Maejo University, Chiangmai, Thailand.

Tomich, T.P., J. Kussipalo, K. Menz and N. Byron. 1997. *Imperata* economics and policy, *Agroforestry Syst.* 36:233–261.

Tomich, T.P., van Noordwijk, M., Vosti, S., and Whitcover, J. 1998. Agricultural development with rainforest conservation: Methods for seeking best-bet alternatives to slash-and-burn, *J. Agric. Econ.* 19: 1–2, 159–174.

Torquebiau, E. 1992. Are tropical home gardens sustainable? *Agric. Ecosyst. Environ.* 41:189–207.

Turkelboom, F. and van Keer, K., Eds. 1996. *Land Management Research for Highland Agriculture in Transition,* Mae Jo University, Thailand, and Catholic University of Leuven, Belgium, 53 pp.

van Noordwijk, M., Tomich, T.P., Winahyu, R., Murdiyarso, D., Partoharjono, S., and Fagi, A.M., Eds. 1995a. Alternatives to Slash-and-Burn in Indonesia. Summary Report of Phase 1. ASB-Indonesia Report No. 4, International Centre for Research in Agroforestry, Southeast Asian Regional Research Program, Bogor, Indonesia. 151 pp.

van Noordwijk, M., van Schaik, C.P., de Foresta, H., and Tomich, T.P. 1995b. Segregate or Integrate: Nature and Agriculture for Biodiversity Conservation, International Centre for Research in Agroforestry, Bogor, Indonesia, 16 pp.

van Noordwijk, M., Hairiah, K., Partoharjono, S., Labios, R.V., and Garrity, D.P. 1997. Food-crop based production systems as sustainable alternatives for *Imperata* grasslands? *Agroforestry Syst.* 36:55–82.

Wells, M. and Brandon, K. 1992. People and Parks: Linking Protected Area Management with Local Communities, The World Bank, Washington, D.C.

12 Crop Yield Variability Due to Erosion on a Complex Landscape in West Central Ohio

Edie Salchow and Rattan Lal

CONTENTS

INTRODUCTION

The integrated effects of landscape position, soil properties, and climate determine the erosion–productivity relationships of erosional and depositional phases. The objectives of this study were to determine, using readily available technology: (1) the relative yields of four erosion phases located on complex slopes, and their variability; (2) the relative area of each phase and the resulting area-weighted average yield; and (3) the compensatory effect of each phase on losses by other phases. Yield reductions due to erosion depend on soil properties, location in the landscape, the relative area of eroded soils on the farm, and on weather. Yields of corn and soybeans on four erosional phases were measured on small plots located in three fields in west-central Ohio. A simple and inexpensive survey allowed calculation of relative areas of four erosional phases. The degree to which yield losses on one erosion phase were offset by gains on other phases was determined for the years 1992 through 1997. The yields on the severely eroded phase were the most variable.

One goal of site-specific crop management (SSCM) is to maximize yields on spatially variable land. Georeferenced measurements, including grid soil sampling, "on the go" yield monitoring, and/or infrared photography, identify areas of high

0-8493-0702-3/00/$0.00+$.50

and low yields and allow correlation of yields with soil and site properties. Eroded soils on complex slopes are an important example of spatial variability where SSCM technologies might be applied. Under uniform management, crops yields along eroded complex slopes varied by as much as 17% for corn and 44% for soybeans in southeastern Nebraska (Jones et al., 1989) and by as much as 40 to 50% for corn in southwestern Ontario (Battiston et al., 1987; Aspinall et al., 1989). Yield reductions occur because erosion, especially when accelerated by tillage, changes soil properties which already differ according to the catena concept (Hall et al., 1982; Frye et al., 1987; Kreznor et al., 1989; Khakural et al., 1992, Brubaker et al., 1994; Govers et al., 1994; Lobb et al., 1994). Pennock et al. (1994) reported that these differences do not necessarily translate into consistent yield differences, however, which makes it difficult to predict erosion–productivity relationships on complex landscapes.

One question to be answered is whether yield reductions along eroded slopes are significant to overall field average yields. Khakural et al. (1992) reported that, under uniform management (high amounts of residue), reduced yields on poorly drained, low-lying, depositional areas were more than balanced by enhanced yields on well-drained, eroded areas. Ebeid et al. (1995) reported that low-lying, depositional areas did relatively better then eroded areas in dry years. Therefore, while maps showing areas of high and low yields may differ year to year because of weather variability, field average yields may remain relatively constant due to compensating effects.

A second question is whether compensating effects represent a temporary or a long-term mechanism of yield stability. The erosion–productivity simulation model of Perrens et al. (1984) predicted higher average productivity for typical soils on complex slopes of northern Indiana and Illinois, despite yield reductions on slightly (SL), moderately (MOD), and severely (SEV) eroded areas, due to yield gains on depositional (DEP) areas. However, these gains lasted only for a period of 20 years, after which eroding subsoil covered DEP areas and reduced their yields.

A third question concerns the relative area of each erosion phase. Daniels et al. (1989) considered yield losses on severely eroded areas to be of minor economic importance, since such areas made up only 11% of the total area of farm fields in the southern Piedmont; approximately 15% of the cultivated land area in southwestern Ontario is composed of moderately to severely eroded shoulder slopes (Kachanoski et al., 1992).

The SSCM technologies which are well suited to answering questions about the relative productivity, areal extent, and significance to total yield of various areas in a field, and the relationships among soil and site factors and yields, are expensive and unavailable to some producers and researchers. Extrapolation of results from studies of isolated plots, which differ only in soil depth, to a complex landscape is too simplistic because the integrated effects of soil property variation, weather, erosion/deposition, runoff/run-on, and percolation/subsurface flow occurring on the landscape are neglected (Cassel and Fryrear, 1989; Jones et al., 1990; National Research Council, 1993; Olson et al., 1994). An intermediate approach to erosion–productivity assessment would be to measure yield and soil properties on plots with varying degrees of erosion and deposition, located on a surveyed landscape. Since erosion phase and landscape position are highly correlated within a region (Carson and Kirkby, 1972; Kreznor et al., 1989; Pennock and Acton, 1989; Hall and Olson, 1991), the resulting data would provide a basis for estimation of erosion–productivity relationships on the entire farm, and on similar areas.

The objectives of this study were to determine, using readily available technology: (1) the relative yields of four erosion phases located on complex slopes, and their variability; (2) the relative area of each phase and the resulting area-weighted average yield; and (3) the compensatory effect of each phase on losses by other phases.

SITE DESCRIPTION

Three privately operated farms (A, B, and C) were chosen from Clark County, in west-central Ohio, where 12% of the farmland is classified as highly erodible land (HEL) (John Grieser, District Conservationist of Clark County, 1997, personal communication). Each is located near the South Charleston branch of the Ohio Agricultural Research and Development Center (39°58′15″, 83°32′30″).

Farms A, B, and C were planted to corn (*Zea mays*)/soybean (*Glycine max*) rotation, with Farm A in conventional tillage and Farms B and C in conservation tillage. According to the soil survey (Miller, 1997) the soils on the 3.3-ha segment of Farm A under study are an eroded Strawn silty clay loam (fine-loamy, mixed, mesic Typic Hapludalf) and Kokomo silty clay loam (fine, mixed, mesic Typic Argiaquoll). Strawn and Kokomo silty clay loams and Crosby silt loam (fine, mixed, mesic Aeric Ochraqualf) are found on the 2.3-ha segment of Farm B. The soils on the 3.4-ha segment of Farm C are an eroded Miamian silty clay loam (fine, mixed, mesic Typic Hapludalf) on 6 to 12% slopes, Miamian silt loam on 2 to 6% slopes, and Kokomo silty clay loam on low-lying, level areas.

Soils of the Strawn series are located on crests, shoulders, and backslopes with 2 to 6% slopes; Kokomo soils are found on footslopes, shallow, closed depressions, and broad areas with less than 2% slopes; the Crosby series is located on rises and foot slopes with 0 to 2% slopes; Miamian is found on knolls, rises, summits, shoulders, and backslopes (Miller, 1997).

Natural Resource Conservation Service criteria (Soil Survey Staff, 1994) were used to identify three replications of the four erosion phases at each farm; taxonomic classifications and detailed soil horizon descriptions were reported by Fahnestock et al. (1995a). All SL, MOD, and SEV phases were classified as fine, mixed, mesic Typic Hapludalfs, except for one replicate of SL which was classified as fine, mixed, mesic Mollic Hapludalf. The DEP phase soil classification included Mollic Hapludalf, Typic Argiaquoll, Aquic Argiaquoll, Typic Hapludoll, Aquic Hapludoll, and Typic Udorthent over Typic Argiaquoll (Fahnestock et al., 1995a). Cesium-137 analysis showed that variable soil loss and soil loss rates had occurred on each phase (Bajracharya et al., 1997). Fahnestock et al. (1995a,b) reported differences in texture, organic carbon content, pH, infiltration rates, and nutrient status of soils at each replicated phase. Fahnestock et al. (1995b) reported yields and nutrient status of soils and crops in 1992 and 1993.

METHODS

Relative surface elevation was recorded every 15.2 m at Farms B and C, using a dumpy level. At Farm C, clinometer readings were taken every 7.6 m along 10 305 m transects in one direction, and along two perpendicular 122-m transects. Surface plots were generated using Surfer (Golden Software, 1997). The total area of each erosion phase was calculated after (1) locating the 12 plots identified by Fahnestock et al. (1995a); (2) identifying slopes where runoff velocity would most likely result in scouring or deposition, following the reasoning of Carson and Kirkby (1972) and Jones et al. (1990); and (3) marking observed areas of surface gravel or accumulated fine particles. Areas planted to grassed waterways were subtracted from the studied farm areas.

Average yields on each erosion phase from 1992 to 1997 were based on crops harvested from the three replications of each erosion phase, which varied in size from 1.5 to 4 m^2. Analysis of variance was used to determine significant differences between phase yields (SAS Institute, 1994). Yearly area-weighted average yields were calculated by multiplying each phase yield average by the estimated area of that phase and summing the results.

PHASE YIELD AVERAGES AND THEIR VARIABILITY

Table 12.1 shows yield averages, coefficients of variation (CVs) and least significant difference (LSD) values for three replications of the four erosion phases on each farm for the years 1992 through 1997.

For Farm A, there were no significant differences in yield in 1992 (corn), 1993 (soybeans), and 1997 (corn); no data were collected in 1994; the DEP yield was significantly higher than SL, MOD, and SEV yields in 1995 (corn) and 1996 (soybeans). Figure 12.1 shows the distribution of precipitation obtained from the South Charleston research station for the years 1992 through 1997. Yield losses on eroded phases in 1995 and 1996 were attributed to late planting, seed loss from sloping MOD and SEV areas, and surface crusting on level SL areas which reduced infiltration. The level DEP areas, on the other hand, had sufficient drainage to support the crop during the wet spring.

For Farm B, there were no significant differences in yield in 1992 (corn), 1996 (corn), and 1997 (soybeans); the DEP yield was significantly higher than SL, MOD, and SEV yields in 1993 (soybeans), when the high water holding capacity of the DEP phase supported the crop through a dry May; the DEP yield was significantly higher than the MOD yield in 1994 (corn). The level position and deeper soil of the SL soils kept available water at these sites higher than that of the MOD sites. Since the SEV phase had the lowest available water capacity, the higher cation exchange capacity (CEC) due to higher clay content (Fahnestock et al., 1995a) could have been responsible for the relatively high yield on that phase. In 1995, the DEP yield was significantly higher than SL and MOD yields (soybeans) because it provided a relatively even water content in a year with sporadic, intense precipitation. The SEV

TABLE 12.1
Average Yield (Mg/ha) of Three Replicates of the Slightly Eroded Phase (SL), Moderately Eroded Phase (MOD), Severely Eroded Phase (SEV), and Depositional Phase (DEP), with Coefficients of Variation in Parentheses

Farm	Phase	1992 C	1993 S	1994 nd	1995 C	1996 S	1997 C
A	SL	10.1 a	3.4 a		7.8 b	2.7 b	8.9 a
		(17.1)	(57.9)		(19.9)	(3.0)	(12.3)
A	MOD	11.0 a	2.9 a		7.1 b	2.4 b	8.6 a
		(18.2)	(62.6)		(21.6)	(14.2)	(16.0)
A	SEV	8.5 a	2.4 a		5.0 b	1.8 b	8.4 a
		(38.2)	(59.1)		(29.9)	(27.8)	(19.7)
A	DEP	7.2 a	4.2 a		11.1 a	3.8 a	8.9 a
		18.4	(59.5)		(20.0)	(21.6)	(19.5)
LSD		4.1	1.9		3.2	1.0	2.8
Farm	Phase	**1992 C**	**1993 S**	**1994 C**	**1995 S**	**1996 C**	**1997 S**
B	SL	15.0 a	3.4 b	11.2 ab	2.2 b	8.9 a	2.0 a
		(11.4)	(1.3)	(16.5)	(57.0)	(9.8)	(25.1)
B	MOD	14.4 a	3.3 b	11.0 b	2.4 b	9.0 a	2.2 a
		(10.8)	(13.2)	(29.5)	(17.5)	(16.3)	(23.7)
B	SEV	14.7 a	3.6 b	12.1 ab	3.4 ab	7.7 a	1.6 a
		(23.9)	(26.9)	(12.9)	(14.3)	(26.9)	(50.0)
B	DEP	15.8 a	5.7 a	14.9 a	4.1 a	7.3 a	2.4 a
		(6.7)	(11.1)	(6.1)	(6.9)	(7.1)	(17.1)
LSD		4.1	1.2	3.9	1.4	2.1	1.1
Farm	Phase	**1992 S**	**1993 C**	**1994 S**	**1995 C**	**1996 S**	**1997 C**
C	SL	4.0 a	7.7 a	3.4 ab	10.1 a	2.7 ab	8.5 a
		(10.0)	(29.6)	(30.4)	(18.3)	(13.1)	(8.4)
C	MOD	3.6 a	8.9 a	2.5 b	11.8 a	2.2 ab	8.8 a
		(9.6)	(11.4)	(17.7)	(5.2)	(33.2)	(10.3)
C	SEV	2.1 b	4.8 b	2.7 ab	12.9 a	1.9 b	7.9 a
		(11.8)	(24.4)	(8.1)	(39.7)	(20.1)	(25.2)
C	DEP	3.4 a	9.1 a	3.6 a	10.6 a	2.9 a	7.6 a
		(9.4)	(10.1)	(4.9)	(16.1)	(8.8)	(21.3)
LSD		0.6	2.8	1.1	5.4	0.9	2.6

Significant differences in yield between phases of the same farm and year in a column are marked by dissimilar letters. Fisher's LSD is calculated with $\alpha = 0.05$. S = soybeans (13% moisture); C = corn (15% moisture); nd = no data.

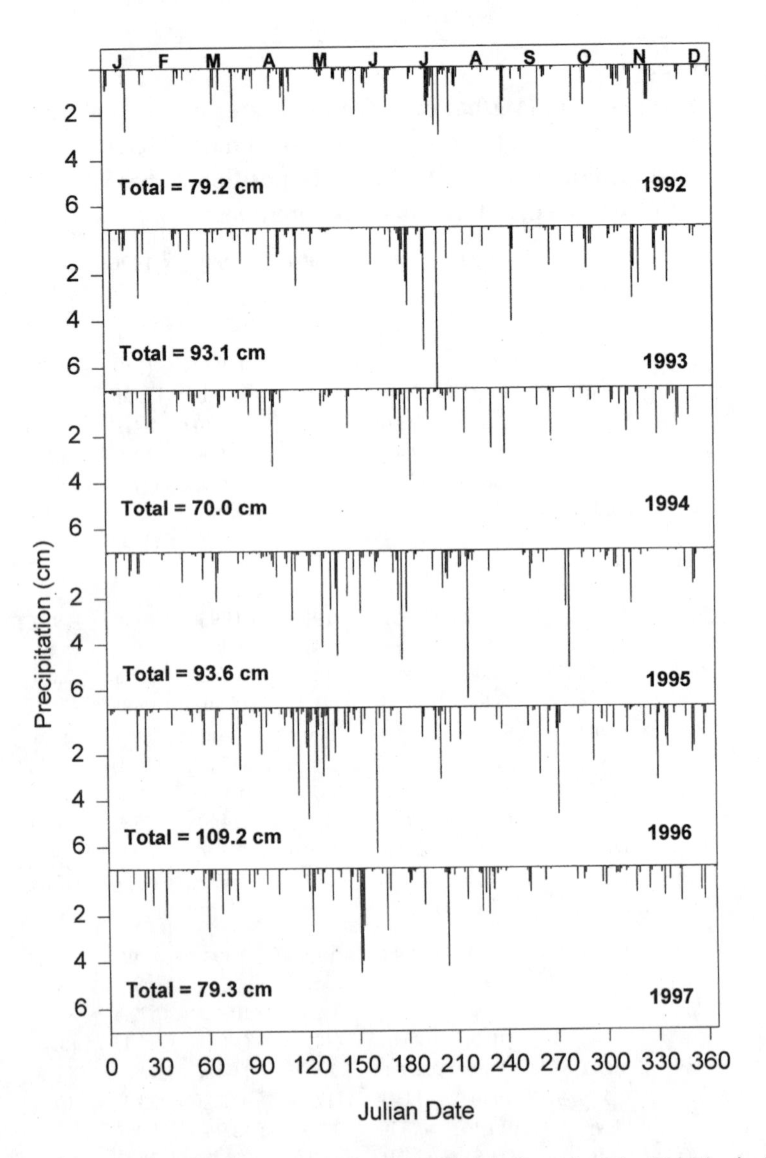

FIGURE 12.1 Daily precipitation record from the Western Branch of the Ohio Agricultural Research and Development Center, South Charleston, Ohio, for the years 1992 through 1997.

sites provided high CEC as well as drainage, while ponded water at the level SL sites reduced plant stand.

For Farm C, the SEV yield was significantly lower than other phase yields in 1992 (soybeans) and 1993 (corn). The SEV areas of Site C are prone to gullying. The DEP yield was significantly higher than the MOD yield in 1994 (soybeans); there were no significant differences in yield in 1995 (corn) and 1997 (corn); the DEP yield was significantly higher than the SEV yield in 1996 (soybeans).

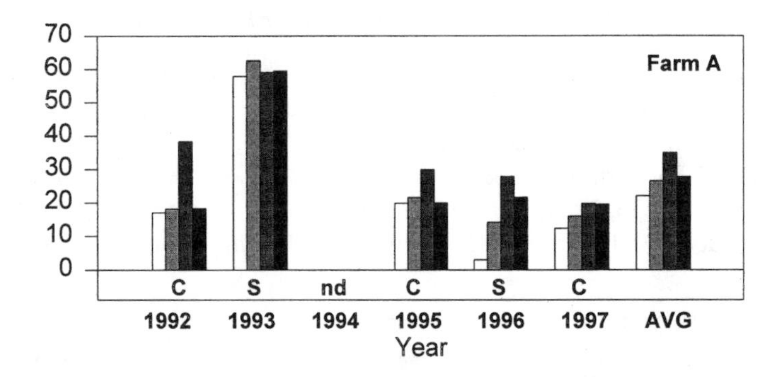

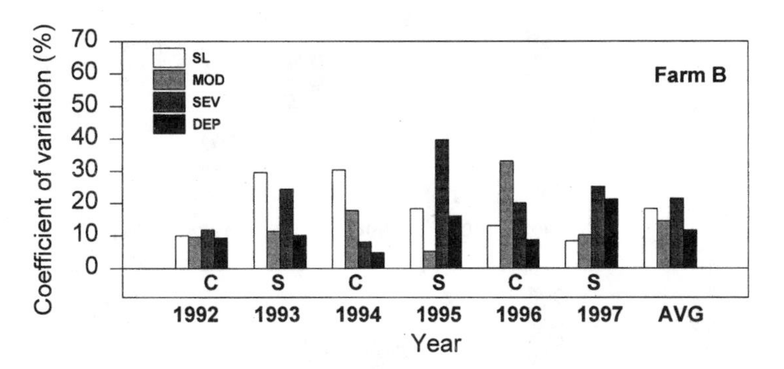

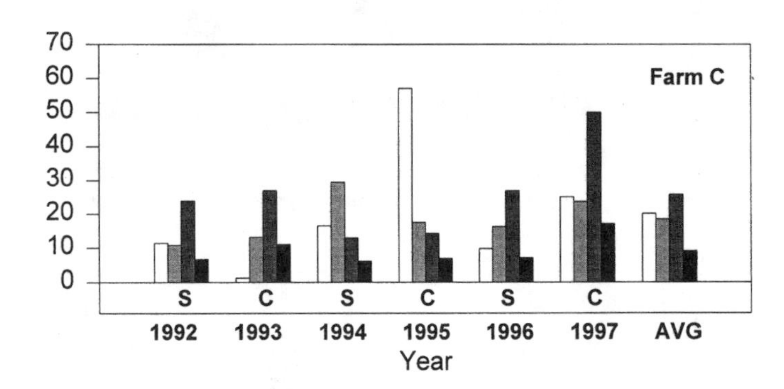

FIGURE 12.2 Coefficients of variation (%) of the average yield of three replications of each phase. AVG represents the average CV for slightly eroded (SL), moderately eroded (MOD), severely eroded (SEV), or depositional phase (DEP) over the years 1992 through 1997. S = soybeans, 13% moisture content; C = corn, 15% moisture content; nd = no data.

Figure 12.2 shows the coefficients of variation (CVs) of the phase average yields each year. Some variability in this study could be attributed to differences in surroundings in the landscape among replicate plots of each phase, which affected drainage. The SEV phase exhibited the largest 7-year average value of CV at all three farms.

PHASE AREAS AND AREA-WEIGHTED AVERAGE YIELDS

The area-weighted average yield represents the combined effect of erosion-phase plot averages (three replicates) and the estimated area of each phase in each study area. For Farm A, the percentage areas of SL, MOD, SEV, and DEP were 29, 39, 14, and 18%. For Farm B, the percentages were 13, 39, 16, and 32%. The percentages for Farm C were 36, 29, 8, and 27%. Generally, area-weighted yields between 1992 and 1997 decreased with time (Figure 12.3), following a countywide trend related

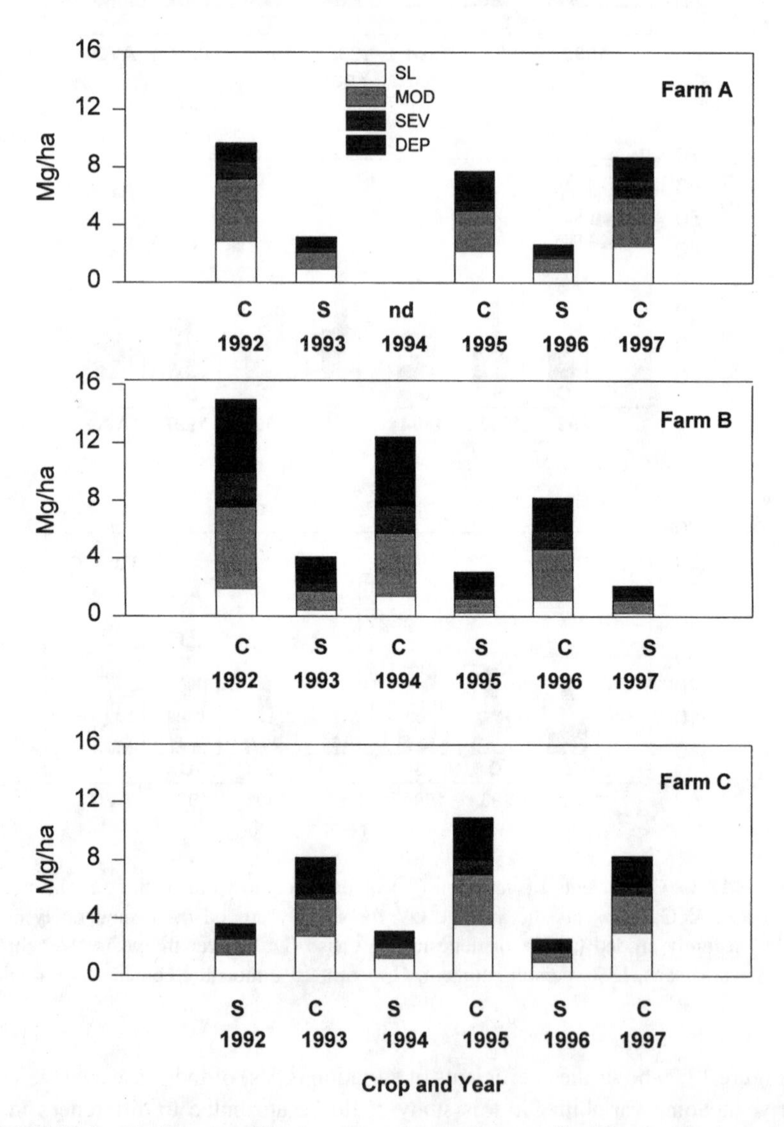

FIGURE 12.3 Total area-weighted average and relative contribution of each phase for the years 1992 through 1997. S = soybeans, C = corn, nd = no data.

to amount and distribution of annual precipitation (Alan Armstrong, owner of Farm B, 1997, personal communication).

TABLE 12.2
Area Weighted Average Yield (Mg/ha) as a Percentage of Average Yield of the Slightly Eroded Phase (first line) and as a Percentage of the Highest-Yielding Phase for That Farm in That Year (second line)

Farm	1992	1993	1994	1995	1996	1997
A	96.2	95.5	nd	99.3	97.4	97.6
	88.1 M	76.1 D		69.6 D	70.4 D	97.6 SL
	C	S		C	S	C
B	99.5	119.6	111.4	138.4	92.5	108.5
	94.7 D	72.7 D	83.3 D	75.2 D	91.8 M	87.7 D
	C	S	C	S	C	S
C	89.1	107.0	91.2	108.5	94.4	98.0
	89.1 SL	89.8 D	87.6 D	85.0 SV	88.8 D	93.9 M
	S	C	S	C	S	C

SL = slightly eroded; M = moderately eroded; SV = severely eroded; D = depositional phase. S = soybeans (13% moisture); C = corn (15% moisture); nd = no data.

For illustration of the effect of erosion on yields, Table 12.2 shows the area-weighted average yields expressed as a percentage of the slightly eroded phase and also of the erosion phase which produced the highest yield that year. For Farm A, area-weighted averages ranged from 95 to 99% of the SL yield averages; for Farm B, from 92 to 138%; and for Farm C, from 89 to 108%. As a percentage of the highest yielding phase, which in 11 out of 17 cases was the DEP phase, the area-weighted averages ranged from 70 to 97% for Farm A, from 72 to 94% for Farm B, and from 85 to 93% for Farm C.

COMPENSATING EFFECTS

One way to highlight erosion–productivity relationships is to compare the proportion of the area-weighted average contributed by each phase to the proportion of the study area contributed by each phase. In the case of Farm B in 1993, 10.6% of the crop was produced on 12.7% of the field area which was SL, for a difference of –2.1, which is graphed in Figure 12.4; 31.5% of the crop was produced on 39.3% of the field area which was MOD, for a difference of –7.8; 14.1% of the crop was produced on 16.2% of the field area which was SEV, for a difference of –2.1. These differences from the area-weighted average were made up for by the DEP area, which produced 43.8% of the crop on 31.8% of the area, for a difference of 12.0%. Therefore, the DEP area compensated for losses at the three eroded phases at Farm B

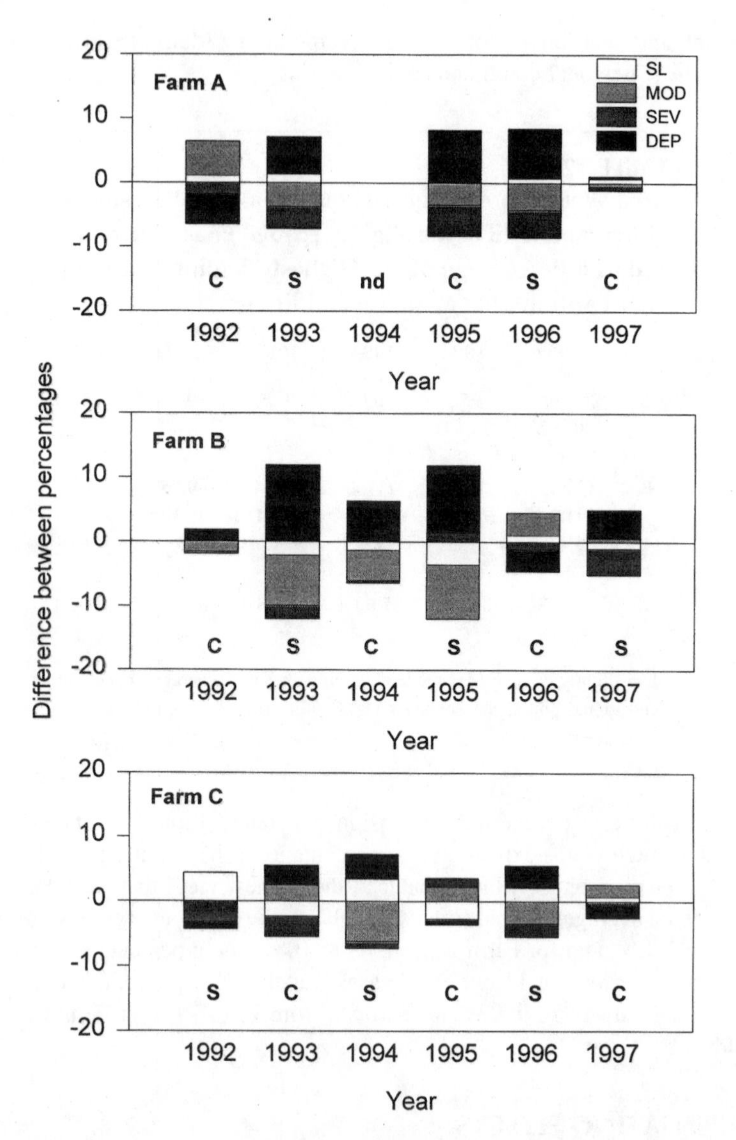

FIGURE 12.4 Difference between percentage of total field area contributed by each phase and percentage of area-weighted average contributed by each phase.

in 1993. This also occurred in 1992 and in 1994 at Farm B and in 1995 at Farm A. In all other years, yields from SL (in 1993, 1996, and 1997 at Farm A; in 1994 and 1996 at Farm C) or from MOD (in 1995 and 1997 at Farm B; in 1993 at Farm C) helped to compensate for losses at the two remaining phases. In 1 out of 5 years at Farm A, in 1 out of 6 years at Farm B, and in 3 out of 6 years at Farm C, the DEP phase contributed less than its share to the area-weighted averages, and was compensated by other phases.

DISCUSSION

The data presented in this study represent the average of three replicates of each erosion phase. Each replicate varied slightly in site characteristics and, therefore, in soil characteristics and drainage. The averages were considered to be representative values and were used to calculate area-weighted average yields. However, the high CVs of the phase yield averages demonstrated the ranges in integrated attributes due to position in the landscape. Better predictions of yield might result using the approach taken by McCann et al. (1996), who grouped management elements according to similarity in yield, plan curvature, slope length, and other site characteristics.

The average values, considered as a nested sample, could be used to predict the effects of erosion on yields in areas of similar topography and soils. It will soon be possible to obtain digitized maps of soils overlaid onto topographic quads with 5 or 10 ft (1.52 or 3.05 m) contour intervals (USDA, 1997), to which erosion classes can be assigned. Lanyon and Hall (1983) interpreted soil and topographic maps in this way, in order to predict areas of slippage; Khakural et al. (1996a) augmented measured values with data from soil surveys in order to predict erosion–productivity on topographic maps.

Two other approaches, parametric equations or simulation models, can be used to scale the data from this study to the regional level (West and Bosch, 1998). The measurements of soil physical and chemical properties obtained from all erosion phases from the three farms, coupled with the elevation data, could be developed into workable relationships (pedotransfer functions) between yields and soil and site properties, similar to those of Khakural et al. (1996b), which related corn and soybean yields to depth to carbonates, surface P and K content, relative elevation, slope gradient, aspect, and profile curvature.

Because the compensating effects shown in this study were related to weather pattern, predictions might be improved by taking the approach of Bruce et al. (1990), who included soil water tension at two depths between days 214 and 280 in their regression equations, to account for drainage and precipitation effects.

One limitation of this study was the subjective estimation of erosion phase areas. The Water Erosion Production Project (WEPP) model could be used to validate the areas of eroded phases on the three farms, as well as to predict changes in relative areas with time, and yields on each phase (Lane et al., 1993).

One problem in this study was the difficulty experienced by several researchers in locating replicates of the erosion phases every year. A simple and inexpensive solution would be to mark each phase with metal tape, buried below the depth of tillage. Such tape can be detected with an inexpensive metal detector and is used by utility companies for marking the location of buried power lines.

REFERENCES

Aspinall, J.D., R.G. Kachanoski, and H. Lang. 1989. Tillage-2000 Soil Conservation: 1989 Annual Report, Ontario Ministry of Agriculture and Food and University of Guelph, Guelph, 24 pp.

Bajracharya, R.M., R. Lal, and J. Kimble. 1998. Use of rad. fallout CS 137 to estimate soil erosion in 3 farms in West Ohio, *Soil Sci.* 163:133–142.

Battiston, L.A., M.H. Miller, and I.J. Shelton. 1987. Soil erosion and corn yield in Ontario. I. Field evaluation, *Can. J. Soil Sci.* 67:731–745.

Brubaker, S.C., A.J. Jones, K. Frank, and D.T. Lewis. 1994. Regression models for estimating soil properties by landscape position, *Soil Sci. Soc. Am. J.* 58:1763–1767.

Bruce, R.R., W.M. Snyder, A.W. White, Jr., A.W. Thomas, and G.W. Langdale. 1990. Soil variables and interactions affecting prediction of crop yield pattern, *Soil Sci. Soc. Am. J.* 54:494–501.

Carson, M.A. and M.J. Kirkby. 1972. *Hillslope Form and Process*, Cambridge University Press, London, 475 pp.

Cassel, D.K. and D.W. Fryrear. 1989. Evaluation of productivity changes due to accelerated erosion, in *Proceedings of Soil Erosion and Productivity Workshop,* University of Minnesota, St. Paul.

Daniels, R.B., J.W. Gilliam, D.K. Cassel, and L.A. Norton. 1989. Soil erosion has a limited effect on field scale crop productivity in the southern piedmont, *Soil Sci. Soc. Am. J.* 53:917–920.

Ebeid, M.M., R. Lal, G.F. Hall, and E. Miller. 1995. Erosion effects on soil properties and soybean yield of a Miamian soil in western Ohio in a season with below normal rainfall, *Soil Technol.* 8:97–108.

Fahnestock, P., R. Lal, and G.F. Hall. 1995a. Land use and erosional effects on two Ohio alfisols: I. soil properties, *J. Sustain. Agric.* 7:63–84.

Fahnestock, P., R. Lal, and G.F. Hall. 1995b. Land use and erosional effects on two Ohio alfisols: II. crop yields, *J. Sustain. Agric.* 7:85–100.

Frye, W.W., R.L. Blevins, and L.W. Murdock. 1987. Onsite impacts of management of highly erodible land, in *Alternative Uses of Highly Erodible Agricultural Land,* Agricenter International, Memphis, TN.

Golden Software. 1997. Surfer, version 6. Golden, CO.

Govers, G., K. Vandaele, P. Desmet, J. Poesen, and K. Bunte. 1994. The role of tillage in soil distribution on hillslopes, *Eur. J. Soil Sci.* 45:469–478.

Hall, G.F. and C.G. Olson. 1991. Predicting variability of soils from landscape models, in *Spatial Variabilities of Soils and Landforms,* Soil Science Society of America Special Publication No. 28, Madison, WI.

Hall, G.F., R.B. Daniels, and J.E. Foss. 1982. Rates of soil formation and renewal in the USA, in *Determinants of Soil Loss Tolerance,* ASA Special Publication Number 45, ASA, Madison, WI.

Jones, A.J., L.N. Mielke, C.A. Bartles, and C.A. Miller. 1989. Relationship of landscape position and properties to crop production, *J. Soil Water Conserv.* 44:328–332.

Jones, A.J., R.A. Selley, and L.N. Mielke. 1990. Cropping and tillage options to achieve erosion control goals and maximum profit on irregular slopes, *J. Soil Water Conserv.* 45:648–653.

Kachanoski, R.G., E.G. Gregorich, and R. Protz. 1992. Quantification of soil loss in complex topography, in R.G. Kachanoski, M.H. Miller, D.A. Lobb, E.G. Gregorich, and R. Protz, Eds., Management of Farm Field Variability, SWEEP/TED Report, University of Guelph, Guelph, Ontario, 156 pp.

Khakural, B.R., G.D. Lemme, T.E. Schumacher, and M.J. Lindstrom. 1992. Effects of tillage systems and landscape on soil, *Soil Tillage Res.* 25:43–52.

Khakural, B.R., P.C. Robert, and D.J. Mulla. 1996a. Predicting corn yield across a soil landscape in west-central Minnesota using a soil productivity model, in P.C. Robert, R.H. Rust, and W.E. Larson, Eds., *Precision Agriculture, Proceedings of the 3rd International Conference,* June 23–26, Minneapolis, MN. ASA/CSSA/SSSA, Madison, WI, 1222 pp.

Khakural, B.R., P.C. Robert, and A.M. Starfield. 1996b. Relating corn/soybean yield to variability in soil and landscape characteristics, in P.C. Robert, R.H. Rust, and W.E. Larson, Eds., *Precision Agriculture, Proceedings of the 3rd International Conference,* June 23–26, Minneapolis, MN, ASA/CSSA/SSSA, Madison, WI, 1222 pp.

Kirkby, M.J. 1978. *Hillslope Hydrology,* Wiley, London, 309 pp.

Kreznor, W.R., K.R. Olson, W.L. Banwart, and D.L. Johnson. 1989. Soil, landscape and erosion relationships in a northwest Illinois watershed, *Soil Sci. Soc. Am. J.* 53:1763–1771.

Lane, L.J., M.A. Nearing, J.M. Laflen, G.R. Foster, and M.H. Nichols. 1993. Description of the U.S. Department of Agriculture water erosion prediction project (WEPP) model, in *Overland Flow: Hydraulics and Erosion Mechanics,* A.J. Parsons and A.D. Abrahams, Eds., UCL Press, London, 440 pp.

Lanyon, L.E. and G.F. Hall. 1983. Land-surface morphology: 2. Predicting potential landscape instability in eastern Ohio, *Soil Sci.* 136:382–286.

Lobb, D.A., R.G. Kachanoski, and M.H. Miller. 1994. Tillage translocation and tillage erosion on shoulder slope landscape positions measured using ^{137}Cs as a tracer. *Can. J. Soil Sci.* 75:211–218.

McCann, B.L., D.J. Pennock, C. van Kessel, and F.L. Walley. 1996. The development of management units for site-specific farming, in P.C. Robert, R.H. Rust, and W.E. Larson, Eds., *Precision Agriculture, Proceedings of the 3rd International Conference.* June 23–26, Minneapolis, MN, ASA/CSSA/SSSA, Madison, WI, 1222 pp.

Miller, K.E. 1997. Soil Survey of Clark County, Ohio. U.S. Department of Agriculture, Natural Resources Conservation Service, Washington, D.C. In cooperation with the Ohio Department of Natural Resources, Division of Soil and Water Conservation, Ohio Agricultural Research and Development Center, The Ohio State University Extension, Columbus, OH, and the Board of County Commissioners, Clark County, OH. Unpublished report.

National Research Council. 1993. *Soil and Water Quality: An Agenda for Agriculture,* National Academy Press, Washington, D.C. 516 pp.

Olson, K.R., R. Lal, and L.D. Norton. 1994. Evaluation of methods to study soil erosion-productivity relationships, *J. Soil Water Conserv.* 49:586–590.

Pennock, D.J. and D.F. Acton. 1989. Hydrological and sedimentological influences on boroll catenas, central Saskatchewan, *Soil Sci. Soc. Am. J.* 53:904–910.

Pennock, D.J., D.W. Anderson, and E. de Jong. 1994. Landscape scale changes in indicators of soil quality due to cultivation in Saskatchewan, Canada, *Geoderma* 64:1–19.

Perrens, S.J., G.R. Foster, and D.B. Beasley. 1984. Erosion's effect on productivity along nonuniform slopes, in *Erosion and Soil Productivity, Proceedings of the National Symposium on Erosion and Soil Productivity,* Dec. 10–11, 1984, ASE, New Orleans, LA.

SAS Institute. 1994. SAS/STAT User's Guide. Version 6. SAS Inst., Cary, NC.

Soil Survey Staff. 1994. *Keys to Soil Taxonomy,* 6th ed. U.S. Government Printing Office, Washington, D.C.

USDA, 1997. Implementation of the SSURGO QIT Recommendations, Status Report. U.S. Department of Agriculture, Natural Resources Conservation Service, Washington, D.C., 34 pp.

West, L.T. and D.D. Bosch. 1998. Scaling and extrapolation of soil degradation assessments, in R. Lal, W.H. Blum, C. Valentine, and B.A. Stewart, Eds., *Methods of Assessment of Soil Degradation,* CRC Press. Boca Raton, FL, 558 pp.

13 A Landscape/Lifescape Approach to Sustainability in the Tropics: The Experience of the SANREM CRSP at Three Sites

William L. Hargrove, Dennis P. Garrity, Robert E. Rhoades, and Constance L. Neely

CONTENTS

0-8493-0702-3/00/$0.00+$.50

INTRODUCTION

The Sustainable Agriculture and Natural Resource Management Collaborative Research Support Program (SANREM CRSP), funded by U.S. AID, aims to develop principles and methodologies for achievement of sustainable ecosystem management at the watershed and landscape scale. In contrast to much of the earlier literature in environmental, agricultural, and ecological science which portrays humans as the despoilers of ecosystems, the SANREM approach acknowledges the role of people as integral actors who must be directly engaged in sustainable development if it is to be successful. The SANREM approach, therefore, uses an intersdisciplinary, farmer-participatory research, training, and information exchange strategy to bridge between scientific knowledge and local expertise in the resolution of problems. A landscape ecology approach is used to describe and understand the complex internal, external, and interactive processes within and between the individual ecosystems of a toposequence transecting two or more agroecological zones. Coupled with the landscape notion is that of "lifescape," or the human dimension, which includes economic, cultural, and social aspects in interaction with the physical and biological dimensions of ecosystems. Interventions appropriate to the local farm community members are designed and evaluated in concert with the ultimate users who are involved in all steps of the research process. The approach includes emphasis on collaboration between biological and social scientists and development practitioners in the U.S. and host countries. It also strives for interinstitutional collaboration that involves private and public institutions from each country as well as the U.S.

The SANREM project is considered one of the first of its kind, a pioneering effort to solve complex natural resource and sustainable agriculture problems involving a wide range of stakeholders who may have conflicting interests among themselves. These stakeholders include local communities, international research and development organizations, U.S. universities, host country universities and government agencies, local government and nongovernment organizations (NGOs). Also, there is a need to address experimentally a hierarchical range of spatial and temporal scales across a landscape. In attempting to meet these challenges, we have many successes, many failures, and many lessons learned. This chapter outlines the experiences of SANREM at the three core sites in the Philippines, Burkina Faso, and Ecuador, and presents critical lessons learned in implementing a large landscape-scale research and education program.

GENERAL APPROACH

THE LANDSCAPE/LIFESCAPE

The concept of the landscape is illustrated in Figure 13.1. From this perspective, landscapes are much more than just the topography across which animals, plants, soils, water, and other materials move. In this view, the term *landscape* is used to refer to spatial patterns of the biological and physical processes and *lifescape* to refer to the human dimension relative to that spatial template. Given that ecosystem interactions take place in real time and space and involve both nature and people,

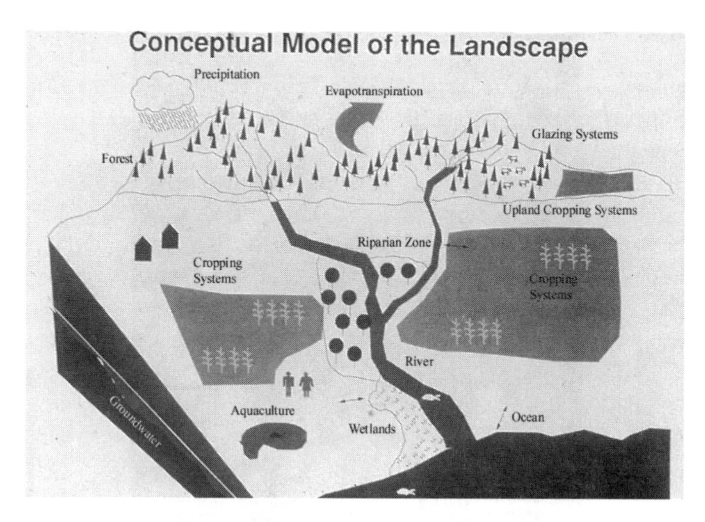

FIGURE 13.1 Concept of landscape.

the landscape/lifescape is the most appropriate scale for applied research aimed at the conservation of biodiversity and at identifying and enhancing sustainable strategies of livelihood for the populations that inhabit and interact in the environments thus defined.

A landscape/lifescape approach requires that the research encompasses hierarchical analytical units both spatially and sociodemographically, from experimental plots, to entire fields, to agroecological systems, to watersheds, and beyond; likewise, from farmers, to families, to communities, to ethnic groups, to national and international institutions and organizations. SANREM CRSP focuses on two broad ecological zones: the humid tropics and semiarid/subhumid tropics, which offer characteristics that make them of particular interest in studying sustainability issues. Within these zones, sequences of climatic, topographic, biological, vegetative features and mosaics of rural and urban activities define various agroecological areas, which in turn comprise the landscape (see Figure 13.1). The SANREM CRSP participatory process of problem diagnostic and prioritization of research questions and research projects jointly by local people and scientists has emphasized interrelations and interactions within agroecological zones as well as across the entire landscape. The Participatory Landscape/Lifescape Appraisal (PLLA) and priming activities to jump-start concrete projects with immediate benefits to local people have helped to deepen the understanding of landscape ecology among both SANREM CRSP investigators and local people by stimulating reflections and representations of sustainability issues and linkages across agroecological zones. This was facilitated by the utilization of indigenous terminologies and ethnoclassifications to develop a characterization to the landscape. Finally, throughout the entire process for reviewing and implementing work plans at all levels (from local communities to rigorous scientific peer review), efforts have been directed to link and consolidate the data generated for various agroecological zones toward an integrated perspective of the landscape ecology.

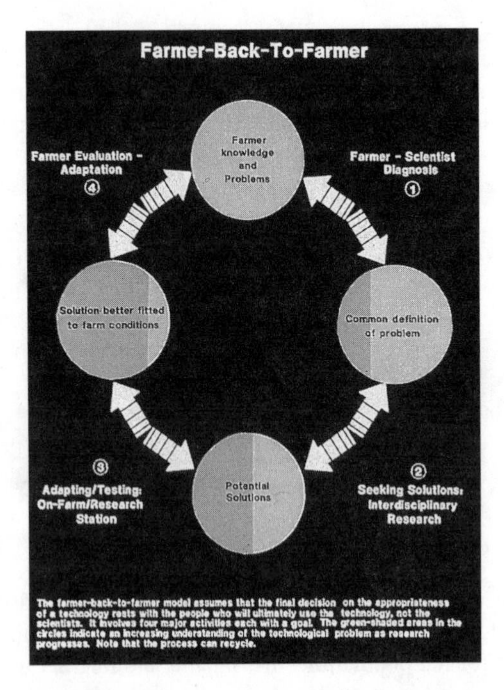

FIGURE 13.2 Farmer-back-to-farmer model.

Participation

SANREM is committed to doing research that is in synchrony with, builds upon, and is useful to the real-life experience of local people. All too often the local community members and actual natural resource managers are omitted from the dialogue and direct involvement in natural resource management research and development projects (Chambers, 1983; Chambers et al., 1989). The Farmer-back-to-Farmer model (Figure 13.2) described by Rhoades (1984) offers an effective framework for such an endeavor, requiring that farmers are included in every step of the research process, from problem diagnosis and definition, to the identification and testing of potential solutions, to evaluation and adoption of the findings. A farmer-centered approach, however, should not be mistaken as portraying farmers as a homogeneous category, sharing common goals and interests. Rather, social differentiation — whether based on gender, ethnicity, or class — is recognized and taken into account in the analysis of the constraints to agricultural sustainability. This means going beyond an intellectual commitment to participation, and investing time and attention to identifying and including all sectors of the community, especially those less privileged. A truly participatory research process calls for considerable work to be devoted to developing a common agenda and relationships of mutual trust and commitment between researchers and the diverse groups in the community. It also entails the voluntary association with the project by all the parties involved, a clear understanding of expectations and obligations of all, and transparency in all aspects of the project, including management and financial.

Because of the diverse nature of the contexts and multiobjectives of research across different scale levels, participation in SANREM CRSP varied along a continuum of involvement, from a minimum level, in which permission for program activities was requested from all appropriate local officials and leaders of local groups; to a higher level, in which the program established an identifiable presence in the community; to a more effective and desirable level of participation, in which the community identified and verified the problems to be addressed by the research, and researchers regularly reported their findings to the community in ways that are understandable and usable by them. What degree and sorts of participation are most appropriate at various stages of the research process and how to measure and assess them remain questions open to debate and are the subject of continuing debate and earnest inquiry among the SANREM CRSP research partners. Despite the fact that the degree and intensity of participation will vary across time and space, the SANREM experience shows that commitment to and willingness to provide resources beyond that given by U.S. AID (time, land, cost share) is greatly enhanced due to the feeling of ownership in the project.

DESCRIPTION OF SITES: THE EXPERIMENTAL LANDSCAPES/LIFESCAPES

General Characteristics

All of the SANREM experimental landscapes/lifescapes can be characterized in the following ways:

- Multiple ecosystems managed by a mosiac of stakeholders;
- Natural areas with significant sources of biodiversity;
- Diverse agriculture;
- Significant water resources;
- Critical human and social issues

The humid tropical highland/lowlands and the subhumid, seasonally arid tropical savannahs where SANREM is working in Asia, Africa, and South America have characteristics that make them relevant and representative zones for the study of sustainable agriculture and natural resource management.

In the highland/lowland regions of the humid tropics, these characteristics include:

- Bioreserves of plant and animal genetic diversity;
- Centers of cultural and ethnic diversity;
- High rates of soil loss and sedimentation;
- Significant downstream impacts (pesticide contamination, salinization, siltation, destruction of coastal resources and living aquatic resources);
- Zones of human migration; and
- Significant forest resources.

The "connectedness" in the landscape between uplands and lowlands and the impact of land management on uplands and lowlands have been documented elsewhere for the humid tropics of Southeast Asia (Magreth and Doolette, 1990; Garrity, 1992).

In the subhumid, seasonally arid tropics, the features of special interest include

- Desertification;
- High spatial variability;
- Human and livestock populations that generally exceed the carrying capacity of the land;
- Food supplies that are highly dependent on the vagaries of rainfall and tropical storms; and
- Critical vulnerability to famine.

Over the past 5 years, we implemented our program at three sites, two that are representative of the humid, highland/lowlands (in the Philippines and Ecuador) and one that is representative of the subhumid, seasonally arid savannahs in Burkina Faso. Each of these is described in more detail below.

Site Descriptions

The Philippines

The landscape selected for research and development in the Philippines was the watershed of the Manupali River which drains approximately 80,000 ha and is a tributary of the Pulangi River Basin. The Manupali watershed is located on north-central Mindanao (longitude 125°W, latitude 8°N), 20 km south of Malaybalay and 85 km south of Cagayan De Oro City.

The watershed has moderate to steep slopes, with greater than half of the land area having a slope of >15%, fragile soils, and severe erosion. Soils in the area are well-drained, highly erodible clays of low pH. A rapid increase in small-scale farms, with insecure land tenure and a necessity to provide food for the family, has resulted in unsustainable agricultural practices on the uplands. Farm sizes are typically 0.25 to 19.5 ha with an average size of 3.0 ha. Farm sizes less than 3.2 ha typically are tenant or leasehold farms. The Manupali watershed has a diversity of agricultural crops. In the lowlands, there are large tracts of irrigated wetland rice, sugarcane, and pineapple which utilize high external inputs. In addition, cash market–oriented production of lettuce, potato, cassava, tomato, and cabbage as well as some small vegetable gardens for household use is present in the uplands. The upper reaches of the Manupali watershed are occupied by the Mt. Kitanglad National Park, a primary forest biopreserve.

The Philippine Council for Agriculture and Resources Research and Development (PCARRD) coordinates agricultural and forestry research at the national level. In addition to government agencies such as the Department of Agriculture, Department of Environment and Natural Resources, University of the Philippines at Los Baños, and Central Mindanao University, several NGOs, both local and international, are active in the watershed. The SANREM CRSP has developed strong linkages between these sectors through an explicitly integrative team research approach.

Some of the critical issues that threaten sustainability include

- Deforestation through slash-and-burn agriculture and encroachment on the bioreserve;
- Expanding high-input vegetable production in the uplands of the watershed;
- Water quality deterioration due to soil erosion and runoff containing pesticides and fertilizer nutrients;
- Siltation and declining water quality of the Pulangi IV reservoir.

Ecuador

The Ecuador site was the last site to be identified and implemented by SANREM CRSP. In July 1994, a zone of colonization located in the southern buffer zone of the Cotacachi-Cayapas Ecological Reserve in Northwest Ecuador was chosen. Designated as one of the biological "hotspots" on Earth, the area is generally characterized by an area of in-migration of people, many of whom historically practiced slash-and-burn agriculture to clear the forest and install pasture. Much of the region has been deforested and its land use is undergoing continuous intensfication as farmers and ranchers try to extract profits from the area. The participating institutions in Ecuador include Facultad Latinoamericana de Ciencias Sociales-Ecuador (FLACSO-Ecuador); Universidad San Francisco de Quito (USFQ); Faculty of Veterinary Medicine and Zootechnics, Universidad Central (FMVZ-UC); Sistema de Información y Desarrollo Comunitario (COMUNIDEC); Heifer Project International (HPI); Centro de Datos para la Conservación (CDC); and Coordinadora Ecuatoriana de Agroecologìa (CEA). Four communities were chosen for SANREM focus: Palmitopamba, La Perla, Chacapata, and Playa Rica. U.S. institutions included the Department of Anthropology and Institute of Ecology, University of Georgia, Department of Sociology, Iowa State Univerisity, and Center for Aquatic Resources, University of Auburn.

Some of the critical issues threatening sustainability include

- In-migration of people and continuing deforestation;
- Extractive uses of the bioreserve;
- Slash-and-burn agriculture and low-quality pasture conversion;
- Soil erosion;
- Declining water quality.

Burkina Faso

Burkina Faso was chosen for the subhumid savannah research site because it encompasses several agroecological zones of interest, it has a strong national research organization and university, and the University of Georgia, as well as several other consortium members, has had a long-term collaboration there.

The chosen site is located in the Sudano-Sahelian zone (750 to 800 mm annual rainfall) of Burkina Faso, and includes the village of Donsin (approximately 12°45′N latitude, 0°39′W longitude) and the surrounding area. Donsin is located approximately 100 km northeast of the capital city of Ouagadougou, and requires a 2.5- to 3-h drive. The watershed is approximately 8500 ha and contains approximately 1500 inhabitants. It is a part of the Niger River drainage area. It is an area of people out-migration and one that has national priority in terms of sustainable agriculture.

As is typical in the Sudano-Sahel, the annual rainfall is extremely variable. In 1991 the watershed received 900 mm of rain, while in the previous year (1990), only 450 mm were received. The native vegetation is typical of the West African savannah and the agriculture comprises of millet, sorghum, groundnuts, and livestock, primarily cattle, sheep, and goats. There are small amounts of cotton, rice, and cowpeas also grown in the area.

The soils are derived from igneous materials, primarily granite. The soils are highly weathered, acid, and ferruginous. They are very low in phosphate. The topography ranges from plateaus to the savannah plain to low-lying, ephemeral streams. There are some outcrops of granite within the watershed.

Some of the critical issues threatening sustainability include

- Water runoff and soil erosion;
- Declining water quantity and quality for household use;
- Declining soil fertility;
- Vanishing biodiversity in the "bush" due to extractive uses.

THE RESEARCH PROCESS AND EXAMPLES OF ACCOMPLISHMENTS

Overview of the Research Process

The SANREM CRSP has developed an interactive process to build a strong base for successfully outlining a sustainable research agenda and addressing the issues which have been outlined. The process was modeled after Conway's agroecosystem analysis (Conway, 1985) and includes the following steps/phases:

1 *Site selection and networking*. This includes a reconnaissance of existing information and secondary data.
2. *Descriptive/diagnostic phase*. This includes a participatory PLLA and institutional analysis. The PLLA is a participatory diagnostic survey in the community which gathers community perceptions and information related to the landscape/lifescape.
3. *Preliminary analytical phase*. This includes a stakeholders' workshop and development of hypotheses.
4. *Strategic planning phase*. This results in the framework plan and identification of key indicators. The framework plan outlines the research and education agenda and is the guiding document for project activities.

5. *Research implementation.* Hypotheses are tested and key indicators are monitored.
6. *Analysis, interpretation, synthesis, prediction.* The goal is to improve our understanding of the principles of sustainability.
7. *Training, global synthesis, transfer.* Results, key findings, and methodological lessons learned are shared, and successful models are replicated.

Some examples of activities and accomplishments from implementing this process at three sites are discussed below.

Diagnostic Phase

The most important component of the diagnostic phase of each of our sites was the conduct of a PLLA. The PLLA was developed by our program as a major diagnostic tool in a participatory, landscape-scale research process. The objectives of PLLA are to:

1. Gain an understanding of interrelationships in the agroecosystem;
2. Identify and gain a collective understanding of constraints to natural resource and agricultural sustainability from the local community's perspectives;
3. Initiate and establish community–scientist dialogue;
4. Facilitate the community to identify their natural resources and become more aware of linkages in the landscape;
5. Gain "real-time" experience for diverse partners (international scientists, national scientists, NGOs, local community, etc.) working together and enhancing skills in listening, negotiating, and visioning; and
6. Develop a participatory research agenda.

PLLA is based on tools, strategies, and experiences of participatory rural appraisal, rapid rural appraisal, and farming systems research. It differs, however, in (1) scale, as it moves the description (baseline data) from the farm scale to the landscape and (2) the focus from identification of problems to the relationships on a landscape scale, the interaction of human activity with the biophysical environment, and the long-term sustainability of the landscape/lifescape. The process of PLLA includes team-building and training, reconnaissance of secondary data and ongoing programs, institutional analysis, field work, analysis, and verification. For the field work, a variety of participatory tools can be utilized including (1) informal, open-ended interviews with individuals or small groups, (2) resource mapping, (3) resource flow diagrams, (4) preference ranking, (5) seasonal calendars, (6) Venn diagrams, and others. A variety of information can be collected, depending on the primary purpose of the appraisal. The specific information to be collected dictates the appropriate tools to be used. An important step is verification in which the information collected is presented to the community(ies) for verification.

In our case, different approaches to the PLLA were used at each of the three sites. In the Philippines a more free-form approach was utilized in which open-ended interviews were the basis of the approach. In Burkina Faso, the Methode Acceleree du Recherche Participatif (MARP) provided the framework. In Ecuador, a local NGO, COMUNIDEC, led four communities in an "AutoDiagnostico" (Self-Diagnosis). In each case, the methodology utilized was a proven one used previously in the respective area or region. Community members participated in a variety of ways: (1) individual interviews, (2) small-group activities (2 to 10 individuals), (3) large-group activities (>10 individuals), and (4) whole village or community meetings. Gender- and/or age-differentiated data have been collected by applying the tools to community groups differentiated by gender and/or age.

The SANREM CRSP has used this approach to conduct appraisals in small watersheds (300 to 80,000 ha) in six countries (The Philippines, Burkina Faso, Cape Verde Islands, Ecuador, Costa Rica, and Morocco). The timescale for completion of the team-building, training, reconnaissance, and field work for these small watersheds was from 4 to 8 weeks. In our case, these appraisals formed the basis for developing a participatory research agenda focused on natural resource management. The PLLA is not intended to replace but to augment, ground-truth, and/or contextualize more quantitative demographic and biophysical data that might be available. In addition, we augmented our original data by subsequently conducting more-focused, "thematic" PLLAs on such topics as human health and nutrition, land tenure, livestock, etc.

Some of the characteristics of the technique include:

1. It is rapid and cost-effective;
2. Some of the local people's knowledge, understanding, and perspectives is revealed;
3. Rapport with local people is developed, forming the basis for future activities;
4. It is flexible; the technique can be used to focus on almost any information of local interest;
5. The process can heighten the community's awareness of environmental, social, or other issues;
6. Results can be incomplete, details missing;
7. Results are only as good as the sample;
8. Bias can be introduced by individuals conducting the appraisal.

The PLLA results were used to develop the SANREM research and education agenda. At each site, a framework plan for research and education activities was developed by a team of program participants. This formed the basis for subsequent annual work plans.

Baseline Data

Much baseline data describing the biophysical and socioeconomic conditions were collected for each site and much of these data were placed in a geographic information system (GIS) format at the University of Georgia. All of the data sets cannot be described here, but data on the following biophyscial and socioeconomic factors have been collected for each site:

Biophysical data
- Agroecological zones and topography
- Vegetation and biodiversity surveys
- Soils and land use
- Water quality and supply
- Others

Socioeconomic data
- Household size and composition
- Ethnicity and class
- Livelihood strategies
- Tenurial status
- Farm resources
- Credit and market availability
- Others

Field Research on Natural Resource Management Technologies

One of the important goals of SANREM was to develop and evaluate viable management strategies for achieving sustainability in agricultural and natural ecosystems. In the early stages of SANREM CRSP implementation, we chose to focus on management strategies that sustain productivity on cropped land, that enhance environmental quality, and that are transferrable and have demonstrable applicability to different agroecosystems. The field research that has been conducted on improved management strategies focuses on one or more of the following:

- Improved agroforestry systems
- Improved vegetable production systems
- Integrated pest management
- Soil management for improved water infiltration and fertility (zai)
- Composting and animal manure management
- Improved pasture management

Technologies that are chosen for SANREM CRSP attention are related to the constraints to sustainability identified by the PLLAs and address important issues related to linkages in the landscape.

Community Monitoring

One of the more successful activities of SANREM CRSP has been a project in the Philippines whose goal was to facilitate the formation and development of community-based water quality monitoring teams and raise community awareness of water issues. This project led by Dr. Bill Deutsch of Auburn University has trained local people in basic water quality testing, bacteriological monitoring of drinking water, and data interpretation. The training and water testing activities initiated by the team has led to the constitution of a registered NGO in the community, the Tigbantay Wahig ("Water Watchers"). The group has collected more than 1800 water samples at 29 sites on five rivers in the municipality of Lantapan and at the Pulangi IV Reservoir. This has resulted in a watershed-level database of water quality that is useful for educators, policy makers, and the general public in the area. Preliminary findings indicate that the information on suspended solids and *Escherichia coli* concentrations are not only scientifically valuable, but are good indicators of sustainability because they are simple and inexpensive to measure, they are related to land productivity and human health and thus are of interest to the general public, and they are important to policy makers at the local and national levels. A growing number of citizens are becoming involved in water quality monitoring, and the community-based groups are collecting credible data that are essential for management decisions. The water-monitoring teams of Lantapan provide a valuable community service and model of citizen action that applies to other regions of the Philippines and to other countries including the U.S.

Training and Capacity Building

From its inception, the SANREM CRSP established community-based education, ecological awareness, and information exchange as a priority to complement the formal training at the undergraduate, graduate, and visiting scientist levels more typical of university-based collaborative programs. In the initial phase, through both global and on-site training and workshops focused on familiarizing all program participants with the SANREM CRSP cornerstones and philosophies and equipping collaborators with the tools for participatory research on a landscape scale. Some of the general types of training are listed below.

- Training in participatory research methodologies;
- Indicators of sustainability;
- Field training for technicians and community members;
- Environmental education;
- Institutional strengthening;
- Informing policy;
- Graduate and undergraduate training.

SUMMARY AND CONCLUSIONS

Summary of Accomplishments

SANREM accomplishments in its first 5 years (1992 to 1997) include the following:

- Development of a participatory process methodology at the watershed or landscape/lifescape level that includes local communities;
- Creation of a network of viable partners that includes NGOs, government organizations, universities, and local communities who are trained to examine multistakeholder and multiscalar level problems;

- Interdisciplinary and scientific/local characterization of the landscape/lifescape through collecting baseline data sets and participatory research;
- Development and implemention a demand-driven research agenda;
- Design, evaluation, and demonstrated improved technologies for sustaining landscapes, such as improved agroforestry systems, improved vegetable production systems, etc.;
- Promotion of education, training, and information exchange in sustainability and environmental issues including direct involvement of community members.

Through this process, many lessons have been learned, some of which are summarized below.

Summary of Lessons Learned

1. Direct and active involvement of stakeholders is crucial to establishing research priorities, understanding the biophysical and socioeconomic constraints to sustainability, and identifying how constraints can be addressed, while heightening the awareness of stakeholders regarding their own landscape and sustainability issues.
2. Building a participatory, demand-driven research agenda that is relevant to local people has high transaction costs.
3. A "strategic" approach to collaboration is needed. The comparative advantage of each collaborating institution must be assessed in terms of benefit to that institution and to the overall program.
4. The most significant impacts of SANREM CRSP so far have been on participating institutions, many of whom have adopted participatory methodologies and/or landscape approaches and are better integrating research, extension, education, and development.
5. Sustainable agricultural technologies must integrate indigenous knowledge, stakeholders' goals and perspectives, and western science. This is the advantage of participatory approaches over the traditional researcher-driven technologies.

Key Finding and Conclusion

The SANREM experience has taught those who participate in the process that achieving sustainability is a "negotiated process"; the keys to sustainability do not lie in technologies developed from "Western science" alone. Sustainable agricultural technologies must integrate indigenous knowledge, stakeholders' goals and perspectives, and Western science. This is the advantage of participatory approaches over the traditional researcher-driven technologies.

Sustainability is a balancing act among competing global concerns and goals (articulated by Agenda 21) and the concerns and goals of local people at the community and household level (primary concerns are for livelihood and quality of life). The challenge is to find ways to "bridge the gap" between global and local agendas.

REFERENCES

Chambers, R. 1983. *Rural Development: Putting the Last First,* Longman Scientific & Technical Publishing, London.

Chambers, R., A. Pacey, and L.A. Thrupp. 1989. *Farmer First: Farmer Innovation and Agricultural Research,* Longman Scientific & Technical Publishing, London.

Conway, G.R. 1985. Agroecosystems analysis, *Agric. Admin.* 20:31–55.

Garrity, D.P. 1992. Sustainable land use systems for the sloping lands of southeast Asia, in *Technologies for Sustainable Agriculture for the Tropics,* American Society of Agronomy, Madison, WI.

Magrath, W.B. and J.B. Doolette. 1990. Strategic issues in watershed development, in *Watershed Development in Asia,* J.B. Doolette and W. B. Magreth, Eds., The World Bank. Washington, D.C.

Rhoades, R.E. 1984. *Breaking New Ground: Agricultural Anthropology,* International Potato Center, Lima, Peru.

14 Model-Based Evaluation of Land Use and Management Strategies in a Nitrate-Polluted Drinking Water Catchment in North Germany

Kurt Christian Kersebaum

CONTENTS

INTRODUCTION

Nitrogen leaching from agricultural land has become evident for pollution of water resources in many regions of the world. Groundwater pollution has to be seen in a watershed context because in the long term it affects the quality of surface waters as well. Simulation models linked to geographic information system (GIS) provide a useful tool to assess the effects of different agricultural practices on water quality. A model was used to evaluate actual and alternative management practices in a nitrate-polluted drinking water catchment of northern Germany. Simulations are able to describe nitrate leaching for a given land use with sufficient accuracy. Scenario

0-8493-0702-3/00/$0.00+$.50

simulations show that adoption of fertilizer supply to the demand of the crops and optimization of crop rotation can significantly reduce nitrate pollution. Nevertheless, the high vulnerability of soils for leaching requires the conversion of parts of the arable land into low-fertilized pasture to meet the drinking water standard.

Nitrate contamination of groundwater resources is a common problem in many countries with intensive agricultural production. The threshold for drinking water within the European Union is at present 50 mg $NO_3\ l^{-1}$ (11.3 mg NO_3–N l^{-1}), the guideline is 25 mg $NO_3\ l^{-1}$ (5.65 mg NO_3–N l^{-1}). Nitrate concentration in many locations of Europe is at or exeeds the drinking water standard (Owen and Jürgens-Gschwind, 1986; ECETOC, 1988). In Germany, 72% of the public water supply originates from groundwater or spring water (Statistisches Bundesamt, 1996). In a monitoring of groundwater resources in Germany, 15% of the investigation sites showed increased nitrate concentrations between 25 and 50 mg $NO_3\ l^{-1}$, and for 11% an excess of the threshold was observed (Umweltbundesamt, 1997) whereby higher concentrations were found mostly in the shallow aquifers. However, protection of groundwater resources from non-point-source pollution with chemicals like nitrate is relevant to sustain drinking water resources. Groundwater contamination over a long-term also becomes evident for surface waters.

For Germany, the percentage of nitrogen inputs from non-point sources to surface waters is estimated to originate 66% (Hamm, 1991) and 71% (Umweltbundesamt, 1997), respectively, from exfiltrating groundwater, whereas erosion accounts for only 12% (Bruckhaus and Berg, 1990) and 10% (Umweltbundesamt, 1997), respectively. Although in the U.K. and the U.S. surface runoff is recognized as the most relevant pathway (e.g., Heathwaite 1995), in the flat landscapes of North Germany, as well as in the Netherlands and Denmark, groundwater exfiltration plays the major role (Van der Molen et al., 1998). Because of improved standards for waste water treatment, the percentage of non-point-source pollution with nitrogen to surface waters has increased to about 60% of the total input (Umweltbundesamt, 1997).

Worldwide, agriculture plays a major role as a contributor to non-point-source pollution to soil and water resources (Humenik et al., 1987). High amounts of mineral fertilizers and high local concentrations of animal production are responsible for the nitrogen balance in many Western European countries showing a high surplus varying from 87 to 367 kg N ha^{-1} (Isermann, 1990). For Germany, Bach and Frede (1998) recently calculated a surplus of 111 kg N ha^{-1} in 1995. A high portion of this surplus is expected to be leached out with seepage water. The contribution of agriculture to the non-point-source contamination of German groundwater resources with nitrogen is estimated to be about 80% (Isermann, 1990). For surface water a contribution from agriculture of 48 and 55% of the total nitrogen input is calculated for, respectively, Germany (Isermann, 1990) and the European Union (Isermann and Isermann, 1995).

The transport of nutrients with the river streams also affects the coastal areas of the sea. In the German Bight of the North Sea, the nitrogen concentration has increased during the 1980s by a factor of 3 (Umweltbundesamt, 1997). Lidgate (1987) estimated that about 60% of the nitrogen input might originate from agriculture, alhough the development of land-use strategies and ecologically oriented agricultural management systems to protect groundwater resources is evident also for river watersheds and coastal regions to prevent eutrophication.

Several strategies to reduce nutrient losses to groundwater and surface water were developed (e.g., Van der Molen et al., 1998). There is an urgent need to predict effects of alternative forms of land use and management practices on water quality to support decisions at the political as well as at the management level. Simulation models may play an important role in that way. In a watershed context often models are used that concentrate on surface runoff processes and their impact on river systems (e.g., Young et al., 1987). For subsurface waters it is more important to describe the processes of nitrogen transport and transformation within the root zone and the vadose zone. Several simulation models of different complexity exist to describe nitrogen dynamics in the soil–plant system. An overview and comparison of some models can be found in CEC (1991), de Willigen (1991), and Dieckkrüger et al. (1995). Most of these models are dedicated to work at the point or field level. In some cases these models were applied on a watershed or regional scale (e.g., Kragt et al., 1990; Vereecken et al., 1990; Styczen, 1995; Kersebaum, 1995).

To assure that models are accepted by decision makers, there is a need to validate them in a well-investigated area preferably in a subcatchment of a watershed. Therefore, we used data of a small drinking water catchment which is part of the watershed of the river Wümme. In such drinking water catchments the conflict between farmers who are interested in a high production and water suppliers who have to meet the drinking water standards is very pronounced.

INVESTIGATION AREA

The investigated area in the north of Germany is the catchment of a water supply plant for the neighboring community of about 15,000 inhabitants. The area consists of 1.35 km^2 which is the official protection zone and an "extended zone" of 1.45 km^2 (Figure 14.1a), which was later identified as part of the well catchment system based on simulations with the groundwater model MODFLOW (McDonald and Harbaugh, 1988). The small size of the catchment, as well as the shallow depth of the aquifer with a short travel time from the root zone to the groundwater, makes it suitable for study of the effects of land use. The landscape is relatively flat with an elevation range of 25 to 35 m above sea level, which means that surface runoff can be neglected.

Since the early 1970s the nitrate concentrations in the four wells of the water plant increased over the present threshold of 50 mg NO_3 l^{-1}. During recent years the steady increase has been stopped and the concentration ranges from 75 to 137 mg NO_3 l^{-1} with an average of 99 mg NO_3 l^{-1}. In fact, the water supply station has to mix with imported water to meet the drinking water standards.

The high concentration of nitrate has been attributed to the extensive arable land use (Figure 14.1b) with a high surplus of nitrogen. Based on pumping experiments, the average travel time across the protection zone is calculated to be about 6 years (Walther et al., 1990). The average travel time from the root zone to the aquifer is 5 to 6 years; therefore, the observed concentrations should be correlated with the management practice 4 to 6 years ago. The following study covers the period from 1990 to 1996.

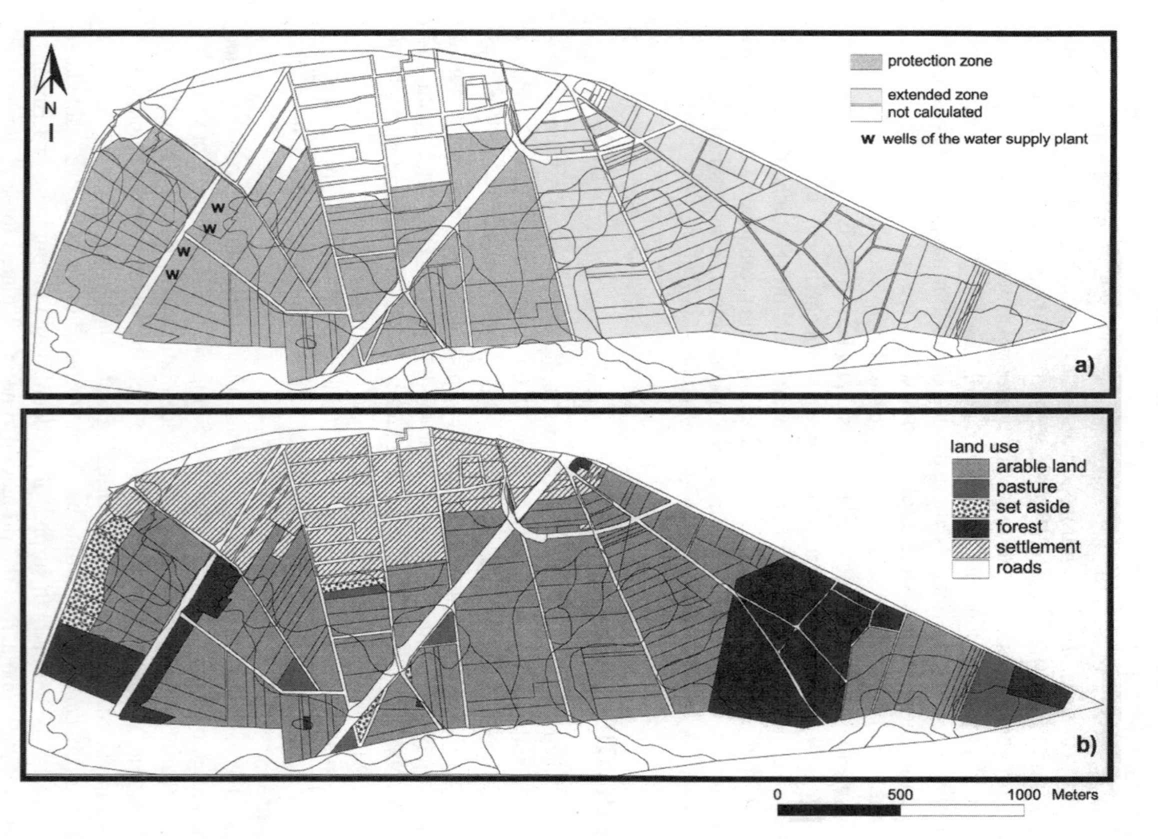

FIGURE 14.1 (a) Drinking water catchment with protection zone, wells of the water plant and polygons from the intersection of the field and the soil map. (b) Land use of the drinking water catchment.

OBJECTIVES

The specific objectives of the study addressed the following questions using a simulation model:

What is the extent of the nitrate contamination of the groundwater for the land-use and management practice of the last 6 years?
To what degree can the water quality be improved by optimizing environmentally sound agricultural practices such as fertilizer supply and crop rotation without fundamental changes of the cropping system?
What changes in land allocation (arable vs. pasture) need to take place to meet the drinking water standards?

Futhermore, model results for the present situation should be compared with observed values to evaluate the validity and applicability of the model under practical conditions without calibration.

DATABASE

During the first years, observations were focused on the official protection zone. During the period from 1989 to 1992 detailed data about the management of each field (crop, sowing and harvest date, yields, fertilization, and normally used crop rotation) were elicited by a questionnaire. Field-specific investigations of the mineral nitrogen content in the root zone (0 to 90 cm) according to the N_{min} method (Scharpf and Wehrmann, 1975) and organic matter content of the plow layer (0 to 30 cm) were carried out in fall 1990, which were used as initial values for simulation. Mineral nitrogen contents in 0 to 90 cm were also measured for all fields of the protection area two to three times per year (spring, summer, and autumn) from 1991 to 1995. The measurements represent mixed samples from 16 points of each field.

Between 1993 and 1996 the main field crops in that area were mapped, and a survey was made about the crop-specific average nitrogen fertilization. Because there was no information on catch crops during this time, their use was estimated for each field based on the typical rotations given in the questionnaire. The extended area was first mapped for field crops in 1996. Typical crop rotations were used to assess the field crops between 1990 and 1995 assuming that they were managed in the same way as in the protection zone.

The soils in the area were mapped and characterized by the State Office for Soil Research of Lower Saxony based on AG Bodenkunde (1982). The soils are mostly sandy with some small areas (<0.2%) of peat soils. The values for the field capacity of the mineral soils vary from 116 to 211 mm in the upper 100 cm. The soil map was digitized and intersected with the digital map of land use using the GIS Arc/Info (ESRI, 1992). Together with daily weather information from a neighboring station of the German National Weather Service, the attributed data of each polygon were used as input for the simulation model.

Between 1987 and 1990 a network of 20 piezometers was installed. At eight of these observation wells with a diameter of 125 mm, water quality was investigated in two to four different depths. During 1989 and 1995 average nitrate concentrations in the upper part of the aquifer (6.5 to 12 m below ground) were observed ranging from 20 mg NO_3 l^{-1} at the downstream border of the forest area in the extended zone, 90 mg NO_3 l^{-1} at the eastern border between the protection zone and the extended zone, and 110 mg NO_3 l^{-1} at one of the wells of the water plant (see Figure 14.3b later).

MODEL DESCRIPTION

The model HERMES used here is a functional deterministic simulation model for advisory purposes. That means that the input requirements are restricted to those data which are usually available from farmers or public sources. The model considers soil water dynamics, net mineralization and denitrification, nitrate transport, and nitrogen uptake by plants. It is described here just briefly. More-detailed information can be found in Kersebaum and Richter (1991) and Kersebaum (1995).

The soil water dynamics is simulated using a modified capacity model which also allows upward movement in the soil. Capacity parameters like field capacity and wilting point were derived according to AG Bodenkunde (1994) from the texture class of each horizon. Potential evapotranspiration is calculated from the daily vapor pressure deficit and plant-specific time-variable coefficients (Haude, 1955; Heger, 1978). Water uptake by plants considers the simulated leaf area index and root distribution over depth as well as the soil water content distribution to calculate actual evaporation and transpiration. Capillary rise to the root zone is estimated by an integrated table considering the texture class and the distance to the groundwater (AG Bodenkunde, 1994). The distance to the groundwater is calculated every time step starting from the lowest layer which falls below 70% of available water, which is given as the boundary condition of the daily capillary rise fluxes of the table.

Net mineralization considers two fractions of potentially mineralizable nitrogen with different decay coefficients that are dependent on soil moisture and temperature. The potentially mineralizable nitrogen is calculated from the organic matter content and the residues of the previous crop. At harvest, a plant-specific portion of the simulated nitrogen uptake is recycled automatically to both fractions if the residues are not removed from the field. Denitrification is described using a Michaelis–Menten kinetic, dependent on soil nitrate content, water-filled pore space, and temperature. The transport of nitrogen is simulated by the conventional convection–dispersion equation. N uptake by plants is described by a generalized crop growth model using plant-specific parameter sets to give reasonable time courses and amounts of nitrogen uptake. Plant growth is reduced by water or nitrogen deficiency. The time step of the model is 1 day. The soil compartments are 10 cm thick.

The input requirements are daily weather data, soil profile description — texture class, organic matter content, groundwater level — and management data — crops, sowing and harvest date, handling of residues, yield (optional), date and amount of nitrogen fertilization or irrigation. For those years (1992 to 1996) when exact management data were not available, typical crop-specific schedules for sowing, harvest, and fertilizer distribution were defined.

The model application is restricted to the agricultural land. That means that nitrogen leaching under forest was estimated roughly. Annual nitrogen deposition in forests was estimated at about 40 kg N ha^{-1} for the area of Lower Saxony (Meesenburg et al., 1994). By regarding values of nitrogen leaching under different forest areas within Europe in dependence of their specific nitrogen deposition (Dise and Wright, 1995), an assumption of 20 kg N ha^{-1} annual loss seems to be reasonable, which means that the average nitrate concentration would be 35 mg NO_3 l^{-1}.

RESULTS

For the simulation period 1990–91 simulations could be carried out using more-detailed information on crop management of each field. Therefore, simulated mineral nitrogen contents in the soils in autumn 1991 were compared with observed values. The comparison of single observations on the fields gives only poor correlations ($r = 0.52$). Nevertheless, there are some uncertainties within the information coming from the farmers, especially for the amount and application dates of manure. Another problem is that observations represent mixed samples of a field, whereas simulations sometimes calculated different values for different parts of the field depending on the soil type. By looking at the average mineral nitrogen contents after different crops (Figure 14.2A), a much better correlation ($r = 0.87$) between measurements and simulations was obtained.

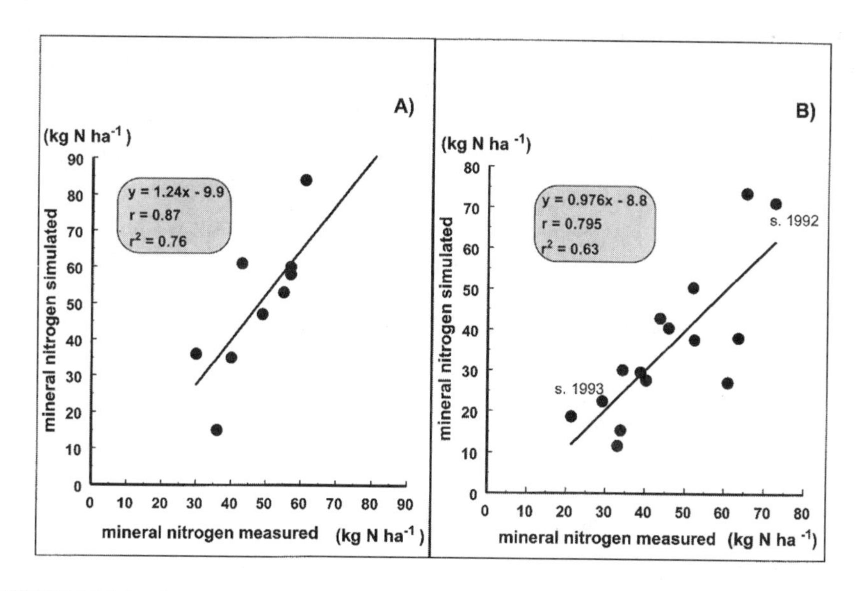

FIGURE 14.2 Comparison of average measured and simulated mineral nitrogen contents of all fields within the protection zone (A) after different field crops in autumn 1991 and (B) at different dates (spring, summer, autumn) from 1991 to 1995 (s. 1992 = summer 1992, s. 1993 = summer 1993).

TABLE 14.1
Percentage of Land Use of the Drinking Water Catchment for the Present Situation and Three Scenarios

		Scenarios		
			Conversion Arable Land to Pasture	
Crop	Present Situation	Optimized Management	Category 1	Category 1 + 2
Winter cereals	19.4	32.2	27.6	22.0
Summer cereals	15.9	10.6	8.8	7.2
Corn	19.4	13.3	11.2	10.2
Sugar beets	6.7	6.6	6.0	4.7
Potatoes	12.6	11.3	9.9	7.5
Grass	1.4	1.4	1.4	1.1
Pasture	3.2	3.2	13.8	26.9
Set aside	1.6	1.6	1.5	0.6
Forest	19.8	19.8	19.8	19.8
Total	100	100	100	100
Fields with catch crops	19.7	38.6	32.3	18.5

A comparison of the average mineral nitrogen contents observed at three dates of each year within the protection zone with simulated values (Figure 14.2B) shows that the general trend of mineral nitrogen content in the root zone of different years and seasons is well reflected by the model simulations.

The percentage of the different crops and the portion of arable land where catch crops were used to prevent leaching during winter is shown in Table 14.1. There are only a few small fields with pasture and only 20% is forest area. The highest percentage of the arable land is used for corn which is used mostly as silage. Growing catch crop is only applicable if the previous crop is harvested before the middle of August; otherwise, the growing season would be too short. It was estimated that catch crops were grown on less than 20% of the arable land.

The average seepage during the period from 1990 to 1996 is estimated as 324 mm/year with a range from 250 mm under forest to 380 mm on the arable land with the lowest water-holding capacity. Within the period, the annual seepage (calculated from October to October of the next year) varied between 138 mm in 1995–96 and 478 mm in 1992–93. The average seepage of each polygon was used to calculate how often on average the soil solution in the root zone has been exchanged during 1 year as an indicator for the vulnerability of leaching. Beside the field capacity of each layer, the specific effective rooting depth for each soil and the actual land use (arable, pasture, forest) has to be considered to calculate the field capacity of the root zone of each site. The exchange frequency for each site is calculated dividing the specific annual seepage by the field capacity of the root zone.

The lowest exchange frequency of less than one time per year is calculated for the forest area due to the higher rooting depth and the lower seepage (Figure 14.3a). The frequency for arable sites range from 1.8 to 4.84 because field capacity and rooting depth in the mostly sandy soils are fairly low. These are usually the areas with the highest nitrogen losses and the highest concentrations. The spatial distribution of the average nitrate concentrations in the seepage during the investigation period is shown in Figure 14.3b. The average nitrate concentrations (from 1989 to 1995) in the upper part of the aquifer are also shown for the observation wells and the wells of the water supply plant. Figure 14.4 shows a probability plot of the simulated nitrogen concentrations of the polygons and the resulting mixed concentration for the corresponding area. While the concentrations of those areas that are used by forest or pasture range from 17 to 35 mg NO_3 l^{-1}, the arable land ranges from 60 to 144 mg NO_3 l^{-1}. This forms a mixed concentration for the whole area of 86 mg NO_3 l^{-1}. Due to the higher portion of forest in the extended zone, the average concentration here is a little less (81 mg NO_3 l^{-1}) than in the protection zone, which has been calculated to be at 91 mg NO_3 l^{-1}. A comparison of observed concentrations to simulated nitrate leaching from the root zone has to consider the travel time to the aquifer. From the seepage amounts, it can be calculated that in 1995 only the nitrate leached before 1992 had reached the groundwater. Regarding only this period, the area weighted mean was 108 mg NO_3 l^{-1}, which corresponds well to the observed concentrations in the aquifer. Within the simulation period, the calculated nitrate concentrations vary from year to year depending on the specific weather conditions and fluctuations in the percentages of different crops. Highest concentrations (135 mg NO_3 l^{-1}) were estimated for the leaching period 1992–93 because a drought summer in 1992 led to bad nitrogen efficiency and high residual mineral nitrogen after harvest (see Figure 14.2B) followed by a very wet year which caused nitrogen leaching even during the growing season 1993. This results in low mineral nitrogen contents in summer 1993 (see Figure 14.2B) so that in the following wet period nitrate concentrations were lowest with 44 mg NO_3 l^{-1}.

SCENARIOS

To assess how far an environmentally sound optimization of agricultural practices can improve the water quality without fundamental changes of the cropping system, the following steps for a scenario calculation were defined:

- Adoption of the fertilizer amounts to the demand of the plants based on average yields.
- Change of the site-specific actual rotation to achieve a nearly complete crop cover over the year by cropping with winter cereals after late-harvested crops and cultivation of winter-tolerant catch crops between summer-harvested crops and spring-sown crops.

The resulting land-use distribution is shown in Table 14.1. While the percentage of winter cereals increased, the percentage of corn and summer cereals decreased. Corn fields, especially, were observed and simulated to have high mineral nitrate contents because high amounts of slurry were usually applied to corn. The portion of fields grown with catch crops increased to nearly 40%.

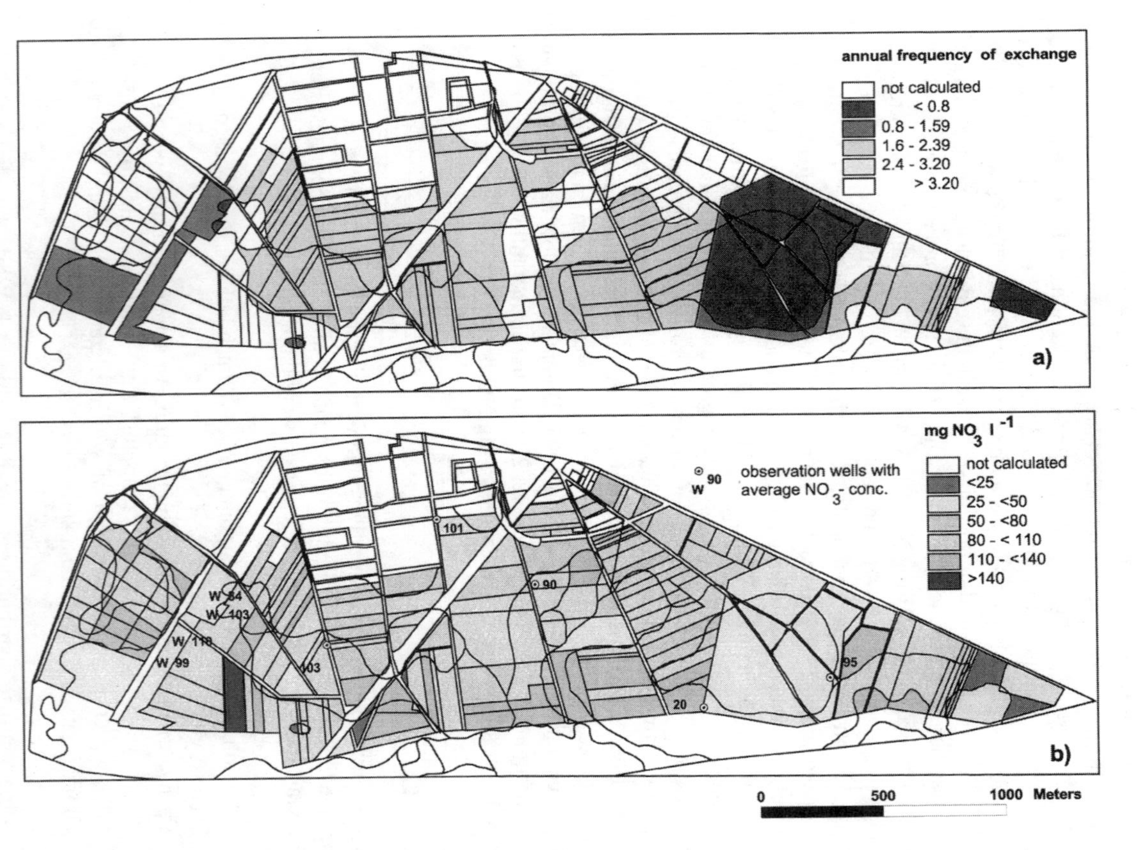

FIGURE 14.3 (a) Simulated exchange frequency of the soil solution in the specific root zone. (b) Spatial distribution of simulated annual nitrate concentrations of seepage water (average 1990 to 1996) with measured nitrate concentrations in the upper aquifer at several observation and water supply wells (average 1989 to 1995).

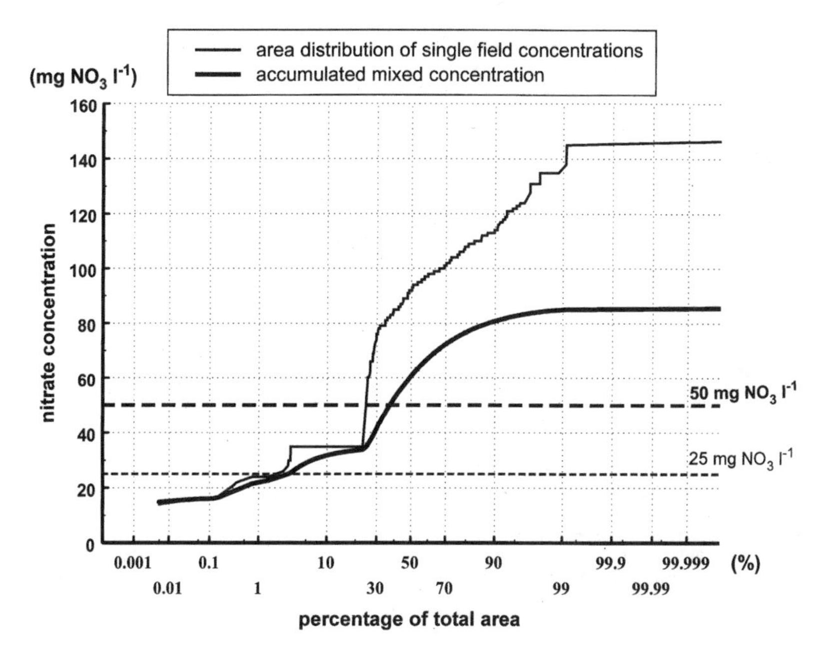

FIGURE 14.4 Area distribution of the simulated nitrate concentrations in the catchment and corresponding mixed concentration for the present land use (averages 1990 to 1996).

Figure 14.5 shows the results of the simulation for the scenario "optimized management." Compared with the present land use where aproximately 30% of the area shows nitrate concentrations in the seepage water of more than 100 mg NO_3 l^{-1}, only 2% of the area is estimated to be still slightly above this value. The resulting mixed concentration would be 63 mg NO_3 l^{-1} for the protection zone, 55 mg NO_3 l^{-1} for the extended zone, and 59 mg NO_3 l^{-1} for the total area. Because this would not be enough to meet the drinking water standard, scenarios were developed to calculate where changes in land allocation (arable vs. pasture) need to take place. The preferences to change arable land into pasture are derived from the properties of the soil map units regarding:

- Their exchange frequency of the soil solution;
- Their specific amount of nitrate loss at a given land use compared with the other soil map units;
- Their spatial situation regarding the wells of the waterwork and the protection zone, because only in this zone can regulations be applied.

In the first scenario, an area of 27 ha neighboring the wells, which had the highest priority according to the above-mentioned criteria, was converted into pasture. This results in a mixed nitrate concentration of 54 mg NO_3 l^{-1} for the total catchment. Therefore, a second scenario was calculated in which an additional 34 ha of the protection zone was converted into pasture. Figure 14.6 shows that now 90% of the area would have nitrate concentrations below 65 mg NO_3 l^{-1}, which would

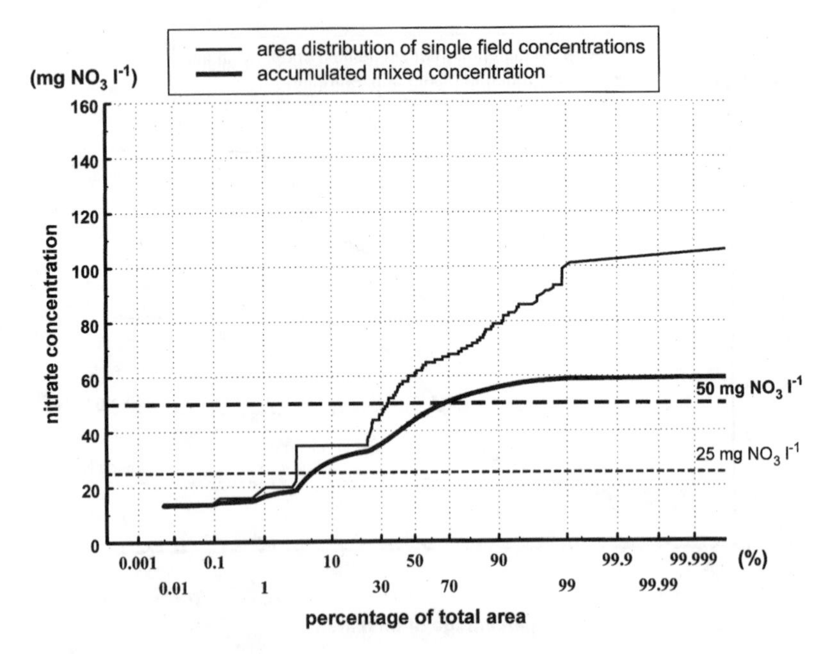

FIGURE 14.5 Area distribution of the simulated nitrate concentrations in the catchment and corresponding mixed concentration for the scenario "optimized management" (averages 1990 to 1996).

be sufficient to meet the drinking water standards because the resulting mixed concentration of the total area would now be 46 mg NO_3 l^{-1} (39 mg NO_3 l^{-1} in the protection zone).

DISCUSSION

To assess the effects of present and probable land use and management practice in a drinking water catchment a relatively simple management model was applied. Compared with the so-called state-of-the-art models, the model input requirements are restricted to those which are usually available from farmers or soil maps. Model comparisons (de Willigen, 1991; Diekkrüger et al., 1995) have shown that results of such management models show a similar quality of the results to complex state-of-the-art models. However, they have their limitations, which should be taken into consideration if applied.

The model was applied without calibration for the area under practical conditions. This means that the data situation especially for the crop management is not comparable with investigation sites. Under these conditions, the results of the model are satisfactory, although the correlation to single field measurements of mineral nitrogen content in the root zone was poor. This might be an effect of some generalizations, e.g., the use of "representative" soil profiles and parameters or standard nutrient contents of manure. Of course, there are some processes like ammonia volatilization which are described very roughly so that relevant N losses to the air

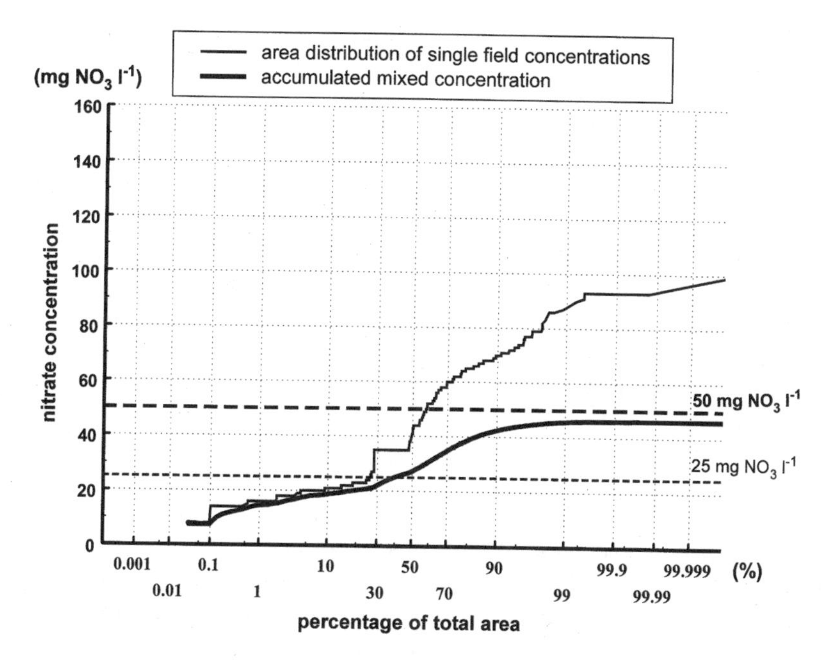

FIGURE 14.6 Area distribution of the simulated nitrate concentrations in the catchment and corresponding mixed concentration for the scenario "conversion category 1 + 2" (averages 1990 to 1996).

might cause errors when organic or mineral fertilizers are applied. Nevertheless, the model seems to be able to reflect the trend of residual nitrate after harvest between different crops and the effect of the specific weather conditions in different years on mineral nitrogen content of the root zone.

The nitrate concentration of the investigated area simulated over a period of 6 years agrees with the observed situation in the aquifer. The high temporal variability of estimated nitrate concentrations for different years indicates that risk assesments have to be based on more than 3 or 4 years monitoring. For a better interpretation, a link to a groundwater model would be desirable.

The scenario simulations show that there is a considerable potential to reduce nitrate pollution of the groundwater within the present land use by the adoption of the fertilization to the demand of the crop and by optimizing the crop rotation. Nevertheless, significant losses of nutrients can be observed even in regions with a relatively low surplus of nutrients (Van der Molen et al., 1998). This is also the case in the area studied where the poor soils with a low water-holding capacity have a high vulnerability of nitrogen leaching under arable land use. Therefore, it is necessary to convert some of the arable land into pasture to meet the threshold for drinking water. A moderate application of nitrogen to the grassland of 120 kg N ha^{-1} was assumed for the pasture scenario. Simulations indicate a significant increase of nitrate concentrations for pasture if nitrogen amounts of more than 280 kg ha^{-1} were applied. These results are in good agreement with Kolenbrander (1981). For a farming system in the Netherlands, Hack-ten Broeke and de Groot (1998) estimated nitrogen concentrations under grassland with 320 kg N ha^{-1} fertilizer input exceeding those of arable land.

To achieve a reduction of the nitrate concentration for the whole area below the threshold of 50 mg NO_3 l^{-1} the conversion of 23.7% of the area (31.4% of the arable land) from arable land into pasture would be necessary.

In principle, the model can be and has been applied to larger areas (Kersebaum et al., 1995; Kersebaum and Wenkel, 1998). On a larger scale, deficiencies of input data especially for management practices will become more evident. Another problem is that the variabilty of crops in other regions might be much larger and cannot be covered by the model. A main knowledge gap is the simulation of a forest ecosystem. There is also still a need to improve the very rough approach to describe nitrogen fixation by legumes which would be necessary for a better estimate of organic farming systems.

CONCLUSIONS

Results indicate that adoption of fertilizers to demand of the crop as well as the careful design of the rotation can reduce nitrate leaching to the groundwater significantly. Depending on the specific site it might be that even low-input arable systems lead to nitrate concentrations which exceed the drinking water threshold. In those specific areas, agricultural activities have to be balanced by the establishment of low-output areas. Low-input pasture is one possibility to reduce nitrate concentrations below the threshold.

Simulation models can provide a useful tool to select portions of an area for the most effective conversion regarding groundwater quality if they do not require a lot of expensive research for the data input. They can also provide a tool to visualize processes going on in the fields to enhance the understanding and cooperation of farmers.

Additionally, limitations to spatially distributed simulations may be related to deficiencies in the process descriptions or lack of data. Although the last point forces modelers to make generalizations and simplifications to reduce the input requirements of their model, the other point seems to argue in favor of more complex and mechanistic model approaches. Several comparisons of integrated nitrogen models applied for the same data set indicate that results of complex mechanistic models are not better and are sometimes even worse than simpler functional models (de Willigen, 1991; Diekkrüger et al., 1995).

ACKNOWLEDGMENTS

The author wishes to thank the State Office for Ecology of Lower Saxony, Germany, which supported this study and provided most of the database.

REFERENCES

Bach, M. and H.-G. Frede. 1998. Agricultural nitrogen, phosphorus and potassium balances in Germany — methodology and trends 1970 to 1995, *Z. Pflanzenernaehr. Bodenk* 161:385–393..

Bodenkunde, A.G. 1982. *Bodenkundliche Kartieranleitung,* Schweizerbart, Stuttgart.

Bruckhaus, A. and R. Berg. 1990. Anforderungen des Gewässerschutzes an eine ordnungsgemäße Landwirtschaft, Umweltbundesamt Texte 19/90.

CEC (Commission of the European Community). 1991. Nitrate in Soils, Soil and Groundwater Research Report II, Brussels.

Diekkrüger, B., D. Söndgerath, K. C. Kersebaum, and C. W. McVoy. 1995. Validity of agroecosystem models. A comparison of results of different models applied to the same data set, *Ecol. Modelling* 81:3–29.

Dise, N. B. and R. F. Wright. 1995. Nitrogen leaching from European forests in relation to nitrogen deposition. *Forest Ecol. Manage.* 71: 153–161.

ECETOC. 1988. Nitrate in Drinking Water, European Chemical Industry Ecology and Toxicology Centre, Technical Report No. 27, Brussels.

ESRI (Environmental System Research Institute). 1992. Understanding GIS. ARC/INFO *Method Rev. 6 Env. Syst. Res Inst.*, Redland, U.S.A.

Hack-ten Broeke, M. J. D. and W. J. M. de Groot. 1998. Evaluation of nitrate leaching risk at site and farm level, *Nutr. Cycling Agroecosyst.* 50: 273–278.

Hamm, A., Ed. 1991. *Studie über Wirkungen und Qualitätsziele von Nährstoffen in Fließgewässern,* Academia Verlag, Sankt Augustin.

Haude, W. 1955. Zur Bestimmung der potentiellen Evapotranspiration auf möglichst einfache Weise, *Mitt. Dtsch Wetterdienst* 11.

Heathwaite, L. 1995. Sourdes of eutrophication: hydrological pathway of catchment nutrient export, in G. Petts, Ed., Man's Influence on Freshwater Ecosystems and Water Use, Publ. 230, IAHS, Wallingford, 161–175.

Heger, K. 1978. Bestimmung der potentiellen Evapotranspiration über unterschiedlichen landwirtschaftlichen Kulturen. *Mitt. Dtsch Bodenk. Ges.* 26:21–40.

Humenik, F. J., M. D. Smolen, and S. A. Dressing. 1987. Pollution from non-point sources: where we are and where we should go, *Environ. Sci. Technol.* 21:737–742.

Isermann, K. 1990. Share of agriculture in nitrogen and phosphorus emissions into the surface waters of Western Europe against the background of their eutrophication, *Fert. Res.* 26: 253–269.

Isermann, K. and R. Isermann. 1995. Die Anteile des N-Austrages mit dem Sickerwasser aus der landwirtschaftlich genutzten Fläche über die (un)gesättigte Zone in die Oberflächengewässer Westeuropas/EU und Deutschlands an der jeweiligen N-Bilanz der Landwirtschaft (1987/92), in *Proceedings of the 5. Gumpensteiner Lysimetertagung "Stofftransport und Stoffbilanz in der ungesättigten Zone,"* 85–91.

Kersebaum, K. C. 1995. Application of a simple management model to simulate water and nitrogen dynamics, *Ecol. Modelling* 85:145–156.

Kersebaum, K. C. and J. Richter. 1991. Modelling nitrogen dynamics in a plant soil system with a simple model for advisory purposes, *Fert. Res.* 27:273–281.

Kersebaum, K. C. and K.-O. Wenkel. 1998. Modelling water and nitrogen dynamics at three different spatial scales — influence of different data aggregation levels on simulation results, *Nutr. Cycling Agroecosyst.* 50: 315–321.

Kersebaum, K. C., W. Mirschel, and K.-O. Wenkel. 1995. Estimation of regional nitrogen leaching in the Northern-Eastern Germany area for different land use scenarios, in J. F. T. Schoute, P. Finke, F. R. Veeneklaas, and H. P. Wolfert, Eds., *Scenario Studies for the Rural Environment,* Kluwer Academic, Dordrecht, 227–232.

Kolenbrander, G. J. 1981. Leaching of nitrogen in agriculture, in J. C. Brogan, Ed., *Nitrogen Losses and Surface Run-Off,* Nijhoff-Junk Publ., Amsterdam, Holland, 199–216.

Kragt, J. F., W. de Vries, and A. Breeuwsma. 1990. Modelling nitrate leaching on a regional scale, in R. Merckx, H. Vereecken, and K. Vlassak, Eds., *Fertilization and the Environment,* Leuven University Press, Leuven, Belgium, 340–347.

Lidgate, H. J. 1987. Nutrient in the North Sea — a fertilizer industry view, paper presented at International Conference on Environmental Protection of the North Sea, London, Session 2: Nutrients.

McDonald, M .G. and A. W. Harbaugh. 1988. A modular three-dimensional finite-difference ground-water flow model, U.S. Geological Survey Techniques of Water-Resources Investigations Book 6, Chapter A1, 586.

Meesenburg, H., K. J. Meiwes, and R. Schultz-Sternberg. 1994. Entwicklung der atmogenen Stoffeinträge in niedersächsische Waldbestände, *Forst Holz* 49: 236–238.

Owen, T. R. and S. Jürgens-Gschwind. 1986. Nitrates in drinking water: a review, Fert. Res. 10: 3–25.

Scharpf, H. C. and J. Wehrmann. 1975. Die Bedeutung des Mineralstickstoffvorrates des Bodens zu Vegetationsbeginn für die Bemessun der N-Düngung zu Winterweizen, *Landwirtsch. Forsch.* 32/I: 100–114.

Statistisches Bundesamt. 1996. *Statistical Yearbook 1996 for the Federal Republic of Germany,* Metzler-Poeschel, Stuttgart.

Styczen, M. E. 1995. Regional simulations as a basis for the assessment of national nutrient emissions at various input scenarios in Denmark, in J. F. T. Schoute, P. Finke, F. R. Veeneklaas, and H.P. Wolfert, Eds., *Scenario Studies for the Rural Environment,* Academic Publishers, Dordrecht, 157–167.

Umweltbundesamt. 1997. *Daten zur Umwelt — Der Zustand der Umwelt in Deutschland,* Erich Schmidt Verlag, Berlin.

Van der Molen, D. T., A. Breeuwsma, and P. C. M. Boers. 1998. Agricultural nutrient losses to surface water in the Netherlands: Impact, strategies and perspectives, *J. Environ. Qual.* 27: 4–11.

Vereecken, H. M. Vanclooster, and M. Swerts. 1990. A model for the estimation of nitrogen leaching with regional applicability, in R. Merckx, H. Vereecken, and K. Vlassak, Eds., *Fertilization and the Environment,* Leuven University Press, Leuven, Belgium, 250–263.

Walther, W., E. Schmidt, W. Hofmann, and W. Meyer. 1990. Studie Schutzgebiet Scheessel, langfristige Vorgehensweise zur Verminderung der Grundwasserbelastung, in W. Walther, Ed., *Grundwasserbeschaffenheit in Niedersachsen — Diffuser Nitrateintrag, Fallstudien,* Vol. 48, Veröffentlichungen des Instituts für Siedlungswasserwirtschaft, Technische Universität Braunschweig, 365–381.

Willigen, P. de. 1991. Nitrogen turnover in the soil-crop system; comparison of fourteen simulation models, *Fert. Res.* 27: 141–149.

Young, R. A., C. Onstad, D. Bosch, and W. P. Anderson. 1987. AGNPS, Agricultural Non-Point-Source Model. A Watershed Analysis Tool, USDA-ARS Conservation Research Report 35.

15 Quantification of Leached Pollutants into the Groundwater Caused by Agricultural Land Use — Model-Based Scenario Studies as a Method for Quantitative Risk Assessment of Groundwater Pollution

Angelika Wurbs, Kurt Christian Kersebaum, and Christoph Merz

CONTENTS

INTRODUCTION

Groundwater contamination with nutrients, e.g., nitrate, is often related to agricultural land use. Especially on sandy soils, nutrient leaching is a serious problem because of low water and nutrient storage capacity and low denitrification rates. Study of the temporal variations in the N dynamics and the exact hydrochemical

0-8493-0702-3/00/$0.00+$.50

characterization of the groundwater were the basis of the following model. A compound model was developed by linking the deterministic models HERMES, SIFRONT, and FEFLOW, which enables consideration of a complete profile from the root zone to the aquifer. Two farming systems ("integrated" and "organic") were compared with regard to the water and nitrate fluxes. The results of the spatially distributed simulations indicate that the soil and climate conditions have a strong influence on the total amounts of nitrogen leaching, and also affect the distinction between the two management systems.

Agricultural land use is the main non-point-source pollution of groundwater in Germany. About 40% of surface water contamination with nitrogen is caused by exfiltrating groundwater (Flaig and Mohr, 1996). In the glacial landscapes of northern Germany, the impact of groundwater on surface water is highly pronounced. Therefore, the development of sustainable land-use systems to improve or preserve groundwater quality is of great relevance for watersheds.

On the typical sandy soils of North Germany, nitrate leaching is often a serious problem due to low water-holding capacity and low denitrification rates. Additionally, farmers in such areas usually compensate the low productivity of their soils by animal production which often leads to a surplus of nitrogen with the application of organic fertilizers.

An integrated and an organic farming system were compared regarding their effects on reducing nitrate contamination in a typical small water catchment. The main objectives of integrated farming are cost reduction and improvement of quality of both products and production ways, through substituting expensive and potentially polluting inputs, especially fertilizers and pesticides, with both agricultural and ecological knowledge (Vereijken, 1989). Organic farming is a holistic modern farming concept which dramatically reduces external inputs by refraining from the use of pesticides, nitrogen fertilizers, easily soluble mineral fertilizers, and other chemicals. It takes local soil fertility, crop rotations with legumes, area-obtained animal production as the key to the system and aims to optimize quality in all aspects of agriculture and the environment (EG-VO 2092/91, 1991).

To assess land use and management practices in a watershed context it is necessary to determine the nitrate leaching from the root zone as well as the spatial and temporal dynamic in the unsaturated and saturated zone (Cluis, 1995; Srinivasan et al., 1995; Zacharias et al., 1997).

METHODS

The two management systems were established in 1989 on a field of 24 ha (Figure 15.1), where high amounts of slurry were applied during the last decade. Intensive measurements were carried out from 1988 and 1996 in the root zone. Soil samples were taken at three times a year (beginning of vegetation, harvest of the main crop, and end of vegetation period) for 0 to 30, 30 to 60, and 60 to 90 cm depths. Mineral nitrogen contents were analyzed according to Scharpf and Wehrmann (1975) to validate the simulation model HERMES.

The migration behavior of the soil solution was determined using a tracer experiment with chloride. The tracer was applied on the soil surface of an area of

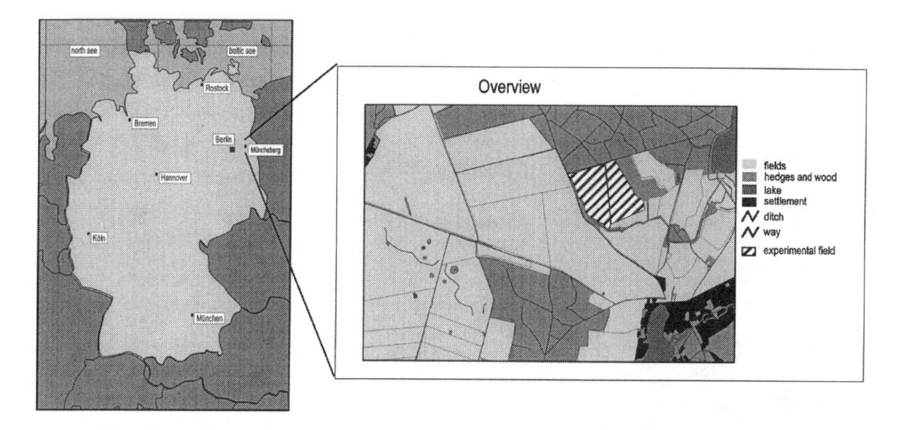

FIGURE 15.1 Location map and overview of the experimental field.

4 m^2 in November 1993 when field capacity was reached. Soil samples were taken in May 1994 in 10-cm increments down to 4.2 m. The water content was determined gravimetrically and the amount of chloride was analyzed with an ion chromatograph (Dionex).

To determine the concentration of nitrate and its variability, samples of the upper 4.2 m were taken from an area of 100 m^2 by a 10×10 m grid because of the obvious soil heterogeneity of the investigation site. In the unsaturated zone, eight drillings were bored to the groundwater level. The sampling of water and sediment was carried out by static cone penetration testing (CPT). In the samples the pH-value, redox potential, and nitrate were measured.

Atmospheric bulk deposition of ammonium, nitrate, sulfate, and other elements was determined with bottle collectors which were especially suitable for event-induced or, with the addition of chemical preservatives, for long-term samplings (Dämmgen et al., 1997).

Between 1987 and 1991 a network of 50 groundwater observation wells was installed at the experimental site on a total area of 40 ha to investigate the geochemical composition of the groundwater at different levels. Since 1992, each observation well has been sampled in spring and autumn. In addition to the physicochemical parameters (pH, pE, electrical conductivity), nitrate, phosphate, and sulfate and cations were analyzed.

The nitrogen dynamics and transport in the root zone, in the unsaturated zone, and in the aquifer were simulated by three models which were coupled by their output and input files (Figure 15.2). The HERMES model (Kersebaum and Richter, 1991; Kersebaum, 1995) simulates nitrogen dynamics in the soil–plant system of the root zone. The model includes nitrogen mineralization, denitrification, transport of nitrate by soil water fluxes, and nitrogen uptake by plants. The water balance is described by a simple capacity approach, and the convection dispersion equation is used for nitrate transport. Net mineralization of two different nitrogen pools is described by first-order kinetics dependent on soil moisture and temperature. Denitrification is calculated by a Michaelis–Menten kinetic dependent on soil moisture,

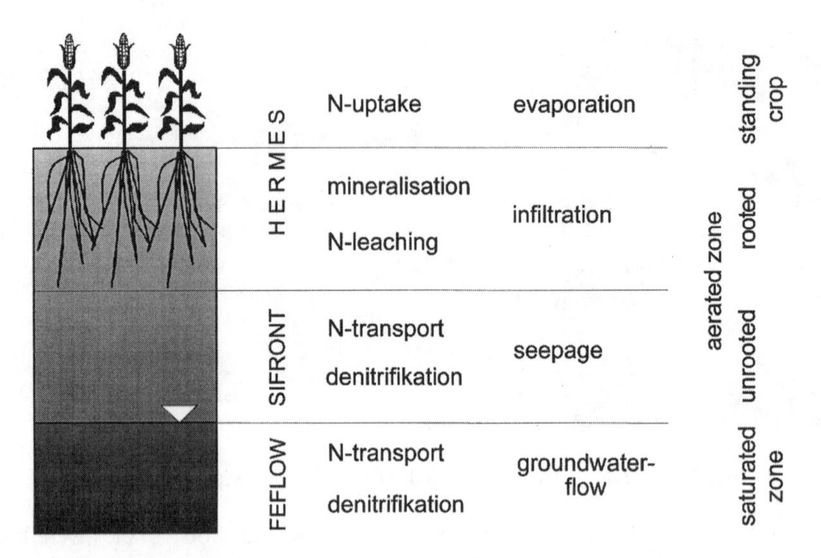

FIGURE 15.2 Parameters of the interlinked deterministic models.

temperature, and nitrate content. Nitrogen uptake by plant is linked to a crop growth-model based on the SUCROS model (van Keulen et al., 1982). The input data required by the model are soil information (texture class, organic matter content), groundwater table depth, daily weather data (precipitation, temperature, vapor pressure deficit, irradiation), and management data (crop, sowing and harvest date, fertilization, yield). The capacity parameters for the water balance model are derived from the soil texture class modified by hydromorphic attributes and soil organic matter content according to Bodenkunde (1994). The time step of the model is 1 day and the soil profile is divided into layers 10 cm deep. The model was already applied for spatially distributed modeling at different scales (Kersebaum and Wenkel, 1998). Because the soil in the study area is very heterogeneous, the simulations were carried out in combination with the geographic information system (GIS) ARC INFO (ESRI, 1992) using a digitized soil map of the field.

The daily output of seepage and nitrogen leaching was used as inputs by the model SIFRONT (Michel, 1991), which simulates nitrate transport and denitrification from the root zone to the groundwater surface. The model describes water dynamics in the unsaturated zone using the Richards equation. Nitrate transport is simulated by the convection–dispersion equation and denitrification uses the same approach as HERMES considering additionally the content of dissolved organic carbon (DOC) to calculate the maximum denitrification potential.

The simulations were carried out for typified subsoil profiles according to the deep drillings. (Voigt and Michel, 1997). Nitrate losses were calculated as well as transport time to estimate the spatiotemporal nitrate output at the interface between the unsaturated and saturated zone. The results for nitrate output is used as input for the groundwater model FEFLOW-3D. FEFLOW-3D (Wasy, 1993), a finite-element subsurface flow system (FEM), represents an interactive, graphics-based modeling system for aquifer-confined or unconfined groundwater flow and contaminat transport (Diersch and Gründler, 1993). Nitrate reduction is described by a first-order kinetic. The simulation model FEFLOW is linked with the GIS ARC INFO (ESRI, 1992) and is able to use different exchange formats. The model uses the initially measured nitrate distribution and the isohypses of the aquifer, the observed concentrations of the groundwater influx at the northeastern border of the area, and the spatially distributed output of SIFRONT simulations (groundwater recharge and nitrate concentration). For the regarded area a network of 19,800 finite elements was constructed in nine calculation layers.

To generalize the results, the model HERMES was used to simulate scenarios for the two management systems based on different soil types (Luvic Arenosol, Hapic Luvisol, and Cambisol) and weather conditions (480 to 550 and 700 to 900 mm annual precipitation, respectively). For each farming system two different crop rotations, typical for the study region, were defined. The precipitation regimes were represented by 8-year weather time series of two different locations in Germany. Each rotation was running eight times with shifted starting dates.

RESULTS

Measurement of bulk deposition showed no relevant atmospheric input of nitrogen compounds (<10 kg N ha^{-1} per annum) onto the study area. Results are in good agreement with corresponding investigations for East German rural environments.

Mineral nitrogen content in the soil (about 130 kg N ha^{-1}) was very high at the beginning of the experiment due to the high amount of slurry applied in the past. Following the changes of the management system, the amount of mineral nitrogen decreased significantly. The comparison of measurements with mineral nitrogen content simulated with HERMES shows coefficients of determination of 0.42 on the organic and 0.56 on the integrated managed plot for the period 1989 to 1996. The correlation, knows that the variability of measurements on the glacial sites is quite high. Results of the grid investigations show coefficients of variation of 0.76 to 1.22 for measured mineral nitrogen content. Schulz (1997) found similar coefficients of variation of 0.33 to 1.48 in a comparable area. Soils formed from glacial deposits are complex, and their thickness and the physical/chemical character vary considerably over short distances (Simpkins et al., 1995). Therefore, the simulation results fit sufficiently with observed data.

During the winter 1993–94 relatively high amounts of precipitation (447 mm) led to a tracer migration of 160 cm on the organic farming plot and 180 cm on the integrated farming plot. This is well reflected by the model which simulates a transport of nitrate to a depth of 180 to 200 cm for both sites.

Results of the spatially distributed simulation of the seepage nitrate concentration during the period 1989 to 1995 are shown for each year in Figure 15.3. Over this period, the area-weighted average amount of nitrate leached from the organic managed part of the field (244 kg N ha^{-1}) was significantly lower as on the integrated part (319 kg N ha^{-1}). Because of a longer soil cover with crops and a higher average water-holding capacity of the soil types on the organic field, the amount of perco-

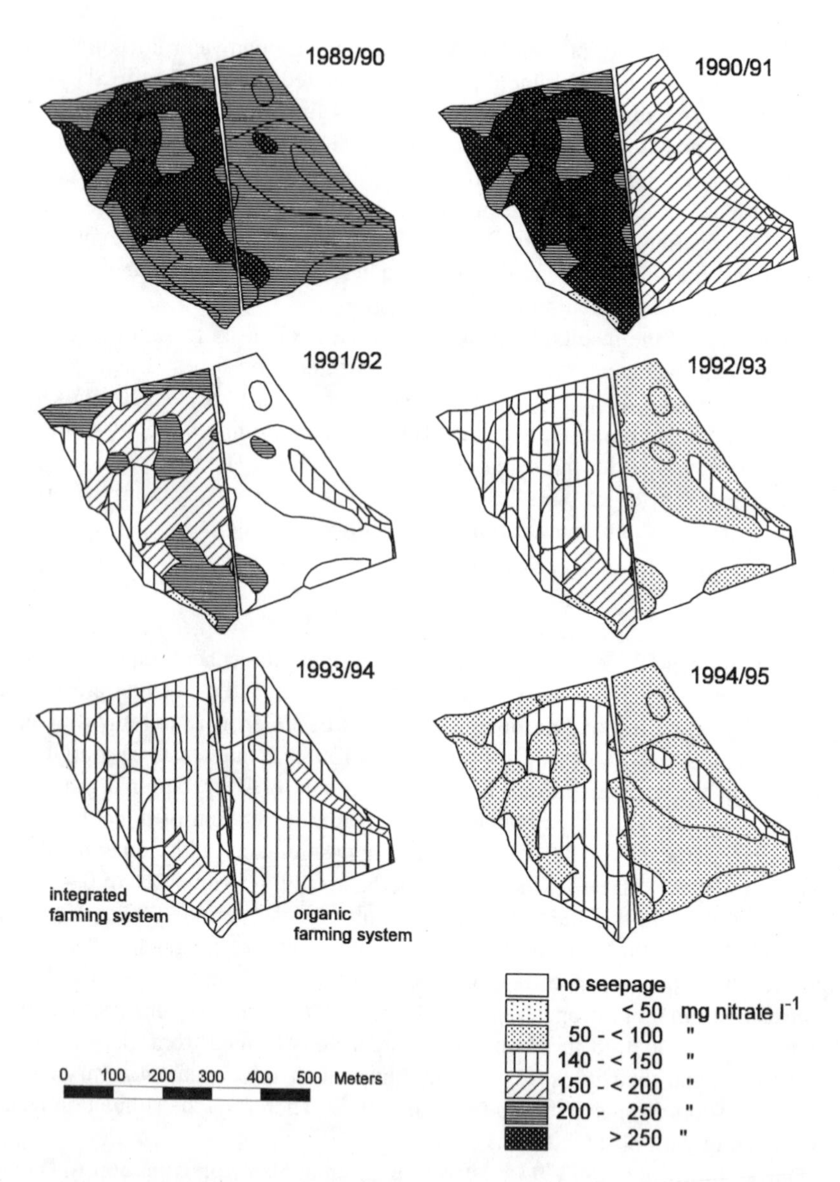

FIGURE 15.3 Simulated nitrate concentration of the seepage water during the period 1989 to 1996.

lation water on this part was also reduced (644 mm vs. 911 mm). Therefore, the differences of the nitrate concentrations between both systems are less distinct (150 vs. 170 mg NO_3 l^{-1}) compared with the loads. Simulations for equal soil profiles showed no significant deviation of the nitrate concentrations between both management systems. The overall averaged nitrate concentration in the seepage water at the bottom of the root zone was calculated to decrease between 1989 and 1995 from 244 to 92 mg NO_3 l^{-1}. By using the simulated seepage and a typical field capacity

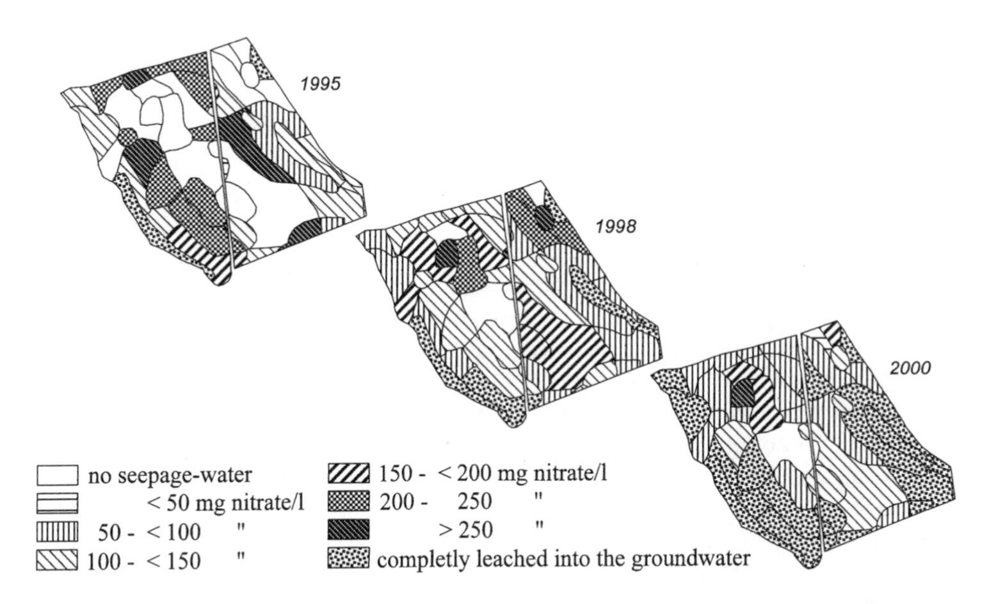

FIGURE 15.4 Prognoses of the nitrate concentration in seepage water reaching the groundwater after 6, 9, and 11 years.

of 140 to 160 mm m^{-1}, an average migration distance of about 0.85 to 0.95 m per annum can be calculated (Gäth and Wohlrab, 1992). This indicates that the migration distance for a period of 6 years would be at 5.1 to 5.7 m. Therefore, the nitrate concentrations below that depth result from the management before 1989. Sampling the interface of groundwater and the unsaturated zone at different depth confirms that prediction. Nitrate concentrations of <100 mg NO_3 l^{-1} were measured at a depth of 4 to 5 m, while below this zone nitrate concentrations increased rapidly to more than 200 mg NO_3 l^{-1}.

The denitrification potential in the unsaturated zone were calculated for typical soil profiles. For a 5-m soil profile percolated with an annual water flux of 100 mm, the model simulates that only 5% of the infiltrated nitrate was degraded by denitrification in a sandy soil, whereas in a loamy sand soil profile nearly 30% of the nitrate was eliminated (Voigt and Michel, 1997). The simulation of nitrate transport and degradation was carried out considering the spatial distribution of soil and subsoil units as well as the different groundwater distances (0.5 to 9.0 m). Figure 15.4 shows the predicted breakthrough concentration at the groundwater level of nitrate leached out of the root zone during 1989 to 1995 at three times (after 6, 9, and 11 years). In major parts of the investigation area none of the nitrate leached in that period reached the groundwater during the first 6 years of simulation because of the high depth of the groundwater table.

To simulate the spatial distribution of nitrate in the groundwater in the period before 1989, different input scenarios were calculated to represent the measured nitrate concentrations. Scenario simulations were used because there was less information about the nitrate input conditions prevailing in the period of 1970 to 1989.

The best results were achieved by a local input of nitrate situated at the norhern part of the integrated site assuming 150 mm per annum seepage and a concentration of 500 mg l^{-1}. These conditions were chosen as a result of information about high amounts of slurry applied at this place. The pollute transport calculation was carried out under the predominating aerobe hydrochemical conditions (redox potential >500 mV). Therefore, appreciable denitrification potential in the saturated zone was ignored during the simulation. Based on the scenario calculations, the period of 1989 to 1996 was simulated with nitrate input calculated by SIFRONT. The time-variable nitrate and seepage amounts from the unsaturated zone were linked to FEFLOW 3D by spatial-distributed conditions of the second order. Figure 15.5 compares the results of the measured and simulated nitrate concentrations in the aquifer in spring 1996. Compared with the calculated results of 1989, there is only a small difference with the simulated distribution of nitrate. The calculation results of SIFRONT indicate that distinct changes of nitrate amounts in seepage will reach groundwater 11 years after changing the agricultural system management. Therefore, appreciable changes of nitrate concentrations can be expected in the year 2000.

The results of the scenarios show the dominating effect of soil type and precipitation on the total amount of nitrate leaching over the simulated period of 8 years. Compared with this the differences between the organic farming system and the integrated system are relatively low under the site conditions of North Germany. Nevertheless, the average annual amount of nitrogen leaching in organic farming systems was estimated to be significantly lower than in integrated farming systems over the period of the crop rotation (Figure 15.6). The nitrate concentration was more affected by the amount of precipitation than by the soil type in both systems. Depending onthe weather conditions and the crop combinations (preliminary/succeeding crops), the nitrogen leaching varies among years. The standard deviation of leached nitrate of the simulated crop combinations was lower in organic farming systems than in integrated systems (Figure 15.7). In dry years the differences between both systems were lower than in wet years. The renunciation of mineral nitrogen and the use of cover crops in the organic farming system are reasons for the reduced leaching. It has to be emphasized that on sandy soils neither the integrated nor the organic system achieved seepage concentrations below the European drinking water standard of 50 mg NO_3 l^{-1} (11.3 mg NO_3–N l^{-1}).

CONCLUSIONS

The application of models allows simulation of the effects of alternative crop rotations and mangement systems on water quality. Nevertheless, the validation of the model applied is an important prerequisite for the reliability of the predictions. Because of the high spatial and temporal variability of the studied variables (e.g., mineral nitrogen content in sandy soils), good correlations between measurements and simulations are difficult to achieve on a field or regional scale. Although a detailed soil map was used for spatially distributed simulations, the spatial variability within soil map units limits the coefficients of determination.

There still are some uncertainties in the model HERMES concerning the processes involved. Mainly, a more accurate approach to simulate nitrogen fixation in

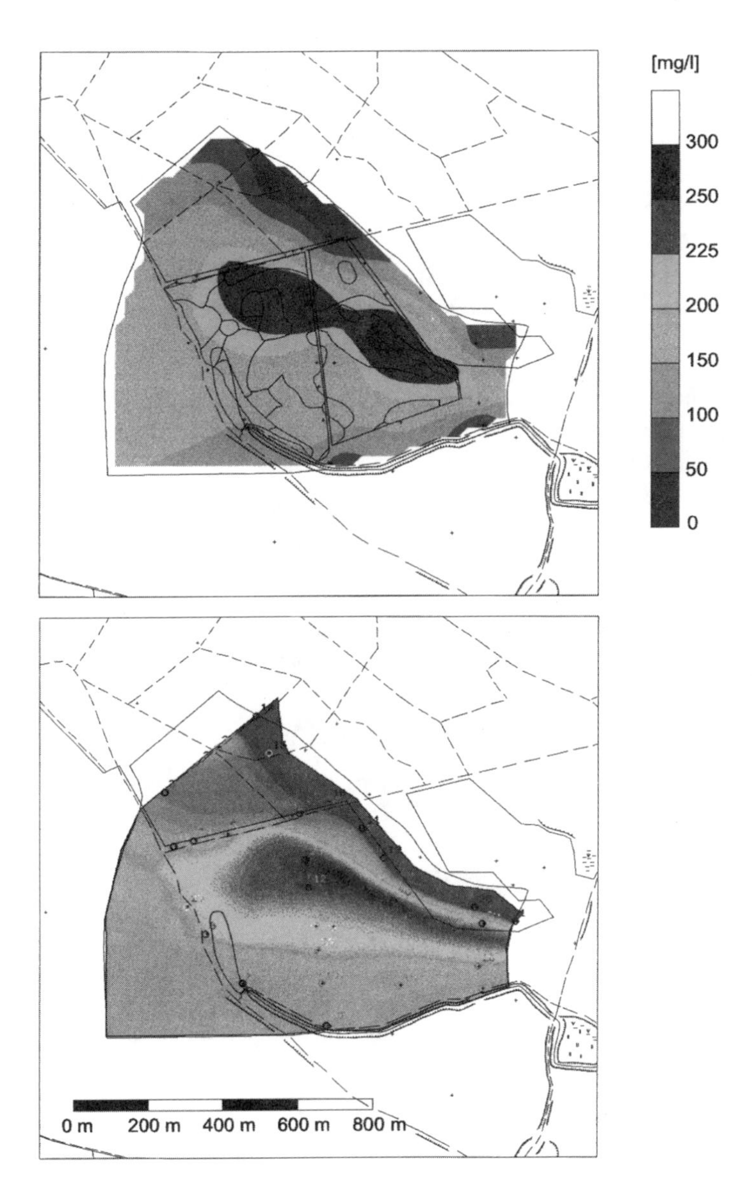

FIGURE 15.5 Distribution of nitrate concentration measured in the groundwater in spring 1996 (at the top) and calculated with the groundwater model FEFLOW considering spatial and temporal dependence of water and substance fluxes from the unsaturated zone (at the bottom).

legumes is required. Nevertheless, the results of the simulation as well as the measurements show a similar trend between the two management systems.

It can be concluded from the simulation results that measured differences between plots in the root zone need to be carefully interpreted because management effects are often masked by changed in soil properties. The scenario simulations indicate

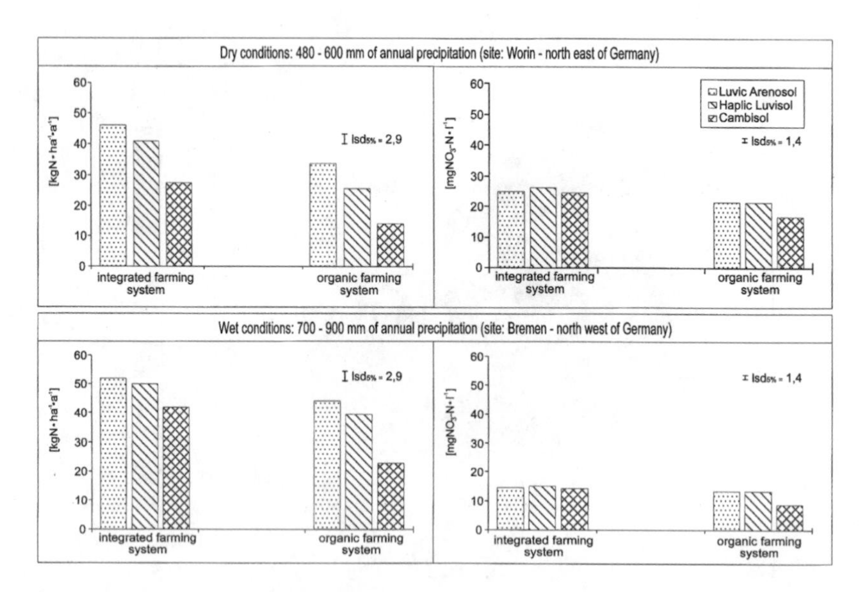

FIGURE 15.6 Average annual amounts of the simulated N leaching from the root zone and N concentration in 2 m depth in integrated and organic farming systems for different soil types and different conditions of annual precipitation.

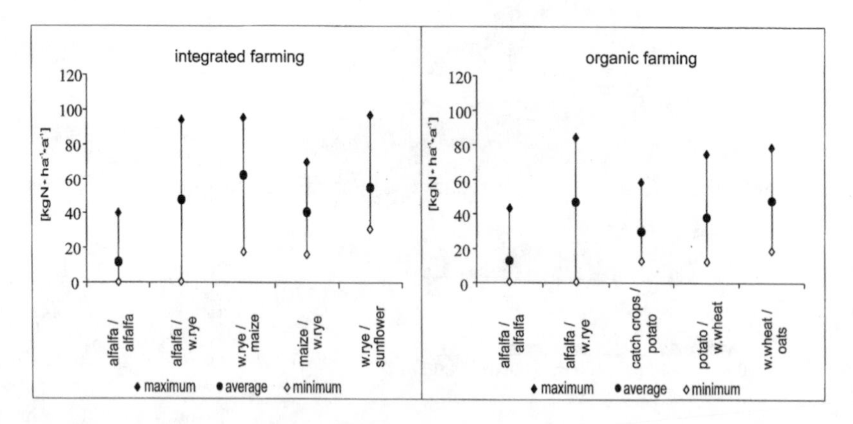

FIGURE 15.7 Simulated nitrogen leaching from some combinations of preliminary/succeeding crops in integrated and organic farming systems.

that soil and climate conditions not only have a strong influence on the total amounts of nitrogen leached, and the distinction between the two management systems. On sandy soils the amount of N leached from the root zone may be reduced by the establishment of organic farming systems under a variety of environmental conditions. In contrast, the nitrate concentration is more dependent on the amount of seepage water than on the farming systems. Only when the annual precipitation exceeds 700 mm is the nitrate concentration of the organic farming system significantly lower than that of the integrated system.

For the input of nitrate into the aquifer the spatial variation of nitrate leaching from the root zone gets an additional variance due to different distances between surface and groundwater level. The resulting differences of travel time will also affect the amount of nitrate degradation in the vadose zone.

Calculations with the groundwater model FEFLOW show satisfactory correlation between the measured and calculated nitrate concentration distribution in the aquifer. The simulations with SIFRONT and with FEFLOW 3D indicate that under the special conditions of the investigation site significant remedial effects for the contaminated aquifer can be expected at the earliest 9 years after the management has changed. Therefore, a distinct improvement of nitrate concentrations in the groundwater can be expected during the next 5 years, which should have consequences for the temporal scale of investigation projects.

In a watershed context the results show that nitrogen leaching can be reduced significantly by using fertilizer at a rate that meets the demand of the plants. Under the typical conditions of climate and soils of northeast Germany the organic farming system shows few advantages compared with the integrated system. The decision for the best management practice has to include site-specific aspects, e.g., climate and soil properties. Concerning the protection of water resources in a watershed, only the regional change to the adopted management systems led to a significant reduction of nitrogen contamination. Therefore, suitable economic and political boundary conditions have to be established based on scenario calculations.

ACKNOWLEDGMENTS

The authors wish to thank the Federal Environmental Agency, the German Federal Ministry of Nutrition, Agriculture and Forestry, and the Ministry of Nutrition, Agriculture and Forestry of Brandenburg state for their financial support of this project.

REFERENCES

Cluis, D. 1995. Specification of loading model for assessing optimal interventions at the watershed scale, in *Water Quality Modeling, Proceedings of the International Symposium*. Orlando, FL, 521–529.

Dämmgen, U., L. Grünhage, and H.-J. Jäger. 1997. The description, assessment and meaning of vertical fluxes of matter withhin ecotopes — a systematic consideration — Enviromental pollution, 96:249–260.

Diersch, H.J. and R. Gründler. 1993. GIS based groundwater flow and transport modeling — the simulation system FEFLOW, in *On Application of Geographic Information Systems in Hydrology and Water Resources, International Conference,* Vienna, Austria.

EG-VO 2092/91. 1991. Verordnung (EWG) Nr. 2092/91 über den ökologischen Landbau und die entsprechende Kennzeichnung der landwirtschaftlichen Erzeugnisse und Lebensmittel. Amtsblatt der Europäischen Gemeinschaft Nr. 198/1 vom 22.07.1991

ESRI (Enviromental System Research Institute). 1992. Understanding GIS. ARC/INFO, *Methd Rev. 6*, Env. Syn. Res. Inst., Redlands, CA, U.S.A.

Flaig, H. and H. Mohr. 1996. Der überlastete Stickstoffkreislauf, *Nova Acta Leopold.* 289(70):168S.

Gäth, S. and B. Wohlrab. 1992. Strategien zur Reduzierung standort- und nutzungsbedingter Belastungen des Grundwassers mit Nitrat, Sonderheft der Deutschen Bodenkundlichen Gesellschaft.

Kersebaum, K.C. 1995. Application of a simple management model to simulate water and nitrogen dynamics, *Ecol. Modelling* . 85:145–156.

Kersebaum, K.C. and J. Richter. 1991. Modelling nitrogen dynamics in a plant soil system with a simple model for advisory purposes, *Fert. Res.* 27:273–281.

Kersebaum, K.C. and K.-O. Wenkel. 1998. Modelling water and nitrogen dynamics at three different spatial scales — influence of different data aggregation levels on simulation results, *Nutr. Cycling Agroecosyst.* 50:315–321.

Keulen, H. van, F.W.T. Penning de Vries, and E.M. Drees. 1982. A summary model for crop growth, in F.W.T. Penning de Vries and H.H. van Laar, Eds., *Simulation of Plant Growth and Crop Production,* Pudoc. Centre of Agricultural Publishing and Documentation, Wageningen, 87–97.

Michel, R.J. 1991. Entwicklung eines Modells zur zeitlichen und örtlichen Verfolgung von Sickerwasserfronten in der Aerationszone und Anwendung zur Beurteilung der vertikalen Wasserbewegung unterschiedlicher Böden im Winterhalbjahr, Abschlußarbeit Postgradualstudium, TU Dresden, 15 S.

Scharpf, H.C. and J. Wehrmann. 1975. Die Bedeutung des Mineralstickstoffvorrates des Bodens zu Vegetationsbeginn für die Bemessung der N-Düngung zu Winterweizen, *Landwirtsch. Forsch,* 32/I:100–114.

Schultz, A. 1997. Informationsbedarf, Komplexität und Aussagegenauigkeit von landschaftsbezogenen Simulationsmodellen, *Arch. Naturschutz Landschaftspflege* 36:107–124.

Simpkins, W.W., T.B. Parkin, T.B. Moorman, and M.R. Burkart. 1995. Subsurface geology: a key to understanding water quality problems, in *Clean Water — Clean Environment — 21st Century, Conference Proceedings,* Kansas City, MO, 251–252.

Srinivasan, R., J.G. Arnold, and R.S. Muttiah. Continental scale hydrologic modeling using GIS, in *Water Quality Modeling, Proceedings of the International Symposium,* Orlando, FL, 231–240.

Vereijken, P. 1989. Experimental systems of integrated and organic wheat production, *Agric. Syst.* 30:187–197.

Voigt, H.-J. and R.-J. Michel. 1997. Ein einfacher Ansatz zur Abschätzung der möglichen Denitrifikation in der Aerationszone, *Mitteilungen Deutsche Bodenkunde Ges.* 85 III:1421-1424.

Wasy Institute of Water Resources Planning and System Research. 1993. Interactive, graphics-based finite simulation system FEFLOW for modeling groundwater flow and contamination transport processes, User's Manual, Wasy Ltd., Berlin, Germany.

Zacharias, S., C.D. Heatwole, and J.B. Campbell. 1997. Spatial trends in the texture, moiture content, and pH of a Virginia Coastal Plain soil, *Trans. ASAE* 40(5):1277–1284.

Socioeconomic Aspects of Watershed Management

16 The Economics of Land Restoration and Conservation in an Environmentally Sound Agriculture

Luther Tweeten and Jeffrey Hopkins

CONTENTS

INTRODUCTION

Some 3 million native Americans resided on what is now the 48 contiguous United States when English settlers founded Jamestown in 1607. Those native Americans intervened only modestly in nature, following a way of life that had sustained their numbers with only transitory deviations for perhaps thousands of years with little resource degradation or depletion. New settlers immediately set to clearing land and cultivating tobacco and other crops, increasing food and fiber output per person and per hectare to sustain a higher population density and living standard than enjoyed by native Americans.

As population increased and standards of living rose, more land was cleared, drained, and plowed for food production. As soils were depleted in the East, settlers

0-8493-0702-3/00/$0.00+$.50

moved West to new lands. By 1890, the frontier was gone and cropping intensity had to be increased to feed everyone. By the 1930s, Americans began to be aware that current rates of natural resource depletion and degradation could not be sustained. At issue is what level of resource use between minor intervention practiced by traditional societies and intensive cultivation practiced by affluent developed economies is economically optimal. The purpose of this chapter is to address that issue and possible policy responses.

The next section briefly inventories the state of soil resources in the U.S. and the world. This is followed by a review of studies of the economically optimal level of soil utilization. The chapter concludes with a synthesis outlining an economic framework to better determine an economic level of soil conservation.

THE SETTING

Dregne and Chou (1992) present comprehensive though crude estimates of the global extent of soil degradation, its annual cost in lost output, proportions of such land reclaimable at favorable benefit–cost ratios, and the value of doing so. Nearly half (47%) of global cropland was judged to be degraded, with percentages varying from 61 in Africa to 16 in North America. An estimated 30% of the 145 million ha of irrigated land was degraded, with the greatest area and incidence in Asia (Dregne and Chou, 1992, p. 276). Fully 73% of rangeland was estimated to be degraded with high proportions in all continents except for "only" 55% in Australia and New Zealand.

Crosson (1996, p. 2) used data from Dregne and Chou to calculate that global soil productivity loss averaged 0.3% per year. Crosson employed a second set of data from Oldeman et al. (1990), who calculated land degradation from judgments of land-use experts in countries throughout the world. Based on a common set of criteria for judging the extent and severity of land degradation, these experts concluded that 1.3 billion ha (27 percent) of the global total annual and permanent cropland and pasture, 4.7 billion ha, was degraded to some extent. Some 84% of the degradation was caused by water and wind erosion. Based on Oldeman et al., Crosson estimated an average annual soil productivity loss of 0.1%. If these estimates of annual soil productivity losses from Crosson are treated as linear rates, global soils will last between 300 and 1000 years. Of course, bottomlands, gently sloped soils, and well-cared-for soils have much greater longevity.

Crosson's estimates have several shortcomings:

- Estimates of soil degradation compiled by Dregne and Chou and by Oldeman et al. are crude judgments that could be improved with a stronger empirical base.
- Crosson's quantification of terms such as "severely degraded" or "slightly degraded" requires arbitrary judgments.
- An approximation of the number of years over which degradation has occurred is necessary to calculate annual rates.
- Comparatively recent changes in tillage practices and technology and ongoing regeneration of soil minerals from parent minerals are not adequately considered.

- The annual degradation rate could change over time. Oldeman et al. argued that the rate of productivity loss may be accelerating; hence the future rate may exceed the past rate of land degradation. In contrast, Crosson (p. 3) contends that "continual population growth in the developing countries and rising demand for food should continue to strengthen farmers' incentives for making [soil conserving] investments"; hence he reasoned that slowing land productivity losses would be "the most plausible future."
- Estimates by Crosson, although crude, are useful in judging overall average trends in soil productivity, but they conceal severe "hot spots" of soil degradation of great concern to soil scientists. Crosson's estimates are necessary but not sufficient to prescribe an optimal level of degradation for society, because downstream costs of soil degradation are not considered. Downstream costs may be significant if we consider the size of the downstream population and the recreation and aesthetic properties of the watershed.

Tweeten (1989, pp. 268, 269), summarizing several studies of the impact of soil degradation, concluded that on average degradation is dropping U.S. land productivity approximately 5% in a century or an average of 0.05% per year. This is half or less the global rate and may fall further in the future if the 1982 to 1992 drop of 25% in wind and water erosion rate continues (U.S. Department of Agriculture, 1994).

Land for producing food is also lost to irreversible development for shopping centers, housing, highways, reservoirs, and the like. Estimates of farmland and cropland loss to nonfarm uses differ widely but on average may be 0.10% per year (Tweeten and Forster, 1993). Adding degradation and development losses, some 0.15% per year of farmland and attendant productivity is lost per year in the U.S. On the other hand, multifactor productivity (crop and livestock output per unit of all production inputs) has grown by 2.0% per year for four decades (Council of Economic Advisors 1997, p. 410), or over 10 times the loss due to land degradation and nonfarm development. Furthermore, private and public rates of return typically average from 20 to 40% on investments in agricultural research and extension to develop new technology, far higher than typical private returns on soil saving structures and practices. While these considerations would seem to argue for a public policy emphasizing research on new production technologies rather than on soil conservation, that strategy has shortcomings noted later in this chapter.

THE PAYOFF FROM RESTORING LAND

Dregne and Chou estimated the annual loss (forgone income) due to soil degradation (which they call desertification) as follows:

	($ billions)
Irrigated land	11
Rain-fed cropland	8
Rangeland	23
Total	42

The analysts estimated that nearly all irrigated land would return the cost of rehabilitation, whereas only 70% of rain-fed cropland and 50% of rangeland could be rehabilitated at favorable benefit–cost ratios. Costs over a 20-year period for rehabilitating desertified land were estimated to be (in 1990 dollars):

	($ billions)
Irrigated land	86
Rain-fed cropland	60
Rangeland	37
Total	213

Benefits over a 20-year period were estimated to be $532 billion in discounted 1990 dollars. Although the benefit–cost ratio for rehabilitation was a favorable 2.5, much of the $213 billion investment in restoration would not be profitable to private firms. The public may be willing to bear part of the cost if justified by externalities such as aesthetic value.

Studies of land restoration costs after strip coal mining give clues to the public perception of the environmental (including aesthetic) value of land. Cost of backfilling and revegetation to restore basic topography and productivity for agriculture in the late 1970s averaged approximately $46,000/ha in Appalachia, $40,000/ha in the Midwest, and $42,000/ha in the West (Committee on Soil as a Resource, 1981, p. 200). These costs of restoring strip-mined land were typically 20 to 40 times the agricultural value per hectare; hence the aesthetic benefits (and to a lesser degree the water quality benefits) of restoration appear to be very high compared with agricultural value of land — assuming heroically that the political process underlying laws requiring restoration properly reflects public choice. Although the magnitudes of those numbers strain credulity, allowing for considerable error still supports the conclusion that society places a high value on land aesthetics.* In addition to aesthetics, the need is apparent for long-term studies of both the technology and economics of soil restoration (Lal, 1995, p. 535).

Another conclusion drawn from the high costs of land restoration is that it is economic to avoid severely degrading land rather than to allow land to degrade severely and then restore it. Because private owners currently do not find it profitable to restore degraded land in many instances, the public either must pay for or require restoration deemed to be in the public interest to preserve water quality and aesthetics.

OPTIMAL SOIL CONSERVATION

The foregoing paragraphs provided an overview of soil degradation issues but did not address the issue of optimal soil conservation. At least three rules have been presented as optimal or sustainable soil-conserving levels. These are to allow land to erode not more than (1) the rate of regeneration of topsoil minerals from parent

* The public interest in preserving and viewing the highly eroded Badlands of the Dakotas seems to belie this conclusion. However, the erosion was not caused by decisions of people. For this and other reasons, the Badlands have unique properties not relevant to restoring other despoiled land.

material, (2) the rate of soil erosion tolerance, or T, and (3) the rate maximizing the net discounted present value of all future production from land.

Topsoil Mineral Regeneration from Parent Material

Soil is often regarded as a renewable resource (Ciriacy-Wantrup, 1968), but in fact soil renewal through genesis is a very slow process. Some of the basic minerals in topsoil from parent material is degraded by entropy each year. On average, soil is regenerated from parent material at an annual rate of 1 ton/ha (conversion of parent material into soil in the A, E, and B horizon) according to Pimental et al. (1995) citing Troeh and Thompson (1993). McConnell (1983, p. 84) states that "natural rebuilding contributes two to five tons per acre [4.5 to 11.2 metric tons/ha] per year depending on soil type, weather, and other variables."

Soil Tolerance Level

Soil conservation policy makes wide use of soil-loss tolerance levels or T values, defined as "the maximum amount of soil loss, in tons per acre per year, that can be tolerated and still achieve a degree of conservation needed for sustained economic production in the foreseeable future with present technology" (U.S. Department of Agriculture, 1974, p. 6). The tolerance level is based on considered judgment of soil scientists (Skidmore, 1982). A typical soil loss tolerance rate is 11 tons/ha.

Several caveats are appropriate:

- Analysts differ not only in estimates of how rapidly soil regenerates through natural processes including incorporation of subsoil minerals into the topsoil but also in estimates of how commercial fertilizers, manures, and the like help to restore soil organic matter, nitrogen, phosphorus, potassium, and many trace minerals.
- T values differ widely by soil type. Deep loess soils can sustain erosion rates well above 11 tons/ha for years with little loss of productivity (Lindstrom et al., 1992). Many glacial till soils can sustain little erosion without losing productivity. As topsoils become thin, the T value decreases and approaches the rate of regeneration from parent material.
- Many nonrenewable minerals systematically added to soils for "sustainable production" are being mined at an unsustainable rate. Reserves of oil used to produce fossil fuels and pesticides are adequate for 50 years, of natural gas used to produce nitrogen fertilizers are adequate for only 69 years, and phosphorus for fertilizers for 240 years at current rates of use (Barton, 1996). Of course, new reserves will be discovered. On the other hand, consumption is likely to rise unless slowed by higher prices. Thus questions of use become a matter of economics.

Economically Optimal Soil Conservation

Figure 16.1 provides a conceptual framework underlying the economics of soil conservation over time. The hypothetical curve on the left shows the initial (time t_0)

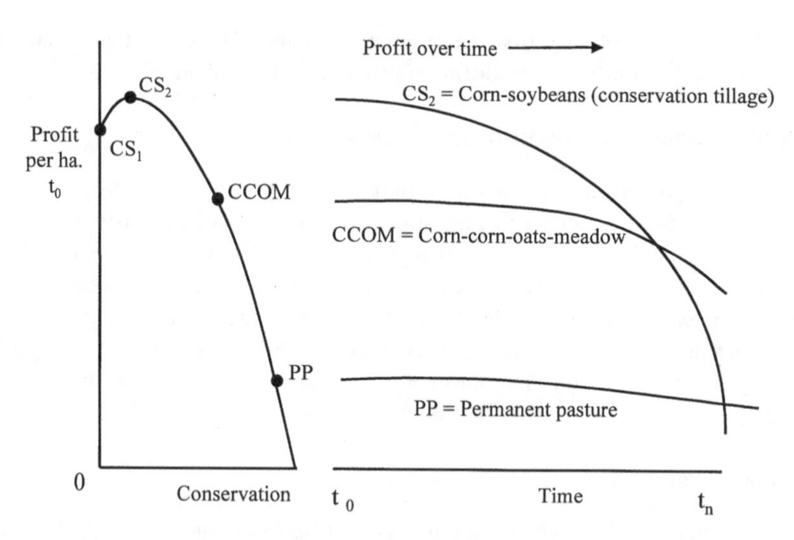

FIGURE 16.1 Profit trade-offs with conservation over time.

tradeoffs between profit (or output) and soil conservation for a typical medium-sized farm in the eastern corn belt. CS_1 is a corn–soybean (CS) rotation using conventional moldboard plow tillage. Using the same rotation but conservation tillage CS_2 with 30% or more surface mulch may simultaneously increase profit, output, and conservation as shown (Hopkins et al., 1996). A corn–corn–oats–meadow rotation (CCOM) protects soil better but sacrifices profit in Figure 16.1. Permanent pasture (PP) is even better for the environment but at a cost in lost short-run profit.

The lines stretching to the right show the profit trajectory over time from year t_0 to year t_n for each cropping system with constant technology. The initial high profit–lower conservation line for CS_2 falls as soil degrades but is everywhere higher than CS_1 (not shown). In contrast, the initial low profit–higher conservation PP rotation profit line falls negligibly over time.

A very conservation-minded operator may sacrifice income to pursue the PP or CCOM rotations in early years. Operators with a long-term perspective are likely to choose such soil-conserving rotations only if they recognize that a corn–soybean rotation erodes land so that productivity falls as depicted over time in Figure 16.1. A profit-minded operator with long-term perspective will choose the rotation offering the greatest present value of net returns — the returns of each year discounted to the present.

The discount rate can influence which rotation has the highest present value and hence is preferred by a profit-oriented investor. A very high discount rate will place a premium on the corn–soybean rotation with a high payoff in early years. If the profit time line for PP always lies below that of other rotations, it will not be preferred whatever the discount rate. If the discount rate is zero, the high profit rotation is the one with the largest area between profit trajectory and the time line in Figure 16.1. An initially preferred cropping system will not lose its advantage over time if the rate of loss of productivity (profit) is less than the discount rate, other things equal.

Ikerd (1997) states that "there is nothing in economics that ensures the regeneration or sustainability of the stocks of resources, either natural or human, which are required to support production of consumer goods and services for current and future generations" (p. 5). In fact, neoclassical economics rather precisely specifies a schedule of flows to be taken from a stock of a mineral. If the annual constant–elasticity demand curve is the same each year in perpetuity and the marginal cost for extraction is constant, the nonrenewable fixed stock resource price will rise at the real rate of interest and the stock will be utilized at the real rate of interest (see Stiglitz, 1976). That real social rate of interest or discount rate is about 3%. It accounts for the premium placed by most people on consuming now rather than later, and the premium on being able to invest now in productive assets to reap a higher payoff in the future, adjusted for risk and uncertainty. Thus, the economically optimal rate of exploitation in this example is 3% of the fixed stock in year 1, 3% of the remaining 97% in year 2, and so on in perpetuity. Several conclusions follow from this simple example:

- A stock resource of value to society is not left unutilized in perpetuity. The issue is how rapidly to utilize it.
- Although a stock resource such as soil is used, it is never used up. Some remains in perpetuity.
- Price rises at the real rate of interest to ration quantity. Assuming demand is unitary elastic in price, a 3% annual reduction in quantity requires a 3% increase in price each year. The reduction in quantity is exactly offset by the increase in price so that total annual revenue from the stock resource is constant in perpetuity, *ceteris paribus*. In perpetuity, the price becomes very large and the quantity used becomes very small.
- By using the above example, the constant-dollar value of the annual flow of output from the stock resource declines each year. For a 3% discount rate, present value of the annual flow in year 100 is $(1/1.03)^{100}$ or 5% of its initial value and in year 200 is 0.3% of its initial value.
- A rational producer maximizing discounted long-term returns on production will reduce soil erosion to its natural replenishing level only when the marginal value of soil becomes very large, soil depth becomes very small, and the social discount rate is zero (McConnell, 1983, p. 88). Whereas in aggregate rising food prices from declining soil quality and crop output will cause producers to cut erosion to negligible levels, the situation for an individual producer is quite different. Depletion of soil on one farm will not raise crop and food prices; hence, operators will not experience the rise in present value of their remaining soil resource required for a rational operator to cut erosion. Thus, the optimal strategy for an individual operator "... even when the social discount rate is zero may include some soil exhaustion" (McConnell, 1983, p. 88). And many costs of soil degradation are borne off site; hence, the on-site decision maker will underinvest in soil conservation even if fully informed and rational.

- Discount rates are higher for producers than for society, discouraging investment in conservation. In the absence of a unique conservation ethic, an economically rational producer maximizing the net present value of the operation will invest in soil conservation if the return on that investment exceeds the discount rate. Discount rates commonly range from about 3% for the public sector to 9% for private investors (Makki et al., 1996, p. 882). As noted earlier, an investment in soil conservation will not pay if the discount rate exceeds the annual rate of decline in value of output (productivity) of eroding soil. Also as noted earlier, on average cropland in the U.S. is losing productivity at a rate of approximately 0.05% per year or even less while global cropland is losing productivity at a rate of 0.1 to 0.3% per year (Crosson, 1996). Even tripling these rates to account for off-site costs normally double the on-site costs leaves productivity losses far short of the discount rate. Hence charging off-site costs to producers would not make erosion control investments economically rewarding to them on average, given current expectations of future supply–demand balances for food.

Many reject the foregoing neoclassical outcome as inconsiderate to future generations who cannot "vote" in elections or the market for conservation. Critics call for lower, even a negative, discount rate to favor future consumption over present consumption. Arguments against that position include the following:

- The present generation hardly enjoys universal affluence — globally over 800 million people are chronically food insecure and 1.3 billion are in abject poverty (FAO, 1996, p. 2). Some 920 million people were food insecure in 1970, and both numbers and proportions of the globally food insecure are declining. Given that recent generations have lived better than their parents, and that this trend is likely to continue, allocation systems that favor future over current consumption through a negative discount rate are inappropriate.
- Future generations might be better off if the current generation spends heavily on high-payoff agricultural or medical research rather than on natural resources.
- A negative or low discount rate would encourage current investment in land-using capital as well as in conservation and other land-saving capital. Hence, a low discount rate might speed land degradation. Use of differential discount rates — a low rate for conservation capital and a higher rate for other capital has been implicitly applied through public financial incentives for conservation structures and practices.

In summary, an environmentally sound agriculture would maximize the discounted net present social value of environmental resources. That outcome occurs where marginal social costs of protecting or enhancing the environment just equal the marginal social benefits of protecting or enhancing the environment. The term *social* means that all incremental costs or benefits are included: those that accrue from production and consumption of private as well as public goods.

Ideally, benefits or costs accruing where markets are poorly developed or non-existent also are included in a social perspective. Examples relevant to soil and water include *existence* and *option* value. Markets may not reflect existence value, defined as benefits people receive and are willing to pay for, such as water quality or soil conservation on land they will never see or use. Option value also may not be included in private market prices but includes the value placed on water quality and soil conservation because people judge they or future generations may benefit from protected resources although current projections do not anticipate the need.

ECONOMICALLY OPTIMAL SOIL CONSERVATION MEASURED FROM SITE-SPECIFIC MULTIPERIOD ANALYSIS

Results of several site-specific economic studies are reviewed below before summarizing conclusions of this chapter.

Xu and Prato (1995) examined the permanent on-site costs of soil erosion based on research plots on Mexico silt loam at the University of Missouri's South Farm in central Missouri producing corn from 1984 to 1989. Based on a logistic model and 35 cm of topsoil, erosion on-site damages were as follows:

	Soil Loss $ per hectare per year (tons)			
Time Horizon	**2**	**3**	**4**	**5**
1 year	0.25	0.38	0.51	0.63
25 years	3.55	5.33	7.12	8.19

Source: Xu, F. and Prato, T., *J. Soil Water Conserv.*, 50, 312, 1995. With permission.

Conclusions were as follows:

- On-site loss from erosion was small. Damage from annual 2.5 cm (1 in.) losses of topsoil was estimated to be $8.06/ha in 1 year and $113.54/ha over 25 years. This annual loss of only 0.2% of land value would be masked and overshadowed by vastly greater year-to-year variation in returns from technology, weather, and prices.
- Erosion damages per year increase as topsoil depth declines. Soil conservation policies that target shallow soils are often more cost-effective than policies targeting deep soils.

Frohberg and Swanson (1976) used a multiperiod programming model to determine an economically optimal soil erosion rate in the Big Blue Watershed, Pike County, IL. Under 1970s conditions, they maximized the net present value of economic benefits over time minus sediment damage and a penalty for losses beyond the 48 years in their planning period.

Topsoil thickness and annual soil loss are shown for period 1 and 3 (16 years apiece) of the 48 year planning period trend under high and low demand for output

TABLE 16.1
Economically Optimal Topsoil Thickness and Annual Soil Loss, Big Blue Watershed, Pike County, IL

Scenario — Demand for Output	Topsoil thickness (cm)			Annual soil loss (tons/ha)		
	Period 1	Period 3	% Change	Period 1	Period 3	% Change
		High (discount rate 8%)				
Slope 4% or less	21.67	19.18	–11	9.81	10.96	11
Slope over 4%	11.97	6.96	–41	26.26	19.24	–27
		Low (discount rate 8%)				
Slope 4% or less	21.67	19.30	–11	9.63	10.37	8
Slope over 4%	11.97	9.42	–21	11.52	16.29	41[a]
		Low (discount rate 2%)				
Slope 4% or less	21.67	19.36	–11	9.43	10.19	8
Slope over 4%	11.97	9.58	–20	10.96	10.35	–6

[a] The result may be in error — the second period soil loss was 11.11 tons/ha, down 4% from period 1.

Source: Frohberg, K. and Swanson, E., Agricultural Economic Research Rep. 161, University of Illinois, Urbana, 1976. With permission.

and for high and low discount rates for the two soil groups (Table 16.1). Several conclusions are apparent from Table 16.1 and other results from Frohberg and Swanson:

- Economically optimal soil loss is within the Natural Resources Conservation Service tolerance value of 9 to 11 tons/ha for flatter soils. Optimal soil loss was rising, however, over time. For the soil group with slope of over 4% the economically optimal soil loss was well over the Natural Resources Conservation Service tolerance levels of 4.5 to 9 tons/ha in several instances but falling (see footnote a of Table 16.1 for a possible explanation of the exception). Thus, an economically optimal soil loss rate is well above tolerance levels of some soils.
- Based on the Frohberg–Swanson model assumptions, it is economically optimal for soil to continue to deplete. Topsoil declined over the 48-year planning horizon by up to 11% on slopes of 4% or less and by up to 41% on steeper slopes.
- A lower demand for farm output does not necessarily slow optimal depletion of soil. High demand and high commodity prices encourage output and hence resource exploitation. But if a demand increment is viewed as permanent, greater investments in soil-conserving practices and structures are warranted.

TABLE 16.2
Impacts of Soil Productivity Losses on Net Present Value and Total Erosion under Alternative Discount Rates for the Four-Mile Creek Watershed

Soil Productivity Loss (corn yield base), kg/centimeters	5.0% Discount Rate		3.5% Discount Rate	
	NPV, $ thou.	Erosion, thousand tons	NPV, $ thou.	Erosion, thousand tons
0.0	12,332	7,392	15,748	7,392
4.4	12,016	7,396	15,003	4,553
10.0	11,754	4,549	14,742	4,395
20.0	11,473	3,977	14,296	3,850
30.0	11,227	3,429	13,982	3,429

Source: Miranowski, J. S., *Am. J. Agric. Econ.*, 66, 66, 1984. With permission.

- A high discount rate reflecting a high premium on current output does not appear to have much influence on soil loss and topsoil thickness. For example, the percentage reduction in topsoil thickness for both soil slope scenarios remains essentially unchanged under the 2 and 8% discount rates. The sharp increase (41%) in soil loss with an 8% discount rate in Table 16.1 is an exception but may be spurious (see footnote a, Table 16.1).
- Reduction in soil productivity had a minor impact on optimal soil loss (Frohberg and Swanson, 1976, p. 26). Despite the assumption of no technological progress, yields changed very little and in some cases increased because of profitable opportunities to apply more nitrogen. The authors conceded, however, that "soil productivity at some point will be reduced more than proportionally with the decrease in topsoil thickness" (p. 27).

Miranowski (1984) employed a multiperiod linear programming model maximizing the net present value of crop production for 50 years (10 5-year periods) with a penalty function in the terminal period to account for even longer-term effects of soil erosion. The model was for the Four-Mile Creek Watershed in Tama County, IA in the heart of the cornbelt. The soils are unusually productive and deep. Corn–soybean (CS), corn–corn–oats–meadow–meadow (CCOMM), and PP rotations were allowed depending on profitability. Discount rates were 3.5 and 5% on net returns over time.

Several observations are apparent from Table 16.2 and other results from Miranowski for the Four-Mile Creek Watershed in Iowa:

- Both the net present value (NPV) from cropping and the erosion rate fall as operators respond to higher yield penalties from erosion. A corn–soybean rotation is at once the most profitable and most erosive rotation; high erosion rates penalize yields and profit of the highest return crops and encourage a shift to less profitable crops in rotation.

- The annual optimal erosion rate when the soil productivity penalty is 4.4 kg of corn per centimeter of soil loss and the discount rate is 3.5% falls from 61 tons/ha in the first 5-year period to 47 tons/ha in the 10th 5-year period (Miranowski, 1984, p. 68). Although the trend indicates progress, both rates far exceed the erosion tolerance rate of 11 tons/ha. Under these same assumptions, net returns fall only 6.7% per hectare between the first and last of the 10 5-year periods. A threshold level of soil loss may be reached when producers experience a sizable loss of profit, but not during the 50-year period of analysis.
- Erosion falls much faster than NPV as the yield penalty rises. Conversely, erosion rises much faster than net present value as the yield penalty falls. Thus, a farmer may be misled by using value of output to judge erosion on the excellent, deep loess soils of Tama County, IA.
- A low discount rate coupled with a heavy (perhaps unrealistically high) soil productivity yield penalty changed the rotation from corn and soybeans to a corn–meadow–meadow–corn rotation (Miranowski, 1984, p. 70).
- If crop prices (especially of hay) are expected to increase over time and if a heavy productivity penalty is paid for erosion, farmers respond with measures to conserve soil. Thus, long-term outlook education programs making farmers aware that such conditions are likely to prevail may be useful to protect soil resources. However, increased commodity prices raised the soil erosion rate in the short run.
- The chisel plow was more profitable than the less-soil-conserving moldboard plow in the study watershed.
- Reduction of topsoil depth raised erosivity, but this conclusion was based on soil science and not the model.

Burt (1981) employed dynamic control theory analysis to determine optimal soil conservation based on topsoil thickness and organic matter under alternative price regimes and rotations for the wheat–pea Palouse area of eastern Washington and western Idaho. Topsoil is rich with an average depth of 46 cm, slopes are long and steep, and precipitation is adequate for high yields of wheat. The soil mantle averages over 30 m of loess except in the bottom of draws where it averages about 3 m and the tops of hills where it averages only 15 cm (Burt, 1981, pp. 83, 84).

With wheat fertilized at a rate that did not limit yield, economically optimal soil organic matter was found to be 3.5% and annual soil loss 11 tons/ha or 0.9 mm/year (Burt, 1981, p. 88). Average topsoil depth would fall from 46 cm to zero in just over 500 years with wheat at $118/metric ton and in just over 400 years with wheat prices at $156/ton. Burt (p. 91) notes that "intensive wheat production with good cultural and fertilization practices is economically justified in the long run … and not a threat to long-run productivity of the soil." However, unlike studies by Frohberg and Swanson and by Miranowski, Burt included no cost for off-site damages through sedimentation and chemical contamination of water.

SUMMARY AND CONCLUSIONS

Several conclusions follow from this review of the economics of optimal soil conservation.

- Global productivity losses from erosion are not well measured and assumptions regarding use of commercial fertilizers and other inputs to maintain productivity are sketchy, but soil erosion at current tonnages appears to be unsustainable and could deplete global soils in 300 to 1000 years. Of course, topsoil on gentle slopes and bottomlands would remain productive.
- Studies are not consistent on whether higher demand and prices for farm output enhances or depresses soil conservation. To encourage substantial additional conservation investment, farm operators and owners must view enhanced output demand as permanent.
- Economic studies assume that future prices are constant, although in fact costs of fertilizers, fuels, and pesticides can be expected to rise with depletion of petroleum, natural gas, phosphate reserves, and soils in a longer period.
- Economic studies of optimal soil conservation over time assume no change in technology. Conservation technology has been improving, and the ability of no-till and conservation tillage to at once maintain profit, output, and soil is one of the real success stories in reducing erosion.
- Soil tolerance (T) levels (averaging about 11 tons/ha as judged by soil scientists) and economically optimal soil conservation broadly coincide based on the few available studies. Both "mine" soil minerals above the rate of soil regeneration from parent minerals. The rate at which parent mineral materials regenerate soil (about 1 ton/ha) appears to be an unduly conservative yardstick for optimal soil conservation.
- Restoring severely degraded farmland is not privately profitable but is publicly justified in some instances. The public places a high value on an attractive landscape as evidenced by extremely high costs relative to agricultural value incurred to restore strip-mined land in industrialized countries. In the case of farmland, however, the best strategy is to follow practices and policies that avoid severe land degradation rather than to degrade first and then restore land.
- Soil productivity loss rates from erosion differ sharply among soils, but on average the rates are too small to be detected by producers. Yield relationships are highly sensitive to rainfall and related stochastic influences, masking changes in the underlying determinants of productivity. That conclusion has numerous corollaries. One is that producers need to be better informed of erosion rates and topsoil thickness, the latter especially when land is sold. Another corollary is that discount or real interest rates for individual producers (likely to be 4 to 9%) and for society (likely to be near 3%) are high relative to productivity losses, hence producers are unlikely to invest optimally in conservation to preserve food output.

- Even if farm owner–operators have low discount rates, are fully informed, and are profit maximizers, they will not conserve soil sufficiently to meet society's needs because they do not pay the cost of off-site damages. Those damages cannot easily be assessed because effluent levels vary widely among farms and have end results ranging from benefits to forests in an alluvial plain to dredging and water treatment costs of over $10/ton of soil deposited in an urban reservoir (see Hitzhusen, 1992).
- A tempting but oversimplified conclusion is that producers should be free to allow their soil to erode (because they pay costs and hence use good judgment regarding on-site degradation) while the public should pay for riparian strips and other measures to control off-site costs. The GAO (1995) reports that reserve land equal to one sixth of the Conservation Reserve Program (CRP), if strategically sited as buffer strips along streams in key watersheds, would be more cost-effective than the current approach. To be sure, riparian strips and the like will be used to control water quality and maintain aquatic habitat more than they will be used to control soil erosion. Because water quality is a more reliable measure of downstream costs than is on-site productivity loss from soil erosion, water quality is attracting more public education and policy focus. Soil erosion and water quality are inseparable, however, because soil erosion and loss of organic matter, sediment, chemical fertilizers (except nitrogen), and pesticides go together. Thus, public policy must jointly address on-site and off-site costs of soil loss, in part, because filter strips and wetlands by no means constrain all effluents and because these measures are ineffective for maintaining on-site productivity.
- Although the low on-site payoff to invest in conservation and the high payoff to investment in research to raise productivity of agriculture would seen to justify a strategy of public investment in science and technology rather than conservation, that conclusion is suspect. Continuing investments in education and science are critical, but neglect of soil conservation would be a serious public policy oversight. The public role is justified given that the private market inadequately addresses problems from (1) ignorance of soil losses or measures to address them, (2) higher private than public discount rates, (3) aesthetic value, (4) off-site costs, and (5) uncertainty regarding the future. The latter may be the most intractable issue for public policy. A resource may have consumptive (current use) value but also existence value and option value. The market mostly handles consumptive use value. Based on existence value, the public may wish to preserve aesthetic value of land even if most people will not see lands that would have been blighted.

The option value is the worth to the public of conserving soil because it might be needed sometime even though current projections do not see the need for special measures to conserve. The latter can best be expressed as a game against nature with outcomes or payoffs A, B, C, and D:

Investment in Conservation	States of Nature	
	Abundant	Sparse
Low	A	B
High	C	D

If "nature" is abundant (rapid growth in agricultural productivity and slow population growth) and investment in conservation is low as at A, error is small. Alternatively, if nature is sparse (slow growth in agricultural productivity, rapid growth in demand) and investment in conservation is high at D, again error is small. A more serious error is high investment in conservation and not need it, outcome C. The most serious error of all, however, is a sparse nature coupled with low investment in conservation, outcome B. As people and nations become more affluent, they are unlikely to want to gamble on future food supplies. Thus, in the future as in the past, the public at least in industrial nations is likely to invest in conservation to avoid outcome B, holding down irreversible loss of cropland to development, and maintaining the option value of a land resource available for food production if needed at some time in an unforeseeable future. T values and economic calculations may call for public policies conserving soils for hundreds of years, but society may want policies to preserve soils for thousands of years.

- A public role in private property is justified to protect society's interest in soil conservation and water quality not protected by the market alone. Loss of on-site cropping productivity to the landowner is a less reliable guide to overall erosion cost than is the level and damage from soil runoff. The latter does not influence decisions of the on-site operator. In theory, a Pigouvian effluent tax on farm operators equal to off-site damage would be ideal but is not workable. Hence, other means are warranted such as those used in the U.S.: Conservation Compliance, Conservation Reserve Program, and Natural Resource Conservation Service efforts.

The U.S. experience is instructive about how public policy might address soil erosion problems (see Tweeten and Zulauf, 1997). Although U.S. soil erosion averages 14 tons/ha of cropland, somewhat higher than the 11 tons tolerance level per hectare, most cropland erodes at less than tolerance rates (Pierce and Nowak, 1994, p. 2). For example, based on the 1977 National Resources Inventory, 23% of cropland incurred sheet and rill erosion in excess of 11 tons/ha, and a like percentage incurred wind erosion in excess of 11 tons/ha (U.S. Department of Agriculture, 1981). But because of overlap, only 34% incurred erosion in excess of 11 tons/ha from wind and water combined.

The erosion problem is even more concentrated than indicated above according to the Agricultural Conservation Program Evaluation (U.S. Department of Agriculture 1980) which indicates that only 13% of the cropland base is eroding at over 11 tons/ha. More recent analysis accounting for no-till technology narrows the problem further. Dicks and Coombs (1994, p. 55) estimate that only 7.3 million ha of cropland cannot be cropped at acceptable soil loss tolerances with "best management practices," notably no-till.

Thus, American soil conservation problems seem tractable with appropriate public policy. The CRP in 1992 included only 2.5 million ha of this "noncropable at T" cropland; hence the CRP altered by the 1996 farm bill will need to focus much more narrowly on converting to soil-conserving uses cropland that cannot be cropped at acceptable T values. If commodity program payments expire in year 2003, operators will lack incentives to participate in the Conservation Compliance Program. If the goal is to reduce all erosion to T or less, the Conservation Compliance Program will need to be continued and cover all highly erodible land not covered by CRP. The purpose is to require "best management practices" to keep operators from "taking" from others by allowing erosion on land that can be cropped at an acceptable T without loss of profit or output through practices such as no-till. Operators are only being required to stop "taking" from others and incur little or no loss themselves from compliance; hence, they might be required to participate in the Conservation Compliance Program. This requirement would strongly "encourage" those who cannot crop at T even with best management practices to place fragile cropland in the CRP where participants are monetarily compensated for "taking." And the CRP might be changed into a cropping easement program allowing grazing or haying of land to remove more fully the supply control dimension of land retirement.

Finally, two additional policy issues are noted. The first is that the depletion of nonrenewable stocks of petroleum, natural gas, and phosphate that attends soil erosion may eventually limit food production more than does degradation of soil. Policies for these natural resources are as important to meet long-term global food and fiber needs as policies regarding cropland.

The second is that sound economics of soil conservation and water quality begins with full marginal cost pricing of resources. Not only is it important to slow erosion, but also to stop subsidizing real estate (e.g., mortgage interest and capital gain tax preferences) and community services encouraging urban sprawl and irreversible loss of cropland to nonfarm uses which may be a bigger threat to cropland than is soil erosion.

ACKNOWLEDGMENT

Comments of Brent Sohngen are appreciated.

REFERENCES

Barton, P. 1996. Renewable and nonrenewable resource, paper presented to the American Association for the Advancement of Science meeting in Baltimore, MD, February 10, 1996. Washington, D.C.: U.S. Department of Interior.

Burt, O. 1981. Farm level economics of soil conservation in the Palouse area of the Northwest, *Am. J. Agric. Econ.* 63:83–92.

Ciriacy-Wantrup, S. V. 1968. *Resource Conservation – Economics and Policies,* Berkeley: University of California, Agricultural Experiment Station.

Committee on Soil as a Resource. 1981. *Surface Mining: Soil, Coal, and Society.* National Research Council, Washington, D.C.: National Academy Press.

Council of Economic Advisors. 1997. *Economic Report of the President*, Washington, D.C.: U.S. Government Printing Office.

Crosson, P. 1996. *Perspectives on the Long-Term Global Food Situation,* Issue No. 3, Washington, D.C.: Federation of American Scientists Fund, Summer.

Dicks, M. and J. Coombs. 1994. Evaluating the Conservation Reserve Program, in *1992 National Resources Inventory Environmental and Resource Assessment Symposium Proceedings,* Washington, D.C.: Natural Resources Conservation Service, USDA, 50–69.

Dregne, H. E. and N.-T. Chou. 1992. Global desertification dimensions and costs, in *Degradation and Restoration of Arid Lands*, H. E. Dregne, Ed., Lubbock: International Center for Arid and Semi-Arid Land Studies, Texas Tech University, 249–281.

FAO. 1996. *World Food Summit: Synthesis of the Technical Background Documents,* Rome: Food and Agriculture Organization of the United Nations, November.

Frohberg, K. and E. Swanson. 1976. A Method for Determining the Optimum Rate of Soil Erosion, Agricultural Economic Research Report No. 161, Urbana: Department of Agricultural Economics, University of Illinois, April.

GAO. 1995. Conservation Reserve Program: Alternatives Are Available for Managing Environmentally Sensitive Cropland, Washington, D.C.: U.S. General Accounting Office, February.

Hitzhusen, F. 1992. The economics of sustainable agriculture: adding a downstream perspective, *J. Sustainable Agric.* 2: 75–89.

Hopkins, J., G. Schnitkey, and L. Tweeten. 1996. Impacts of nitrogen control policies on crop and livestock farms at two Ohio farm sites, *Rev. Agric. Econ.* 18:311–324.

Ikerd, J. 1997. Toward an economics of sustainability, paper presented to joint meeting of Agriculture, Food, and Human Values Society and Association for the Study of Food in Society meeting in Madison, WI, June. Columbia: Department of Agricultural Economics, University of Missouri.

Lal, R. 1995. Trends in world agricultural land use: potential and constraints, in R. Lal and B. A. Stewart, Eds., *Soil Management: Experimental Basis for Sustainability and Environmental Quality,* Boca Raton, FL: CRC Press, 521–535.

Lindstrom, M., T. Schumacher, A. Jones, and C. Gantzer. 1992. Productivity index model comparison for selected soils in north central United States, *J. Soil Water Conserv.* 47:491–494.

Makki, S., L. Tweeten, and M. Miranda. 1996. Wheat storage and trade in an efficient global market, *Am. J. Agric. Econ.* 78:879–890.

McConnell, K. 1983. An economic model of soil conservation, *Am. J. Agric. Econ.* 65:83–89.

Miranowski, J. S. 1984. Impacts of productivity loss on crop production and management in a dynamic economic model, *Am. J. Agric. Econ.* 66: 61–71.

Oldeman, R., R. Hakkeling, and W. Sombroeck. 1990. World Map of the Status of Human-Induced Soil Degradation: An Explanatory Note. Wageningen, The Netherlands: International Soil Reference and Information Center and Nairobi, Kenya: United Nations Environmental Program.

Pierce, F. and P. Nowak. 1994. Soil erosion and soil quality: status and trends, in *1992 National Resources Inventory Environmental and Resource Assessment Symposium Proceedings.* Washington, D.C.: Natural Resources Conservation Service, USDA, 1–23.

Pimental, D., C. Harvey, P. Resosudarmo, K. Sinclair, D. Kurz, M. McNair, S. Crist, L. Chpritz, L. Fitton, R. Saffoury, and R. Blair. 1995. Environmental and economic costs of soil erosion and conservation benefits, *Science* 267:1117–1123.

Skidmore, E. L. 1982. Soil loss tolerance, in *Determinants of Soil Loss Tolerance,* Madison, WI: ASA, SSSA, Chap. 8.

Stiglitz, J. 1976. Monopoly and the rate of extraction of exhaustible resources, *Am. Econ. Rev.* 66:655–661.

Troeh, F. R. and L. M. Thompson. 1993. *Soils and Soil Fertility,* 5th ed., New York: Oxford University Press.

Tweeten, L. 1989. *Farm Policy Analysis*, Boulder, CO: Westview Press.

Tweeten, L. and L. Forster. 1993. Looking forward to choices for the twenty-first century, *Choices*, Fourth Quarter, 26–31.

Tweeten, L. and C. Zulauf. 1997. Public policy for agriculture after commodity programs, *Rev. Agric. Econ.* 19(Fall/Winter):263–280.

U.S. Department of Agriculture. 1974. Universal Soil Loss Equation, Resource Conservation Planning Technical Note IL-4. Champaign, IL: Soil Conservation Service.

U.S. Department of Agriculture. 1980. National Summary Evaluation of the Agricultural Conservation Program, Phase I. Washington, D.C.: Agricultural Conservation and Stabilization Service, USDA.

U.S. Department of Agriculture. 1981. Soil and Water Resources Conservation Act — 1980 Appraisal, Part I. Washington, D.C.: USDA.

U.S. Department of Agriculture. 1994. Summary Report: 1992 National Resources Inventory, Washington, D.C.: Soil Conservation Service.

Xu, F. and T. Prato. Onsite erosion damages in Missouri corn production, *J. Soil Water Conserv.* 50(May/June):312–316.

17 Economic Analysis of Sedimentation Impacts for Watershed Management

Fred J. Hitzhusen

CONTENTS

INTRODUCTION

This chapter starts by putting sedimentation into a watershed context and clarifying the meaning of financial vs. economic and efficiency vs. distribution analysis. The new institutional economics (NIE), alternative types of environmental economic value, and the appropriate methods to measure these values are developed and related to alternative types of sedimentation impacts. A case study of Valdesia Reservoir in the Dominican Republic and some Ohio research results are presented to illustrate several of the foregoing concepts and their implications for improved public policy. Finally, some general conclusions are developed for economic analysis of sedimentation for watershed management.

Soil erosion and resulting sedimentation problems are major concerns throughout the world. Dregne (1972) was the first to evaluate the general state of land degradation worldwide. Much of southwest Asia, large areas of China, India, and Southeast Asia, Northern Africa, Central America, and Mexico were found to suffer from severe land degradation. Dregne concluded that desertification "will blunt crop production increases in the subhumid and semiarid regions of the United States and

0-8493-0702-3/00/$0.00+$.50

Mexico, as well as in the Central American mountains." In South America, "land degradation is most severe on the cultivated lands of the Andes Mountains" and "water and wind erosion have damaged some Argentine farmland." Dregne also emphasized the fragility of soils in the Amazon Basin, where deforestation and encouragement of development have been intensifying in recent years.

The off-site impacts of land degradation or soil erosion are illustrated in the reservoir sedimentation literature by such authors as Allen (1972), Robinson (1981), and Das (1977). For example, the severe effect that sedimentation can have on the economic life and power production capacity of a reservoir is dramatically illustrated by the rapid sedimentation of the Anchicaya hydroelectric project in Colombia reported by Allen.

The Anchicaya project was initiated in 1944 and the first power was produced in 1955. Planners had assumed that the sediment load of the river was not a significant concern. This may have been true prior to the project, when there was no highway and very little deforestation. Concurrent with construction of the dam, however, improved access to the watershed promoted a rapid influx of hillside farmers. The resultant clearing of large portions of the watershed rapidly made it clear that sedimentation was going to be a serious problem for the reservoir.

Allen (1972) quotes Davis in discussing the futility and economic infeasibility of mechanical removal of deposited sediment. Their conclusions are the same "... watershed control is the only practical solution." For Anchicaya, the culprit was clearly deforestation; one badly deforested subbasin contributed only about 30% of the flow but an estimated 80% of the sediment input to the reservoir. As a result, the dam was reduced to a run-of-river power plant, with no effective storage capacity, after only 12 years.

Rapp (1977) and Robinson (1981) suggest that this experience is by no means restricted to Latin America and Linsley and Franzini (1979) argue that it is not limited to developing countries. National aggregate off-site cost estimates for soil erosion establish the magnitude of the problem in the U.S. Clarke et al. (1985) estimated total annual off-site costs for all erosion sources to range from $3 to $13 billion with a point estimate of $6.1 billion of which $2.2 billion was attributed to cropland erosion (in 1980 dollars). Damage to recreational uses accounted for the largest share of costs — comprising nearly 33% of total costs, and boating was the largest recreation subgroup. Other high-impact receptors or users and their percent of total costs included municipal and industrial (14.8), water storage facilities (11.3), dredging (8.5), and preservation values (8.2). Cropland erosion was the largest source at 38% of total erosion. A reanalysis of these results by Ribaudo (1986) suggested a $7.1 billion point estimate.

Later analysis of the Conservation Reserve Program (CRP) by Ribaudo et al. (1989) found that annual water quality benefits from the first 23 million acres enrolled in the CRP totaled $2.05 billion or an average of $89/acre. This suggests that Ribaudo's 1986 point estimate should be adjusted downward by about $2 billion. However, both the Clark and Ribaudo analyses omit several categories of downstream impacts (e.g., dredge spoil disposal, delays to commercial shipping, biological impacts, etc.), which may make their estimates very conservative. A more recent and comprehensive attempt by Pimenel et al. (1995) to estimate the economic on-site and off-site costs of soil erosion in the U.S. places these costs at $44 billion annually. However, there are some serious methodological problems that will become more evident in the following conceptual sections of this chapter and evidence of overestimation by this team of ecologists attempting to do economic analysis.

Clearly, any threat to the longevity of existing and future water resource developments could have serious repercussions and warrants careful evaluation and planning. The primary and most immediate threat to hydroelectric and other water storage projects is sedimentation. A major cause is excessive erosion from watershed overuse or misuse. In many developing countries, the general pattern of heavy land use leading to high erosion rates results from a combination of ill-defined property rights, traditional shifting agriculture, and population growth. In reservoir watersheds, this pattern may be preexisting, or the reservoir project may, as with the Anchicaya reservoir described above, provide the access route for transient farmers into a previously relatively undisturbed area. Another contributor to watershed erosion and sedimentation is deforestation. This may initially be by large-scale legal (or illegal) logging, but a deforested state is often maintained by subsequent agricultural use including extensive livestock grazing in many areas. Other important contributors to sedimentation include stream bank cutting and what many have called "civil engineering erosion" from improper location, construction, and maintenance of roads, culverts, bridges, and channels.

Allen (1972) argue that it is usually much more feasible and less costly to keep sediment out of reservoirs, harbors, etc., utilizing a watershed unit of analysis than it is to cope with it after it has been deposited. The watershed as an hydrologic unit is a useful way of understanding and managing surface water movement and sedimentation from a physical science perspective. It also provides a sufficiently inclusive unit of analysis or accounting stance for more comprehensive (the focus of the following sections) economic analysis of sedimentation. Thus, the focus of the following conceptual sections, the Valdesia case, and the Ohio evidence is primarily on economic analysis of sedimentation management from a watershed perspective.

FINANCIAL, ECONOMIC, DISTRIBUTION, AND NIE ANALYSIS

What passes for "economic" analysis of various sediment management strategies varies widely and can be placed within an "accounting stance" continuum regarding both space and time. One end includes private individual- or firm-oriented, engineering-type financial analysis utilizing current market or administered prices of inputs or outputs. At the other end are societal and intergenerational efficiency concerns and income distribution analysis, including consideration of both weighted and unweighted income distribution impacts. In between lies a series of adjustments or shadow pricing methods to account for full social opportunity cost, willingness to pay, elasticity of supply and demand, unemployed factors, externalities, economic surplus, compensated income, and overvalued currency considerations.

Margolis (1969) suggests the reason private market prices may not reflect full social benefits or costs with the following:

> … there are many cases where exchange occurs without money passing hands; where exchanges occur but they are not freely entered into; where exchanges are so constrained by institutional rules that it would be dubious to infer that the terms were satisfactory; and where imperfections in the conditions of exchange would lead us to conclude that the price ratios do not reflect appropriate social judgments about values. Each of these cases gives rise to deficiencies in the use of existing price data as the basis for evaluation of inputs or outputs.

Costs generated from engineering data and future revenues based on current market prices can be misleading, particularly if one is concerned with societal costs and benefits. These "costs" generally do not represent full opportunity costs or highest use value of all factors of production such as the value of the farmer's time in implementing soil erosion and related sediment control during the busy spring planting season. Alternatively, the opportunity cost of labor may be less than the wage rate in those situations involving underemployed or unemployed labor. Costs based only on engineering data may also omit major technological externalities from soil erosion such as flood damage, water pollution, and ditch, harbor, and reservoir sedimentation.

"Revenues" may also be overstated, particularly in those cases where local currency is overvalued (which is the case in many developing countries) or where a relatively inelastic demand exists for the end product(s). The inelastic demand situation refers to those cases where an increase in supply results in a disproportionate decrease in price. Ward (1976a and b) discusses the use of shadow exchange rates to adjust for overvaluation of local currencies and the Bruno criterion to evaluate the foreign exchange saved or earned by alternative erosion control strategies. This is particularly relevant in evaluating erosion control strategies in developing countries. For example, these countries are frequently dependent on hydroelectric power as a major domestic energy source. Sedimentation of reservoirs reduces power output and may lead to increased importation of oil which requires foreign exchange.

Gittinger (1982) argues for distinguishing between financial and economic analysis, where financial analysis refers to net returns to private equity capital based on market or administered prices. Financial analysis indicates incentives facing market participants. Financial analysis also treats taxes as a cost and subsidies as a return; interest paid to outside suppliers of money or capital is a cost, whereas any imputed interest on equity capital is a part of the return to equity capital. By contrast, Gittinger sees economic analysis as concerned with net economic returns to the whole society, frequently based on shadow prices to adjust for market or administered price imperfections. In economic analysis, taxes and subsidies are treated as transfer payments; i.e., taxes are part of the total benefit of a project to society and subsidies are a societal cost. For purposes of this chapter, financial and economic analysis will refer to private and social concepts of economic efficiency analysis, respectively.

This financial vs. economic distinction is important, but the complementarity of these analytical approaches is equally relevant. Financial analysis provides information on the profitability of a given soil erosion reduction practice (e.g., cover crops or terraces) to individual entrepreneurs or investors and thus gives an indication of the incentive structure and potential adoption rate. Economic or social cost–benefit analysis attempts to determine profitability from a societal standpoint, taking into consideration externalities or environmental costs, pricing of under- or unemployed factors, currency evaluation, etc. The appropriateness of these analytical alternatives depends on the question one is asking. Generally speaking, it is relatively straightforward to assign values to the cost and revenue streams in financial analysis; market prices suffice. However, this is substantially more difficult in full social cost–benefit analysis.

Social costs and benefits or gains and losses from an economic perspective refer to the aggregation of individual producer and consumer measures of full willingness to accept or pay compensation. Individual preferences count in the determination of social benefits and costs and are weighted by income or more narrowly by market power. Since most policy changes involve economic gainers and losers, economists have developed the concept of potential Pareto improvement (PPI) to add up gains and losses to get net benefits. Simply stated, the concept holds that any policy change is a PPI or an increase in economic efficiency if at least one individual is better off after all losers are compensated to their original or before the policy change income positions. As Dasgupta and Pearce (1979) and others argue, the compensation need not actually occur but must be possible.

These measures of social costs and benefits are often not fully reflected in current market prices (or in government-regulated prices) as in the case of crop production in an area with high rates of soil erosion. This divergence results from several factors. First, government subsidies of inputs and/or outputs can lead to levels of input use and outputs in agriculture which are not economically efficient or environmentally sustainable, particularly in the case of agricultural sediments. Second, because there are consumers willing to pay more and producers willing to sell for less than prevailing market or regulated prices, they receive what economists call consumer and producer surpluses. Third, technological externalities in agricultural production exist to the extent that, external to the production and consumption of the resulting output, individuals, households, or firms experience uncompensated real economic losses (or gains) from soil erosion, agricultural chemicals, or other residuals. Finally, there may be willingness to pay (WTP) to keep future economic options such as hydroelectric generation open (see Veloz et al., 1984) or WTP for existence value of plant or animal species threatened by water pollution which are not reflected in the market or government-regulated prices of agricultural inputs and/or outputs.

Figure 17.1 illustrates the concepts of economic surplus, technological externality, and "deadweight" loss. For example, at market price P consumer surplus is equal to area PEC and producer surplus is equal to area PEA. One might think of P as the market price at equilibrium E where marginal private cost MPC of the farmer is equal to demand or marginal willingness to pay. Farm output at Q also includes the joint production of soil erosion and surplus chemicals which impose social costs on downstream watershed residents represented by P* and E*. One might also think of Q as producing residuals (sediment and chemicals) that exceed environmental assimilative capacity, i.e., an inefficient level of output. These external social costs are fully internalized at P′E′Q′ which results in consumer surplus equal to area P′E′C and producer surplus equal to area P′E′B. The area E′E*E represents the dead weight loss at output Q from the presence of the soil and chemical externalities.

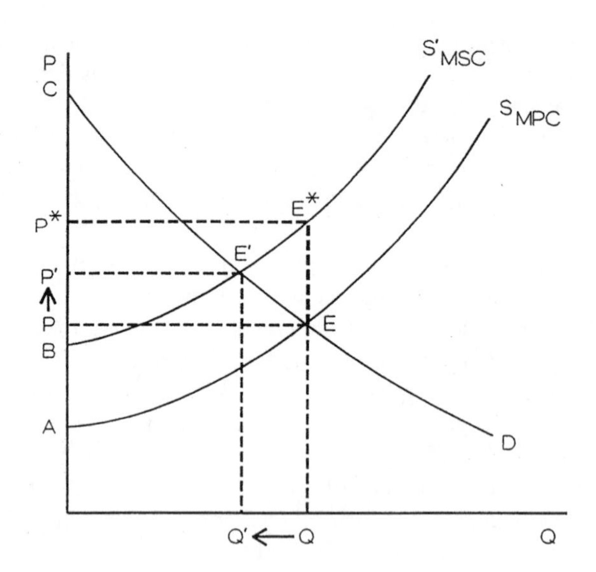

Technological Externality Defined (Dasgupta & Pearce, 1978)

1. Necessary Condition
   Physical interdependence of production and/or utility functions
2. Sufficient Condition
   Not fully priced or compensated

S = marginal private (e.g., upstream farmer) cost function
S' = marginal social (e.g., watershed) cost function
D = demand or marginal benefit function
Q = output quantity
P = price/unit of output

FIGURE 17.1 Soil erosion as an externality.

Once all quantifiable cost and benefit streams have been given prices or shadow values, one must decide on an appropriate rate of discount or time value and a criterion for evaluating and ranking the economic efficiency of alternative soil-conserving strategies or programs. A long-standing controversy on the appropriate discount rate centers primarily on those who support various private opportunity cost vs. social time preference measures. Baumol (1969) has written a classic article dealing with the discount rate controversy.

The alternative efficiency criteria include (1) the ratio of benefits to costs, (2) the net present value, (3) the internal rate of return, (4) the payout period, and several other lesser-known criteria related to optimal time phasing of projects, and the optimal utilization of scarce foreign exchange. Several authors, including Dasgupta and Pearce, Gittinger, and Ward have explored the decision criteria issue in depth. These criteria and the resulting ranking of alternative sediment management strategies are heavily influenced by the nature of future benefit and cost streams, the ratio of future operating costs to initial capital outlay, and the nature of the capital or budget constraint.

Analysts should also be concerned with the equity or income distribution impacts of alternative sediment management strategies. For example, Korsching and Nowak (1982) argue that low-income farmers may be disproportionately impacted by the costs of many soil erosion or sediment control strategies. This can result from their farming of a higher proportion of erodible soils, possessing fewer savings to cover initial investment demands, and having farms too small to capture any major scale economies in erosion control practices.

Economists use several alternative methods for handling income distribution impacts including (1) explicit weighting of net benefits by income class, group, or region (see Blue and Tweeten, 1996, for an example of weighting of costs and benefits among income groups by the marginal utility of income); (2) provision of alternative weighting functions and their distributional consequences to decision makers; (3) estimation of nonweighted net benefits by income class, group, or region; and (4) a constrained maximum or minimum targets approach which maximizes economic efficiency subject to an income constraint, or vice versa (see Ahmed and Hitzhusen, 1988).

When the foregoing concepts of financial, economic, and distribution analysis are combined, it frequently results in more useful analysis for public policy. Actual implementation of public policy is more in the domain of the new institutional economics (NIE). This field (see Satish et al., 1992; Ostrom et al., 1993); focuses on transaction costs (e.g., information, regulation, monitoring, etc.) and alternative mechanisms for collective choice including nongovernmental organizations (NGOs), which have become increasingly important in watershed management. The following section develops the key concepts from environmental economics which are complementary to financial, economic, distribution, and NIE analysis and relevant to improved watershed management.

TYPES AND MEASURES OF ENVIRONMENTAL ECONOMIC VALUE

It is possible to develop specific values, measures, instruments, and options for economic assessment of environmental service flows related to sedimentation and watershed management. The service flows include raw material supply, assimilative capacity, amenities/aesthetics, human habitat, and plant and animal biodiversity. Figure 17.2 summarizes this process. *Direct current use value* refers to use of environmental service flows such as hydropower and water. *External values* are those uncompensated costs or benefits (externalities) from upstream production or consumption processes that are borne or received now or in the future but not reflected in current prices to producers or consumers. *Option value* refers to the willingness to pay to delay the use of something until some future time while *bequest value* refers to a willingness to preserve something for the use of future generations. This is related to the notion of forgone benefits to future users from current exhaustion of a finite resource without any close substitutes. *Existence value* is the willingness

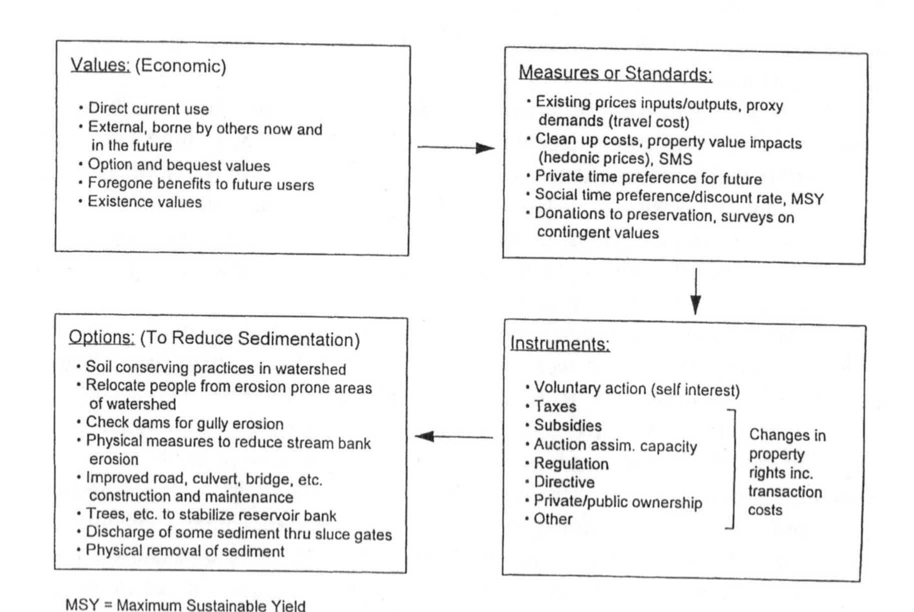

FIGURE 17.2 Monetizing environmental service flows and implementing change/reform.

to pay for preservation of plant and/or animal species without regard for their use by humans (see Hoehn and Walker, 1993; Dixon et al., 1994).

Economists use a variety of measures or methods to infer or discover the foregoing values. Sometimes it is possible to observe values directly in existing prices, and in other cases it is necessary to infer values from prices of closely related complementary goods. In the first case, reduction in commercial fish catch from agricultural pollution (externality) of a reservoir can be measured in lost fishing revenues. However, any reduction in sport fishing in the same lake would require assessment of any decrease in expenditures on goods and services related to boating and sport fishing activity, i.e., the development of a travel cost or proxy demand function method (see Macgregor et al., 1991).

The value of an externality can also be conservatively estimated in some cases by replacement, cleanup, or avoidance costs, such as reservoir dredging and water treatment related to soil sediments and agricultural chemicals. In addition, the impact of an externality on private property values can frequently be estimated by hedonic pricing which is a method for statistically decomposing the sources of value or demand in a property market to allow independent estimation of an environmental amenity or disamenity (see Hitzhusen et al., 1995). Contingent valuation refers to a survey method which estimates willingness to pay values directly from respondents for some change in an environmental service flow (see Randall et al., 1996). This method is the most comprehensive for simultaneously estimating all of the types of economic value outlined in Figure 17.2, but it requires careful development to avoid strategic behavior of respondents to mask their true preferences or WTP.

In cases of environmental service flows with a critical zone or threshold, it may be necessary to establish a safe minimum standard (SMS) or maximum sustainable yield (MSY). Barbier et al. (1990) provide several examples of this alternative. The objective is to avoid irreversible effects to human health or ecosystems such as in the case of nitrate–nitrogen contamination of groundwater in parts of the U.S. Contamination is considered critical to human health at 10 mg/l which was established by the U.S. Environmental Protection Agency. The T value in the universal soil loss equation (USLE) is an example of a somewhat more flexible or reversible SMS relative to long-run productivity of the soil. One could envision a similar safe standard of sediment inflows into a water reservoir to maintain minimum storage capacity. The MSY of a water aquifer may be a withdrawal rate equal to or less than the annual recharge rate, which becomes critical in cases where the aquifer is covered by a heavy rock overburden. In these cases, excess withdrawal can result in irreversible loss in aquifer capacity.

Once some basic economic estimates have been established for environmental service flows relative to sedimentation, it is possible to select instruments to accomplish more efficient and/or equitable outcomes in watershed management. Economists prefer instruments that provide incentives and allow a range of choices as opposed to command-and-control instruments. Examples include taxes, subsidies, and auctioning of assimilative capacity up to some resource constraint or SMS. Well-defined property rights are a recurring theme of economists, and this is equally true of any changes in property or use rights related to environmental service flows. As an illustration, some primary options for reducing reservoir sedimentation in a watershed context are listed in Figure 17.2. These options impose costs, but also generate benefits in terms of increased flood control, water supply, hydroelectric power, recreation, etc. These benefits of reservoir sedimentation reduction are matched with the appropriate economic measures in Appendix 17.A and several are illustrated in the following case study from the Dominican Republic and in the Ohio research results section.

CASE STUDY OF VALDESIA WATERSHED/RESERVOIR

To illustrate some of the concepts introduced in the preceding sections, this section summarizes an earlier evaluation of a soil conservation project for a hydroelectric watershed, dam, and reservoir located in the Dominican Republic (DR). After some background information about the study area has been presented, the method used to estimate reservoir sedimentation under different assumptions regarding land resource management upstream as well as the approach for estimating the benefits and costs of soil conservation are outlined. A more complete description of this research can be found in Veloz et al., (1984) and Veloz (1982.)

VALDESIA WATERSHED AND RESERVOIR

Like many other developing countries, the DR has looked to hydroelectric development to reduce fuel oil imports. Many promising dam sites are found in the Cordillera Central, where most of the country's major rivers originate. Elevations vary in that region from 100 to 3000 m above sea level, and precipitation in many areas exceeds 1500 mm/year.

As late as the 1950s, most of the Cordillera Central was forested. But after the death in 1961 of the country's longtime dictator, Rafael Trujillo, who owned most of the country's standing timber, many forested areas were clear-cut. Peasants then settled on the newly cleared land. The CRIES (1980) study shows that by the late 1970s, most of the region had been converted to rangeland and cropland. High rates of soil loss are now a problem throughout the Cordillera Central, according to Hartshorn (1981).

The Valdesia Dam was one of several projects initiated in order to decrease the dependence of the DR on imported oil. By developing Valdesia and one other site during the 1970s, de La Fuente (1976) states that the DR reduced the share of electricity generated at oil-burning facilities from over 90% to less than 80% by 1981. However, the Direccion General de Foresta (DGF, 1976) suggested that sedimentation was a continuous concern at the reservoir. Even before a major hurricane swept over the island in 1979, annual sediment yield in the reservoir watershed was estimated to be 1.4 million metric tons. Further compounding the problem was the existence of separate government bureaucracies for agricultural production for export, soil conservation, protection of forests, and hydroelectric generation, which appeared to have limited incentives to cooperate.

Study Methodology

Although the Dominican government had determined that the Valdesia reservoir watershed would have a watershed management program in the near future, a specific plan for reducing soil erosion there had not yet been developed. Without a more-detailed plan, it was possible to characterize in a general way the land uses and land management techniques to be encouraged under the project. Consultation of the literature and with professionals who have worked in the DR suggested the following guidelines for four slope classes:

Slope Class A (3 to 20% slope). Continue crop farming on existing cropland. Encourage mulching and contour farming. Renovate all rangeland with a slope of 13 to 20%.

Slope Class B (21 to 35% slope). Encourage mixed cropping/agroforestry enterprises on existing cropland. Renovate all rangeland.

Slope Class C (36 to 50% slope). Convert all existing cropland to agroforestry. Renovate all rangeland.

Slope Class D (over 50% slope). Reforest all existing cropland and rangeland.

To estimate the impacts of project implementation on sedimentation and watershed management, a 1:50,000 scale topographic map of the 85,090 ha watershed was consulted. The watershed was divided into 10-ha cells, each of which was placed into one of the four slope classes. By contrasting existing land use based on the CRIES results with recommended land use, it was found that 57% of the watershed would not be affected by project implementation while 11% would have to undergo a change in land use. In the remaining 32%, soil conservation goals would be met by mulching, contour farming, and range renovation.

Reductions in soil loss resulting from compliance with project guidelines were estimated by applying the USLE (see Wischmeier, 1976; Veloz, 1982) using soil survey data and the estimates of slope that were used to categorize land into the four slope classes. In general, this approach yields conservative estimates of erosion inasmuch as the USLE explains only gross sheet and rill erosion and not gullying, stream bank scour, and other types of soil loss.

Average sedimentation rates were calculated by multiplying an estimated sediment delivery ratio from Onstad et al. (1977) by estimated erosion rates. Given current land use and land management practices and evidence from Espinal (1981), it was estimated that average yearly accumulation of sediment in the reservoir is 921×10^3 metric tons, which is 59% of the rate obtained by dividing total sediment in the reservoir by the reservoir age. However, the latter rate includes the silt deposit by Hurricane David in 1979 as well as the bed load of sediment. Thus, the observed rate probably overstates the long-term sedimentation rate. Implementation of the project throughout the watershed would reduce sedimentation by an estimated 86%, to 130×10^3 metric tons a year.

Results of Analysis

Sedimentation was assumed to have a negligible impact on electricity production as long as it remains possible to operate the hydro plant. However, without any change in land resource management, it was estimated that the generator intake would be clogged with silt after 19 years (it was actually clogged in less than 15 years). A 25% reduction in erosion, however, would extend the lifetime of the dam by at least 6 years since a 25-year time horizon was used (see Veloz, 1982).

The group affected most by project implementation would be hillside farmers. Researchers familiar with the study area supplied information about farm income without the project. Budgets for soil-conserving activities were obtained from the U.S. Agency for International Development (U.S. AID), the Centro Agronomics Tropical de Investigacion y Ensenanza (CATIE), and other sources.

Economic evaluation of the watershed management project outlined above strongly suggests that the net benefits associated with soil conservation depend on the accounting stance. The results of private-level analysis reported in Table 17.1

TABLE 17.1
Private-Level Analysis of Valdesia Watershed Management Project

Slope Class	Hectarage	PNPVi	PNPVi/hectare
A	7,640	DR$12,500,000	DR$1,635
B	17,800	−550,000	−30
C	8,600	−1,925,000	−225
D	2,350	−1,100,000	−470
Total	36,390	DR$ 8,925,000	DR$ 245

indicate that some farmers would benefit from project implementation, whereas others would lose, assuming that all use a real discount rate of 5% and a 25-year time horizon. Those with land in Slope Class A would, on average, benefit; the net returns per hectare they would realize by adopting improved tillage practices would exceed status quo net returns per hectare. However, full project implementation would force farmers who work on steeper land (Slope Classes B, C, and D) to make a change in land use which is, on average, not profitable for them.

Social-level analysis of the on-site economic effects of soil conservation (Table 17.2) yields results that are similar to the results of private analysis of those

TABLE 17.2
Results of On-Site Economic Analysis

Efficiency	Discount Rates			
Criterion	8%	10%	12%	20%
SNPV[a]	1,850,000	1,460,000	1,150,000	420,000
SB/SC	1.49	1.43	1.38	1.19
Bruno	3.2	3.7	4.4	13.74
SIRR	31.4	0		

[a] Shadow exchange rate = 1.50; shadow wage rate = 0.5.

same effects. In the former analysis, labor is shadow-priced at less than the minimum wage because of high under- and unemployment in the DR. Also, the project's exported outputs and imported inputs are evaluated using the parallel market exchange rate rather than the official exchange rate, which overvalues the local currency. Even after these adjustments have been made, social-level analysis yields the same two conclusions as private-level analysis, assuming the same discount rate, time horizon, and accounting stance. First, as an entire group, farmers would benefit from project implementation. Second, farmers with land in the higher slope classes, who would be required to switch from traditional agriculture to agroforestry or reforestation, would lose because of project implementation.

Broadening the accounting stance to incorporate the off-site effects of erosion yields substantially increased estimates of the net benefits of soil conservation. Table 17.3 shows the net present values of extensions in the lifetime of the Valdesia Dam obtained by conforming to project guidelines in selected slope classes. For example, given project implementation in Slope Class A, the additional external benefits of reducing erosion in Slope Class B (see Table 17.3) would greatly exceed the private costs of doing so (see Table 17.1).

Proper conceptualization and analysis of sedimentation from an economic perspective is extremely important if optimal watershed-level corrective measures are to be implemented. A core concept is the private (financial) vs. social (economic) perspective or accounting stance with respect to both space and time. A social or economic accounting stance is primarily concerned with both on-site and off-site costs and returns or returns to the total society over the long run. Several forms of

TABLE 17.3
Social-Level Analysis of Off-Farm Impacts of Valdesia Watershed Management Project

Slope Classes Where Project Guidelines are Followed	Dam Lifetime, Years	Off-site Benefits of Compliance with Guidelines
None	19	0
A	20	DR$1,755,000
A + B	25+	9,350,000+
D	25	9,350,000

corrective or shadow pricing may be necessary in doing the social cost–benefit estimates; this complicates the analysis and results in greater data needs.

Both the private and social perspectives or accounting stances are important as illustrated in the DR case application. The private-level analysis demonstrates the on-site profitability of alternative soil-conserving or sediment-reducing practices to hillside farmers and shows those on steeper slopes as net losers. The social-level analysis includes both on-site and hydropower-related off-site impacts and includes corrective or shadow prices for both labor and overvalued local currency. This analysis shows social benefits of soil-conserving practices to be substantially higher than the private costs in the Valdesia watershed and reservoir in the DR; i.e., full compensation of farmer losses would still result in relatively large net benefits. However, the authors neglected to apply the insights of the NIE in terms of compensating those bureaucracies that perceived higher transaction costs and/or loss of funds from the proposed project.

The tools to determine soil loss rates and sediment transport in less-developed countries are, by and large, rudimentary. Thus, it is hardly surprising that available estimates of the external costs of erosion (e.g., the costs of lost hydroelectricity, of reductions in irrigation water, of flood control, etc.) are imperfect. Similarly, in order to understand better how the hillside farmer makes the decision of whether or not to utilize land in a soil-conserving fashion, better information about the farmer's institutional environment and about production functions for both traditional and less erosive agricultural activities is needed. Finally, better demographic information about the populations that inhabit environmentally sensitive areas would improve our understanding of how watershed management problems evolve over time.

With more research on the on-site and off-site impacts of reductions in soil loss, it will be possible to perform economic analysis that is more comprehensive than the analysis described in this report. However, the case application estimates of the external benefits of soil conservation are conservative since only one type of benefit — the extension in dam lifetime — was quantified and the per hectare returns association with traditional agriculture have probably been exaggerated. Thus, the results of the Valdesia case study probably understate the net benefits. Even so, there is a need for more research that allows for the design and implementation (including the NIE insights) of better strategies for dealing with pressing soil erosion and sedimentation problems in developing countries throughout the world.

SOME OHIO EVIDENCE

Resource economists at the Ohio State University have evaluated the downstream economic impacts of soil erosion in Ohio on several receptors including natural lakes and man-made water reservoirs. For example, Macgregor (1988) used the Ohio Department of Natural Resources State Park Lakes data on lake characteristics, visitations, and dredging to estimate the boater value losses and dredging costs due to sedimentation in 46 state park lakes. Macgregor et al. (1991) report that sediment deposition was significantly and negatively related to boater visitor days, and this coefficient was multiplied by a mean value ($23.92) for a motorized boating recreational day to get boater value loss. These findings indicate an average boater value loss in the 46 lakes of $0.49/ton of sediment, but the values ranged from $0.008 to $11.95/ton of sediment. This emphasizes the need for targeting of soil conservation funds based on off-site economic impacts. The average cost was $1.29 to dredge 1 ton of sediment in 11 state park lakes where dredging was being done in 1987 and costs ranged from $1.20 to $1.46/ton. This dredging is funded by boater license fees and a portion of the gasoline tax. Ironically, farmers are exempt from this tax on fuels used in the farm operation.

Forster et al. (1987) estimated the relationship between water treatment costs and soil erosion in 12 public water treatment plants in western Ohio, depending on surface water storage reservoirs. Independent variables other than soil erosion used in the regression analysis included treatment plant size, storage time of untreated water, and turbidity improvement due to water treatment. Results indicate that a 10% reduction in annual gross soil erosion results in a 4% reduction in annual water treatment costs. The average increase in water treatment costs per ton of sediment delivered was $0.32 at the 12 treatment plants. For all Ohio communities, it is estimated that annual water treatment costs would decline by $2.7 million with a 25% reduction in soil erosion.

Hitzhusen et al. (1995) have done an economic evaluation of state park lake dredging and sedimentation impacts on lake recreators (particularly boaters) and lakeside property values. A majority of the lakes studied are man-made water reservoirs. The earlier cross-sectional travel cost results from Macgregar were updated and combined with a two-stage hedonic pricing model used to estimate the impacts of rates of sediment inflow and accumulation as well as dredging on lakeside property values. The sedimentation economic impacts on property values were statistically significant and generally larger than the impacts on boaters. Some simulation analyses also suggested that lake residents and recreational users were implicitly willing to pay more for upstream erosion control than for dredging.

Appendix 17.B attempts to summarize the foregoing Ohio research results on the off-site or downstream economic impacts of soil erosion. The results are placed in the context of a hypothetical 500-acre row crop farm which is assumed to be upstream from each of the downstream receptors, e.g., state park lakes, harbors, drainage ditches, and water treatment plants. It is also assumed that the 500-acre farm has average gross erosion of 3 T under conventional tillage, which is probably an upper-bound estimate and an average sediment delivery ratio of 10%, which is probably a lower-bound estimate. The T value or sustainable gross erosion value for average Ohio conditions is assumed to be 5 ton/acre/year. The values in each cell of the table are low, mean, or high economic values per ton of sediment delivered and low, mean, and high values per year for the hypothetical farm. Hitzhusen (1992) provides a more-detailed discussion of these results, but it is clear that the magnitudes of these external impacts are not trivial and they vary considerably between sites.

CONCLUSIONS

Several important conclusions flow from the foregoing conceptual and empirical analyses on economic assessment of sedimentation for watershed management. First, it is critical to see sedimentation in the context of watershed management if long-term, cost-effective solutions are to be realized. Second, it is necessary to distinguish carefully among financial, economic (or social benefit–cost), and income distribution analyses. Each makes different assumptions about the accounting procedure for calculating the level and distribution of economic welfare. All three perspectives may be appropriate in the analysis of a given watershed sedimentation project or issue depending on the questions one is asking.

A third set of conclusions relates to the rapidly expanding conceptual and empirical work on measuring environmental service flows. The values of these service flows are usually difficult to observe directly in market prices. Thus, a rather substantial theory, method, and empirical literature has evolved utilizing replacement, avoidance, and cleanup costs, expenditures on related private goods (e.g., travel cost and hedonic pricing), and direct survey or contingent valuation methods. The empirical literature suggests that these non and extra or related market values are frequently larger than those values observed directly in markets. The results from the Valdesia reservoir in the DR and the Ohio as well as other research results on downstream economic impacts of sediments support this contention.

All of the above suggest the need for a substantial increase in general resource economic literacy among noneconomists. Considerably more financial, economic, and income distribution assessment of sedimentation, including efforts at combining efficiency and equity, is also necessary if efficient and fair recommendations are to be made for upstream watershed management. Finally, some of the insights from the NIE on transaction costs and alternative mechanisms for collective choice may be helpful in actually implementing policies for improved watershed management. For example, the proposed Valdesia watershed management project would likely have had a higher probability of being implemented if the perceived higher transaction costs and/or loss of funding of the impacted government agencies had been estimated and compensated.

ACKNOWLEDGMENT

Helpful comments were received from Prof. Luther Tweeten, OSU.

REFERENCES

Ahmed, H. and F. Hitzhusen. 1988. Income Distribution and Project Evaluation in LDCs: An Egyptian Case, paper presented at International Association of Agricultural Economics meeting, Argentina, August.

Allen, R.N. 1972. The Anchicaya Hydroelectric Project in Colombia: design and sedimentation problems, in *The Careless Technology,* Natural History Press, Garden City, NY, 1030 pp.

Barbier, E. B., A. Markandya, and D.W. Pearce. 1990. Sustainable agricultural development land project appraisal, *Eur. Rev. Agric. Econ.* 17, 181–196.

Baumol, W.J. 1969. On the discount rate for public projects, in *The Analysis and Evaluation of Public Expenditures: The PPB System*, Vol. 1, U.S. Congress, U.S. Government Printing Office, Washington, D.C., 489–504.

Blue, N. and L. Tweeten. 1996. The Estimation of Marginal Utility of Income for Application to Agricultural Policy Analysis, Department of Agricultural Economics, The Ohio State University, Columbus.

Clark II, H.E., J.A. Havercamp, and W. Chapman. 1985. *Eroding Soils: The Off-Farm Impacts,* The Conservation Foundation, Washington, D.C.

Council on Environmental Quality. 1980. *The Global 2000 Report to the President*, Vol. 2, *The Technical Report,* U.S. Government Printing Office, Washington, D.C.

CRIES. 1980. *Land Cover/Use Inventory for the Dominican Republic through Visual Interpretation of Land Sat Imagery.* CRIES/USDA/AID/Michigan State University, Santo Domingo, Dominican Republic.

Das, D.C. 1977. Soil conservation practices and erosion control in India — a case study, in *Soil Conservation and Management in Developing Countries,* FAO Soils Bulletin 33, Food and Agriculture Organization, 11–50.

Dasgupta, A.K. and D.W. Pearce. 1979. *Cost-Benefit Analysis: Theory and Practice,* Macmillan, London.

Davis, C.V. 1969. *Handbook of Applied Hydraulics*, 3rd ed., McGraw-Hill, New York.

de La Fuente, Santiago, S.J. 1976. Geografia Dominicana, Thesis, Editora Colegia Quisqueyana, S.A., Santo Domingo, Republica Dominicana, 272 pp.

DGF (Direccion General de Foresta). 1976. Estudio de Conservacion de Suelos de la Cuenca Hidrografica del Rio Nizao: Parte I, Subprograma Forestal Pidagro. Santo Domingo, Republica Dominicana.

Dixon, J.A. et al. 1994. *Economic Analysis of Environmental Impacts*, Earthscan Publications, London.

Dregne, H.E. 1972. Impact of Land Degradation on Future World Food Production, ERS-677, U.S. Department of Agriculture, Washington, D.C., 33 pp.

Espinal, M.H. 1981. *Informe Final Sobre los Levantamientos Batimetricos del Embalse de Valdesia*, CDE, Santo Domingo, Dominican Republic.

Forster, D.L., C.P. Bardoes and D. Southgate. 1987. Soil erosion and water treatment costs, *J. Soil Water Conserv.*, 42:5, 367–369.

Gittinger, J.P. 1982. *Economic Analysis of Agricultural Projects*, 2nd ed., Johns Hopkins University Press, Baltimore, MD, 505 pp.

Glymph, L.M. 1973. Summary: sedimentation of reservoirs, in *Man-Made Lakes: Their Problems and Environmental Effects*, Geophysical Monograph Series, Vol. 17, American Geophysical Union, Washington, D.C., 847 pp.

Hartshorn, G., Ed. 1981. *The Pominican Republic, Country Environmental Profile, a Field Study.* U.S. Agency for International Development, Washington D.C., 109 pp.

Hitzhusen, F.J. 1992. The economics of sustainable agriculture: adding a downstream perspective, *J. Sustainable Agric.*, 2(2), 75–89.

Hitzhusen, F.J. et al. 1995. Economics and political analysis of dredging Ohio's state park lakes, in *Water Quantity/Quality Mnagement and Conflict Resolution: Institutions, Processes and Economic Analyses*, A. Divar and E.T. Lochman, Eds., Praeger, Westport, CT, chap. 36, 485–499.

Hoehn, J. and D. Walker. 1993. When Prices Miss the Mark: Methods of Evaluating Environmental Change, EPAT/MUCIA Policy Brief, No. 3, August.

Hufschmidt, M. et al. 1983. *Environment, Natural Systems and Development: An Economic Valuation Guide*, Johns Hopkins University Press, Baltimore, MD.

Korsching, P.F. and P. Nowak. 1982. Environmental criteria and farm structure: flexibility in conservation policy, in *Farms in Transition: Interdisciplinary Perspectives on Farm Structure*, Iowa State University Press, Ames, 169 pp.

Lee, J.G., D. Southgate, and J. Sanders. 1997. Methods of economic assessment of on-site and off-site costs of soil degradation, in *Methods for Assessment of Soil Degradation,* R. Lal et al., Eds., CRC Press, Boca Raton, FL, 475–494.

Leftwich, R.H. 1966. *The Price System and Resource Allocation*, 3rd ed., Holt, Rinehart, Winston, New York, 369 pp.

Lehman, T.S., F. Hitzhusen, and M. Batte. 1995. The political economy of dredging to offset sediment impacts on water quality in Ohio's state park lakes, *J. Soil Water Conserv.,* 50:6,659–662.

Linsley, R.K. and J.B. Franzini. 1979. *Water-Resource Engineering*, 3rd ed., McGraw-Hill, New York, 160–165.

Macgregor, R. 1988. The Value of Lost Boater Value Use and the Cost of Dredging: Evaluation of Two Aspects of Sedimentation in Ohio's State Park Lakes, Ph.D. dissertation, The Ohio State University, Columbus.

Macgregor, R., J. Maxwell, and F. Hitzhusen. 1991. Targeting erosion control with off-site damage estimates: the case of recreational boating, *J. Soil Water Conserv.,* July-August, 301–304.

Margolis, J. 1969. Shadow prices for incorrect or nonexistent market prices, in *The Analysis and Evaluation of Public Expenditures: The PPB System*, Vol. 1, U.S. Congress, Joint Economic Committee, U.S. Government Printing Office, Washington, D.C., 533–546.

Onstad, C.A., C.K. Mutchler, and A.J. Bowie. 1977. Predicting sediment yields, in *Soil Erosion and Sedimentation: Proceedings of National Symposium*, American Society of Agricultural Engineers, St. Joseph, MI, 43–58.

Ostrom, E., L. Schroeder, and S. Wynne. 1993. *Institutional Incentives and Sustainable Development*, Westview Press, Boulder, CO.

Pimentel, D. et al. 1995. Environmental and economic costs of soil erosion and conservation benefits, *Science*, 267, 1117–1123.

Randall, A., D. de Zoyza, and F. Hitzhusen. 1996. Groundwater, surface water, and wetlands valuation for benefits transfer, chapter in Final Report of project An Economic Analysis of Sustainaable Agriculture Adoption in the Midwest; Implications for Farm Firms and the Environment, of the USDA/EPA Research and Education Grants Program.

Rapp, A. 1977. Soil Erosion and reservoir sedimentation — case studies in Tanzania, in *Soil Conservation and Management in Developing Countries*, FAO Soils Bulletin 33, Food and Agriculture Organization, Rome, Italy, 123–132.

Ribaudo, M.O. 1986. Reducing Soil Erosion: Off-site Benefits, USDA, Economic Research Service, Agricultural Economic Report No. 561, September.

Ribaudo, M.O. et al. 1989. CRP: what economic benefits, *J. Soil Water Conserv.,* Sept.-Oct., 421–424.

Robinson, A.R. 1981. Erosion and sediment control in China's Yellow River Basin, *J. Soil Water Conserv.*, 36(3): 125–127.

Satish, S., F. Hitzhusen and K.V. Bhat. 1992. Watershed management in India: in search of alternative institutional intervention, paper presented at International Symposium on Soil and Water Conservation: Social, Economic and Institutional Considerations, Honolulu, Hawaii, October 19–22.

U.S. Agency for International Development (USAID). 1981). Dominican Republic Project Paper: Natural Resource Management, AID/LC/P-079, Washington, D.C.

Veloz, J.A. 1982. The Economics of Watershed Protection and Erosion Control: A Case Study of the Dominican Republic, M.S. thesis, The Ohio State University, Columbus, 139 pp.

Veloz, J.A., D. Southgate, F. Hitzhusen, and R. Macgregor. 1984. The economics of erosion control in a subtropical watershed: a dominican case, *Land Econ.* 61(2): 145–155.

Ward, W.A. 1976a. The Bruno Criterion in Two Numeraires, Course Notes, CN-22, the International Bank for Reconstruction and Development, Washington, D.C., July, 9 pp.

Ward, W.A. 1976b. Adjusting for Over-Valued Local Currency: Shadow Exchange Rates and Conversion Factors, Course Notes, CN-28, the International Bank for Reconstruction and Development, Washington, D.C., December, 12 pp.

Wischmeier, W.H. 1976. Use and misuse of the universal soil loss equation, *J. Soil Water Conserv.* 31(1): 5–9.

APPENDIX 17.A

Benefits and Economic Measures of Reduced Reservoir Sedimentation

Benefits of Reduced Reservoir Sedimentation	Measures of Economic Benefits
1. Increased flood control	Reduced downstream economic loss to property, life (?)
2. Increased water supply	
• Agriculture	MVP[a] of water in agriculture
• Industry/business	MVP[a] of water in industry
• Households	Basic needs + WTP[b] for water
3. Increased hydroelectric power	Forgone higher costs of alternative electric sources, e.g., oil
4. Increased recreation: fishing (?), swimming, boating	Travel cost and contingent valuation measures of WTP[b]
5. Increased adjacent reservoir property values	Hedonic pricing of property market

[a] MVP = marginal value product of water as under- or unpriced intermediate good in production process.
[b] WTP = willingness to pay of householders and recreators.

APPENDIX 17.B

Farmland Soil Erosion Off-Site Economic Impacts of Sediment: Some Ohio Research Results

Type and Number of Receptor/ Environmental "Sinks" Analyzed	$/ton Sediment			$ Off-Site Impacts/year[a]		
	I_L	I_M	I_H	I_L	I_M	I_H
1. State Park Lakes (n = 46)						
a. Lost boater value (n = 46)	0.007	0.44	11.56	3	218	5779
b. Dredging costs (n = 11)	1.20	1.29	1.46	600	645	730
c. Property values (n = 4/225)	3.04	10.51	16.95	1520	5225	8475
2. Lake Erie Harbors (n = 9)						
a. Dredging costs	2.26	2.90	5.14	1130	1450	2570
3. Ohio River						
a. Dredging costs	1.78	2.52	3.56	890	1260	1780
4. Drainage Ditches (n = 6 counties)						
a. Dredging costs		1.87	—	—	2805[b]	—
5. Water Treatment Costs (n = 12)	0.32	—	—	160	—	

[a] Assumes 500 acres of row crops with average gross erosion of 3 T under conventional tillage. The assumed value for T is 5 ton/acre/year which by definition is sustainable. Sediment delivery ratio of 10% (average).
[b] Sediment delivery ratio = 30%.

18 Land Degradation and Food Security: Economic Impacts of Watershed Degradation

Pierre Crosson and Jock R. Anderson

CONTENTS

INTRODUCTION

At the World Food Summit in Rome in the fall of 1996 it was acknowledged that food security is a problem, primarily in the less-developed countries (LDCs), because of poverty, not scarcity of me resources needed for food production. People are underfed because they do not have enough money to buy the food they need, not because farmers could not produce more. In a study for the Food and Agriculture Organization (FAO), Alexandratos (1995) estimated that the number of undernourished people in the world had declined from 941 million in 1969–71 to 781 million in 1988–90. The percentage of the LDC population undernourished fell from 38 in

0-8493-0702-3/00/$0.00+$.50

1969–71 to 20 in 1988–90. The declines were most marked in Asia, where per capita income in that period increased rapidly. Improvement was more modest in Latin America, where per capita income grew less rapidly than in Asia; and in Africa, where per capita income declined after about 1980, the number of malnourished people increased.

The implication of the World Food Summit argument and the evidence supporting it is that increasing food security around the world will require increasing per capita income, particularly among the poor in the LDCs. However, because so many of the poor in those countries are either directly dependent on agriculture for their income or indirectly dependent on agricultural performance because so much of their income is spent on food, increasing agricultural productivity is a necessary condition for attacking the problem of food insecurity. (It is not a sufficient condition. The poor must receive a major part of the increased income flowing from higher productivity if they are to achieve food security.) Solving the problem of increasing agricultural productivity, in its turn, requires attention to the supplies of resources available to farmers. The supplies of land are particularly critical. This chapter is focused on two dimensions of land supply: (1) the qualitative dimension as it is affected by degradation, that is, the losses of land productivity because of the effects of wind and water erosion and other forms of degradation; (2) land management as it affects soil erosion and subsequent downstream effects on surface water quality.

A final introductory note. The title of this chapter says that it is about watershed degradation. With the exception of the Atacama Desert of Chile, and possibly a few comparable areas around the world, all land in the world receives some amount of precipitation. All such land, therefore, is part of a watershed. Much of the following discussion of the two dimensions of land supply does not explicitly refer to watersheds, but the reference is implicit throughout.

GLOBAL LAND DEGRADATION: WHAT DO WE KNOW?

Despite some claims to the contrary (e.g., Brown and Wolf, 1984) the fact is that prior to the early 1990s almost nothing reliable was known about annual rates of land degradation, let alone its productivity consequences. Rattan Lal, a world-class soil scientist at Ohio State University, emphasized this. Lal was not alone. El-Swaify et al. (1982), authors of what at the time was the most comprehensive published study of soil erosion in the LDCs, concluded that "there is little or no documentation of the extent, impact or causes of erosion" in tropical environments. Nelson (1988), in a report to the World Bank, did a comprehensive review of the evidence of soil erosion and other forms of land degradation around the world. He found the evidence on the extent and severity of the problem to be "extraordinarily scanty." Harold Dregne of Texas Tech University, in another 1988 publication, asserted that estimates of the extent and severity of land degradation were based on "little data and much informed opinion." Writing specifically of soil erosion and its productivity effects, Dregne stated that "there is an abysmal lack of knowledge of where water and wind erosion have adversely affected crop yields." In another report to the World Bank,

Biot and others (1995) emphasized the lack of reliable information about "the physical processes of land degradation, as well as its economic and social causes and consequences" and noted that hard evidence for the assertion that resource degradation is causing declines in crop yields "is often deficient."

The Brown and Pimentel Estimates

The last-quoted statement applies with special force to the estimates of the extent of global soil erosion found in Brown and Wolf (1984) and in Pimentel et al. (1995). After careful analysis of each of these estimates, Crosson (1995a) for Brown and Wolf and (1995b) for Pimentel et al. concluded that in both cases the analytical and empirical underpinnings of the estimates failed to meet even minimal standards of scientific credibility.

The Oldeman, Hakkeling, and Sombroeck Estimates

It was not until 1990 that a set of global estimates of soil degradation meeting those standards was published (Oldeman et al., 1990, conveniently summarized in Oldeman, 1992). This study was based on a survey of the views on land use and degradation of some 200 experts in countries around the world. Respondents were asked to use a common set of criteria in judging the extent and severity of land degradation in their respective areas. Oldeman and his colleagues recognized that some subjective differences in judgment among their respondents were inevitable. Nonetheless, the scientific credentials of Oldeman and colleagues, and of their collaborators, indicate that their estimates provide a promising first approximation to the extent and severity of global soil degradation.

The estimates showed that of the global total of 4.7 billion ha of land in annual and permanent crops and permanent pasture (FAO, 1990), 1.3 billion ha (28%) was degraded to some extent. Water and wind erosion were responsible for 84% of the degraded land. Of the 1.3 billion ha of degraded land, 0.5 billion ha were lightly degraded, 0.6 billion ha were moderately degraded, and 0.2 billion ha were strongly degraded. Lightly degraded land was that on which farmers could restore the lost productivity using their own resources. Moderately degraded land would require investments by public bodies to do the job, and strongly degraded land was beyond repair, its productivity permanently lost for all practical purposes. The estimates were of the *cumulative* loss of productivity over the 45 years from 1945 to 1990.

The Dregne and Chou Estimates

Two years after the Oldeman et al. estimates were published, Dregne and Chou (1992) produced global estimates of degradation-induced losses of land productivity in dry areas of most countries, including all of the big ones. Dry areas are those in arid, semiarid, and dry subhumid zones. The estimates are for three kinds of agricultural land use: irrigated land, rain-fed cropland, and rangeland. Drawing on FAO data, Dregne and Chou found 5.1 billion ha of dryland in the three uses, 88% of it in rangeland, 9% in rain-fed crops, and 3% in irrigated crop production.

Dregne and Chou classified rain-fed and irrigated cropland as slightly degraded (0 to 10% loss of productivity), moderately degraded (10 to 25% loss), severely degraded (25 to 50% loss), and very severely degraded (more than 50% loss). For rangeland the corresponding percentages of productivity loss were 0 to 25, 25 to 50, 50 to 75, and more than 75%. The percentages for rangeland are more ample because range scientists consider rangeland that has lost no more than 25% of its potential productivity to be in good condition. The percentages represent the estimated losses of productivity relative to the potential plant yield of the land in an undegraded state, given the existing technology used, whatever that might be.

Dregne and Chou estimated that some 3.9 billion ha of rain-fed cropland, irrigated land, and rangeland in dry areas of the LDCs have suffered some degree of productivity loss because of land degradation. For each of the categories of severity of productivity loss, Crosson assumed that the loss is at the midpoint of the range given by Dregne and Chou. That is, slightly degraded irrigated and rain-fed cropland has lost 5% of its potential productivity, moderately degraded land has lost 18%, and so on. Crosson then weighted these estimates of productivity loss by the amount of land in each degree-of-severity category in each of the three land uses to calculate the weighted average loss in each use. These averages are irrigated land 11%, rain-fed cropland 14%, and rangeland 45%. Finally, because in terms of lost productivity a 1-ha loss of irrigated land imposes a higher social cost in lost production than a 1-ha loss of rain-fed cropland, which imposes a higher cost than a 1-ha loss of rangeland, Crosson calculated the weighted average loss of the three land uses taken together by weighting the percent loss for each use by its per hectare value of production. According to Dregne and Chou, these values (in prices around 1990) were $625 for irrigated land, $94 for rain-fed cropland, and $17 for rangeland. This calculation showed that the weighted average loss for the three land uses taken together was 12%. This is the *cumulative* loss over some period of time, which Dregne and Chou do not specify. But for most of this land the period must be several decades. Over, say, four decades the average annual rate of degradation-induced loss would be 0.3%.

Crosson Estimates Using Oldeman et al.

As noted above, Oldeman et al. used categories of degradation like those of Dregne and Chou, i.e., lightly degraded, moderately degraded, and so on. However, Oldeman et al. did not estimate the percentage losses of productivity in each of their degree-of-severity categories. Crosson did this (1995b). He assumed a zero productivity loss on the 72% of the land in annual and permanent crops and permanent pasture that Oldeman et al. found to be undegraded. For the 28% of this land that was degraded to some extent he used the Dregne and Chou categories of productivity loss, that is, lightly degraded land has lost 5% of its potential productivity, moderately degraded land has lost 18%, and so on. Weighting these percentages by the amounts of land that Oldeman et al. found in each of the degree-of-degradation categories gave a weighted average loss on the 4.7 billion ha of 4.8%. Over the 45 years used in the Oldeman et al. work, the average annual rate of loss was 0.1%.

This estimate of the degradation-induced loss of productivity using the Oldeman et al. estimates is powerfully conditioned by their finding that 72% of the land in annual and permanent crops and permanent pasture had suffered no productivity loss. Given that, the average loss on the total of such land could not be high even if the losses on the 28% of the land that is degraded were much greater than the percentages Crosson used, i.e., 5% and 18% for lightly degraded and moderately degraded land, respectively, and so on. For example, if the losses on lightly degraded, moderately degraded, and strongly degraded land were 15, 35, and 75%, the weighted average cumulative loss on the total of 4.7 billion ha still would be only 8.3%. Over the course of 45 years the average annual rate of loss would be 0.18%.

In correspondence with Crosson about his use of the work by Oldeman and colleagues, Oldeman did not argue with Crosson's estimate that the average annual rate of degradation-induced productivity loss over the 45 years was no more than 0.1 to 0.2%. He asserted his belief, however, that the trend rate of loss has been, and continues to be, rising. This, of course, would imply that early in the 45 year period the annual rates of loss were less than 0.1 to 0.2%, and that more recently they have been more than that. In this case, using the 45-year average to project losses over the next several decades could seriously underestimate future losses.

There is no compelling evidence to support Oldeman's belief that the degradation-induced losses are on a rising trend. But whatever the case may be in that respect, there is reason to believe that future losses are more likely to fall than to rise. Studies done in Asia (Pingali, 1989) and Africa (Migot-Adholla et al., 1991; English et al., 1994) indicate that the combination of rapid population growth, advances in agricultural technology, and increasing commercialization of agricultural output increases the value of agricultural land, which strengthens property rights in it. The result is to promote farmers' incentives to protect the land against degradation wherever it is in fact under threat. If the pace of agricultural development achieved in the LDCs (except in Africa!) over the last several decades is maintained, then farmers should have increasing incentive to protect the land against degradation; in which case degradation-induced losses should decline.

Comparability of Oldeman et al. and Dregne and Chou

The Oldeman et al. and Dregne and Chou estimates are not directly comparable, not least because Dregne and Chou estimated degradation-induced losses of land productivity only in dry areas and the Oldeman et al. estimates are for all agroclimatic zones. Moreover, the procedures for making the estimates were quite different, Dregne and Chou relying on the literature and their own experience and Oldeman and associates on the informed judgment of experts all around the world. Given these differences, and the poor quality of the data on which both studies had to rely, the results are quite similar: a 12% degradation-induced productivity loss in dry areas over some decades and a 5% loss (estimated by Crosson using the Oldeman et al. data) over 45 years across all agroclimatic zones.

Are the Oldeman et al. and Dregne and Chou Estimates Believable?

These estimates are much lower than what one might expect from much of the literature on the problem of global land degradation. For example, although neither Brown and Wolf (1984) nor Pimentel et al. (1995) give estimates of degradation-induced productivity loss, their very high estimates of the amount of global soil erosion imply substantially higher losses than those that can be derived from the work of Dregne and Chou and Oldeman et al. For reasons given above, we do not believe that the work of Brown and Wolf or of Pimentel et al. can be taken seriously.

However, at least one serious student of the land degradation issue has told us that he cannot accept the very low estimates of loss based on Dregne and Chou and Oldeman et al. At our request, William Larson, a world-class soil scientist recently retired from the University of Minnesota, reviewed the Dregne and Chou and Oldeman et al. studies. Larson readily acknowledged the highly competent and professional underpinnings of both studies. But, he told us, the studies do not provide "the last word," and his personal observations of land degradation around the world, especially in sub-Saharan Africa, have convinced him that land degradation is more severe than indicated in the Dregne and Chou and Oldeman et al. work.

Erosion "Hotspots"

It must be noted that the estimates of degradation-induced productivity loss that can be derived from the Dregne and Chou and Oldeman et al. studies apply at the global level. The low global estimates are entirely consistent with high levels of loss in some areas around the world. Scheer and Yadav (1996) refer to these areas as "hotspots" where "land degradation poses a significant threat to food security for large numbers of poor people, to local economic activity, and to important environmental products and services." Scheer and Yadav identify hotspots in the Indus, Tigris, and Euphrates River basins of south and west Asia; northeastern Thailand and the steeply sloped areas of China and Southeast Asia; the Nile Delta and erosion-prone parts of Nigeria and the Sahel; the irrigated areas of northern Mexico; and the hillsides of subhumid and semiarid portions of Central America. In these, and no doubt other places, the scale of land degradation taking place justifies immediate attention and remedial action. These problems do not as a whole represent a major threat to global agricultural capacity, but for the people who must deal with them they are important.

A Final Point about Degradation-Induced Losses of On-Farm Productivity

We must make a final, important point with respect to the conclusion that on a global average scale land degradation is not a serious threat to agricultural production capacity or food security. Our judgment on this issue most definitely does not mean that we think that land conservation is unimportant. On the contrary, we think that it is quite important. The evidently small degradation-induced losses of land productivity indicate to us that farmers all around the world share our conviction that the resource must be conserved and in fact have taken the steps necessary to achieve that goal. Indeed, they have done such a good job that little if anything remains to be done. To be sure, farmers may not give as much weight to the interests of future generations in protecting the land against degradation as some people think they should. But if the very low losses derived from the work of Dregne and Chou and Oldeman et al. are reasonably on target, then one would have a hard time making a convincing argument that the interests of future generations require that farmers do even more to protect the land against degradation than they at present are.

OFF-FARM CONSEQUENCES OF SEDIMENT FOR SURFACT WATER QUALITY

On a global scale farmers probably have managed and are managing the land in ways that keep degradation-induced losses of soil productivity at a roughly optimal level for them, and almost surely for future generations as well. The overriding reason for this is that wherever farmers have firm, enforceable property rights in the land they have a powerful incentive to protect its productivity. If the land loses productivity, they lose.

Farmers have no comparable incentive to protect downstream users of water against the water quality damages of sediment because the farmers responsible do not bear the costs of the damage. The argument here is that these off-farm sediment damages to water quality almost surely are more costly than the on-farm degradation-induced losses of soil productivity, and the difference in farmer incentives is the main reason.

Discussion of the off-farm costs of sediment damage is necessarily within a watershed context. Lowland and upland areas are linked by the hydrological cycle, and much of the sediment that does damage in lowlands originates on farmers' fields in upland areas.

Off-farm sediment damages are many, and familiar. They include most prominently losses of recreational values because of muddy water, sedimentation of reservoirs and irrigation canals, costs of cleaning the water so that it can be used for domestic, industrial, and commercial purposes, costs of clearing rivers and harbors so they can be used for shipping, and losses of aquatic habitat values when sediment covers fish-spawning areas. Sedimentation of riverbeds may also contribute to increased costs of flooding.

ESTIMATES OF OFF-FARM DAMAGES ARE FEW: THE U.S.

Unfortunately, reliable, comprehensive estimates of the costs of sediment damages are few. Indeed, on a large spatial level there are only two. One was a study by Clark et al. (1985) of the costs of sediment damage in the U.S. Drawing upon a wide variety of data and other information sources, Clark et al. estimated that sediment damages to surface water quality in the U.S. cost somewhere between \$3 billion and \$13 billion per year, with a "best guess" estimate of \$6 billion. The estimates were in prices around 1980. They were estimates of damage costs from all sources of sediment. Clark et al., admittedly with little evidence, attributed one third of the costs to agriculture, which would indicate that the agricultural costs would have been between $1 billion and $4 billion/year. Clark et al. were careful to cite their sources, state their assumptions, and emphasize that much of their data was "soft." Indeed, it was because of these limitations that the range of their costs estimates is so wide.

The U.S. Department of Agriculture (1989) used the erosion productivity impact calculator (EPIC) to estimate the economic cost of on-farm erosion-induced losses of soil productivity in the U.S. over a 100-year period, given early 1980s levels of crop production, technologies, and management practices. The results showed that for the nation as a whole, wind and water erosion would reduce crop yields at the end of the 100 years by less than 3% of what they would be in the absence of erosion. The present value of the yield losses attributable to water erosion alone was calculated by estimating the annual losses over the 100 years, valuing them by 1982 crop prices, then summing the discounted stream of losses, using a discount rate of 4%. This was taken to be the "real," i.e., inflation adjusted, rate of return to agricultural capital. The present value of the losses was $6.2 billion (USDA, 1989, p. 4). The implied average annual stream of losses is $252 million.

This is an estimate of the annual *gross* value of erosion-induced losses of yield, not *net* losses of farm income. The difference between gross and net income is the cost of all the inputs farmers buy to produce crops. The U.S. Department of Agriculture (1989) did not estimate the net losses. However, one can use the relationship between net and gross income in the country as a rough basis for translating estimates of the gross value of erosion-induced yield losses into estimates of net losses. In the early 1990s net farm income in the U.S., not including direct government subsidies, averaged about 20% of gross income (USDA, 1992). Assuming this percentage over the next 100 years, the $252 million annual losses of gross income would imply annual net losses of $50 million.

This loss, however, is attributable only to the cost of yield losses. It does not include the costs of all the things that farmers might do to keep the yield losses so low, e.g., using more fertilizer to replace lost nutrients, investing in terraces, etc. Consequently, the estimated $50 million/year net on-farm cost of soil erosion in the U.S. must be too low, probably substantially so. But if the "true" costs were three or four times the $50 million estimate, they still would be far less than the Clark et al. (1985) estimate of $1 billion to $4 billion/year as agriculture's contribution to the costs of sediment damage.

Because of the different incentives of farmers with respect to the on-farm and off-farm costs of erosion and sediment damage, the conclusion that in the U.S. the off-farm costs are much higher is wholly unsurprising. It is what one would expect.

Estimates of Off-Farm Costs: Java

The only other large-spatial-scale study of the off-farm costs of sediment from farmers' field that we know about was done for the island of Java (Magrath and Arens, 1989). These authors found that, unlike the situation in the U.S., in Java the on-farm costs of erosion-induced losses of productivity were higher than the off-farm costs of sediment damage. The Magrath and Arens study applied only to nonirrigated land in upland areas with very high rates of erosion. This land accounted for 41% of cropland in Java. The extensive and highly productive land in rice was not considered.

The Magrath and Arens study is well done. Its finding that in the region they studied on-farm costs of erosion in lost productivity was higher than the off-farm costs of sediment damage is counterintuitive, given the difference in farmers' incentives with respect to the two kinds of costs. Whether off-farm costs exceed those on-farm elsewhere in the LDCs is not known because estimates of the two kinds of cost are not available for other countries. Eckholm (1976) did a comprehensive review of the then available literature on land and water resource degradation around the world, and found much evidence that in the LDCs sediment was a substantial threat to irrigation systems, municipal water supply systems, and so on. But he made no estimates of the costs of these damages, nor of the on-farm costs of erosion.

More recently, deGraff (1996) asserted that in many countries downstream sedimentation of reservoirs now constitutes the most important form of sediment damage. For many of these reservoirs, rates of sedimentation were seriously underestimated at the time of reservoir construction. deGraff (p. 120) cites a study of 21 reservoirs in India that found that in most cases sedimentation rates were two to three times the expected rate. The consequences for the cost–benefit ratios of these expensive projects could be serious, if not disastrous.

Southgate and Macke (1989) studied the downstream benefits of alternative land management strategies in a watershed in Ecuador where a large hydroelectric facility was located. Sediment deliveries to the reservoir were a major threat to the economic viability of the project. The details of the Southgate–Macke analysis need not detain us here. The most important aspect of their study for our purposes was the description of the strategy the Ecuadorian government had adopted to deal with the sedimentation problem. The strategy had four components (Southgate and Macke, 1989, p. 7):

1. Civil works to keep eroded material out of waterways and upstream from the reservoir;
2. Protection of the forests covering much of the watershed;
3. Reforestation of areas previously in forest;
4. Erosion control on agricultural land in the upper watershed.

Note that erosion control on farmers' fields was only one component of the government's strategy. The others were designed to intercept eroded material before it could be delivered to the reservoir (component 1), avoid increased erosion by protecting currently forested land (component 2), and reduce erosion on denuded land by reforesting it (component 3). Southgate and Macke do not say this, but not only would erosion be controlled on the forested and reforested land, but those areas would also intercept erosion originating on farmers' fields in upland areas.

The four-pronged strategy of the Ecuadorian government for protection of the reservoir makes sense because of a well-known, but not always recognized, fact about sediment: at any given time, much of it is stored somewhere in watersheds where water flows have slowed enough to permit sediment deposition. The fact of sediment storage raises questions about the advisability of a strategy that seeks to control downstream sediment damage primarily by reducing erosion on farmers' fields.

Sediment storage in watersheds gives rise to what Wolman (1977) called large temporal and spatial discontinuities between the occurrence of erosion on the landscape — say, on farmers' fields — and the occurrence of downstream sediment damage. Because of these discontinuities, on-farm soil conservation may reduce erosion from a storm event, but the event likely will mobilize sediment stored in the watershed and move it downstream where the damage it does is no different from before the on-farm conservation measures were adopted.

A study by Trimble (1975) of erosion and sedimentation in the southern piedmont of the U.S. demonstrates the importance of the sediment storage phenomenon. Trimble estimated the amount of erosion and sedimentation occurring in the region from colonial times to about 1970 and concluded that, over that period, only some 5% of the sediment had been delivered to the fall lines of the rivers of the region. In a similar study of the Coon Creek Basin of Wisconsin, Trimble (1981) and Trimble and Lund (1982) estimated that from the 1850s to the 1970s only 6 to 7% of the soil eroded from upland areas and valleys of the region had been exported from the region to the Mississippi. As in the southern piedmont, the rest was stored somewhere in the basin.

Because of sediment storage in watersheds and the resulting temporal and spatial discontinuities between the occurrence of erosion and the occurrence of sediment damage, a strategy of seeking to protect against downstream sediment damage by controlling erosion on farmers' fields may be economically inefficient. The costs of on-farm erosion control occur "today." The benefits of reduced downstream sediment damage may not occur for years, decades, or even centuries if one can believe Trimble's findings. The consequences for a benefit–cost analysis of the erosion control strategy at any plausible discount rate are obvious.

A study by Davie and Lant (1994) of the sediment reduction resulting from the U.S. Convervation Reserve Program (CRP) in two watersheds in southern Illinois supports the argument that the sediment-reduction benefits of on-farm erosion control programs may be long delayed. In the two watersheds 25 to 30% of the land had been put in the CRP beginning in 1986, resulting in very high estimated reductions in erosion. Davie and Lant compared sediment concentrations in the streams 3 to 4 years after the land went into the CRP with concentrations immediately before that. They found a very small, statistically insignificant, reduction in after-the-CRP concentrations. They attributed this both to sediment storage in the watersheds and to the fact that much of the CRP land was 30 to 50 km from the streams.

The fact of sediment storage suggests that often — perhaps usually — the most economical strategy for watershed management of the sediment problem would be to focus on intercepting sediment before it enters sensitive water bodies rather than focusing on reducing erosion on farmers' fields. The exception would be in cases where the water body to be protected is in close proximity to and directly linked by a stream or streams to farmers' fields. Even in cases of close proximity, if the sediment delivered to the water body is by overland runoff it may be more economical to intercept the sediment by use of filter strips or planted trees than by reducing erosion on the land.

The case for a sediment interception strategy probably is strongest in large watersheds where sediment is sporadically moved by storm events over cumulatively long distances. These are the watersheds in which the spatial and temporal discontinuities resulting from sediment storage are likely to be most important and where, consequently, a strategy of controlling downstream sediment damage by reducing erosion on farmers' fields is likely to be least efficient.

In the U.S. the Natural Resources Conservation Service still seeks to deal with the problem of downstream sediment damage primarily by encouraging farmers to adopt on-farm erosion control measures. Clearly, there is a place for this strategy. But in our judgment the case is strong for a sediment interception policy supplementary to, if not in place of, an erosion control policy. The very limited example of the Ecuadorian government's partial adoption of a sediment interception strategy, briefly described above, suggests that that strategy is not just an abstract idea. Perhaps other governments should be considering it as well.

CONCLUSION

Watershed degradation has an effect on food security, but it is indirect and small. It is indirect because the main cause of food insecurity is not natural resource degradation but poverty. In the LDCs increasing agricultural productivity is a necessary, but not sufficient, condition for reducing poverty, thus increasing food security. But, on a global scale, the effect of (land) degradation on productivity has been, and promises to be, small. Over the last four or five decades land degradation has reduced productivity only some 0.1% (Crosson, using data of Oldeman et al.) to 0.3% (Dregne and Chou) per year. And there is reason to believe that conditions strengthening farmers' incentives to invest in soil conservation are spreading in the LDCs, suggesting that low as the annual rates of degradation-induced productivity losses have been in the past, they will be even lower in the future.

So land degradation does not appear to be a threat to food security. It could be, however, that sediment damages to water quality could pose a threat, not directly to food security in the LDCs, but perhaps indirectly by increasing the economic and environmental costs of providing people in those countries with adequate water supplies. The sediment damage threat thus deserves the attention of LDC policy makers concerned with natural resource management. The focus of their attention clearly should be at the watershed level. And in our judgment, as they begin to think about policies to deal with sediment damage, they should give careful consideration to a strategy emphasizing sediment interception, not necessarily to the exclusion of a strategy to reduce on-farm erosion but, at a minimum, as a strong complement to it.

REFERENCES

Alexandratos, N. 1995. *World Agriculture: Toward 2010.* United Nations Food and Agriculture Organization and John Wiley & Sons, New York.

Biot, Y., P. Blaikie, C. Jackson, and R. Palmer-Jones. 1994. Rethinking Research on Land Degradation in Developing Countries, World Bank Discussion Paper, World Bank, Washington, D.C.

Brown, L. and E. Wolf. 1984. Soil Erosion: Quiet Crisis in the World Economy, Worldwatch Paper 60, Worldwatch Institute, Washington, D.C.

Clark, E. II, J. Haverkamp, and W. Chapman.1985. *Eroding Soils: The Off-Farm Impacts,* Conservation Foundation, Washington, D.C.

Crosson, P. 1995a. Future supplies for land and water for world agriculture, in N. Islam, Ed., *Population and Food in the Early Twenty-First Century: Meeting Future Food Demand of an Increasing Population,* International Food Policy Institute, Washington, D.C.

Crosson, P. 1995b. On-Farm Costs of Soil Erosion: What Do We Know? Working Paper 95-27, Resources for the Future, Washington, D.C.

Davie, K.D. and C.L. Lant. 1994. The effect of CRP enrollment on sediment lead in two southern Illinois streams, *J. Soil Water Cons.* 49:407–412.

deGraaff, J. 1996. The Price of Soil Erosion: An Economic Evaluation of Sofl Conservation and Watershed Development, Wageningen Agricultural University, Wageningen, the Netherlands.

Dregne, H. 1988. Desertification of drylands, in P. Unger, T. Sneed, W. Jordan, and R. Jensen, Eds., *Challenges in Dryland Agriculture: A Global Perspective, Proceedings of the International Conference on Dryland Farming,* Texas Agricultural Experiment Station, Bushland/Amarillo.

Dregne, H. and N. T. Chou. 1992. Global desertification dimensions and costs, in H. Dregne, Ed., *Degradation and Restoration of Arid Lands,* Texas Tech University, Lubbock, 249–282.

Eckholm, E.P. 1976. *Losing Ground: Environmental Stress and World Food Prospects*, Norton Publishing, New York.

El-Swaify, S., E. Dangler and C. Armstrong. 1982. *Soil Erosion by Water in the Tropics,* University of Hawaii Press, Honolulu.

English, J., M. Tiffen, and M. Mortimore. 1994. Land Resource Management in Machakos District, Kenya, Environment Department Paper no. 5, The World Bank, Washington, D.C.

FAO. 1990. *The Production Year Book*, FAO, Rome.

Magrath, W. and P. Arens.1989. The Costs of Soil Erosion on Java: A Natural Resource Accounting Approach, Environment Department Working Paper no. 18, The World Bank, Washington, D.C.

Migot-Adholla, S., P. Hazell, B. Biarel, and F. Place. 1991. Indigenous land rights systems in sub-Saharan Africa: a constraint on productivity? *World Bank Econ. Rev.,* 5(1), 155–175.

Nelson, R. 1988. Dryland Management: the Desertification Problem, Environment Department Working Paper no. 8, World Bank, Washington, D.C.

Oldeman, R. 1992. Global extent of soil degradation, in Bi-Annual Report 1991–1992, International Soil Information Reference Center, Wageningen, the Netherlands.

Oldeman, R., R. Hakkeling, and W. Sombroeck. 1990. World Map of the Status of Human-Induced Soil Degradation: An Explanatory Note, International Soil Information Reference Center, Wageningen, The Netherlands, and United Nations Environment Programme, Nairobi.

Pimentel, D. et al., 1995. Economic and environmental costs of soil erosion and conservation benefits, *Science,* 267, 1117–1123.

Pingali, P. 1989. Institutional and environmental constraints to agricultural intensification, *Pop. Dev. Rev.,* 15, 243–260.

Scheer, S. and S. Yadav. 1996. *Land Degradation in the Developing World: Implications for Food, Agriculture and the Environment,* International Food Policy Research Institute, Washington, D.C.

Southgate, D. and R. Macke. 1989. The downstream benefits of soil conservation in third world hydroelectric watersheds, *Land Econ.* 65:38–48.

Trimble, S. 1975. Denudation studies: can we assume stream steady state? *Science* 188, 1207–1208.

Trimble, S. 1981. Changes in sediment storage in the Coon Creek Basin, Driftless area, Wisconsin, 1853–1975, *Science* 219, 181–183.

Trimble, S. and S. Lund. 1982. Soil Conservation and the Reduction of Erosion and Sedimentation in the Coon Creek Basin, Wisconsin, U.S. Geological Survey Professional Paper no. 1234, U.S. Government Printing Office, Washington, D.C.

USDA. 1989. The Second RCA Appraisal: Soil, Water, and Related Resources on Nonfederal Land in the United States, U.S. Department of Agriculture, Washington, D.C.

USDA. 1992. The second RCA appraisal: soil, water and related resources on nonfederal land in the United States, U.S. Department of Agriculture, Washington, D.C.

Wolman, M. 1977. Changing needs and opportunities in the sediment field, *Water Resour. Res.,* 13, 50–54.

19 People, Property, and Profit in Catchment Management: Examples from Kenya and Elsewhere

Mary Tiffen and Francis Gichuki

CONTENTS

INTRODUCTION

Small catchments, such as those used in a Kenyan soil and water conservation program, are better than large ones as project areas, since they are more likely to be social units as well as hydrological ones. Conservation is not worthwhile when

0-8493-0702-3/00/$0.00+$.50

population density is low. It becomes feasible as densities rise over 70/km^2, but it may be necessary first to make markets more accessible. Farmers live by their profits, and profits require flexible responses to new opportunities and reduction of risk. Farmers react with hostility to any threat to their land rights. It cannot be assumed that there is common property and that this can be used to regulate grazing. In Machakos District, Kenya, erosion was widespread in the 1930s. Soil conservation improved in the 1950s as market access improved. Farmers found water conservation profitable. They continued terracing on their own after the official program stopped. Output per hectare in 1990 was 11 times greater than in the 1930s. This was partly due to conversion of pasture to arable land and cultivation of more valuable crops.

WHAT IS A CATCHMENT?

The *catchment area* or *drainage basin* denote the area drained by a stream or river. (These terms are to be preferred to *watershed*, which denotes the ridge dividing catchment areas.) They represent parts of the landscape that are hydrologically connected. From a technical point of view it makes sense to study, plan, and manage areas that are hydrologically linked because of the important role of water in shaping the landscape and influencing the interaction of soil, plant, and atmosphere. Catchment management therefore deals with all lands within the boundaries of the drainage basin with the aim of promoting integrated development to ensure sustainable use and conservation of all the natural resources in the basin. In this chapter we focus on soil and water resources and the socio-economic dimensions that influence the way the land users use and manage the catchment. The main source of sediment and the main beneficiaries of a catchment management project are the most important factors influencing the scope of watershed projects.

What does a catchment mean to the people who live in it? Basically, it represents a collection of properties, mostly belonging to them, sometimes to the state or an outside landowner. Farmers will be well aware of the water flows over their land. They will know that the actions or neglect of people upstream affects their farms, but they normally have no power to influence these people. A catchment is not a social unit. As an example, the people in the upper Zarqa catchment of the King Talal dam in Jordan were not willing to put in costly work for the sake of richer irrigating farmers below the dam, though they would consider what was profitable on their own land (Agrar, 1982). This illustrates the complexity of catchment management brought about by on-site and off-site costs and benefits.

The two most important social units are those of the household and its land, and the units of local government or local society within which people are accustomed to community action. The land is generally owned by particular families, not by the community as a whole. Catchments are most likely to work as a project area if they are small, falling within one local government unit, and where farmers know each other as neighbors. The catchment approach introduced in Kenya in 1987 defines a catchment area as a geographically well-defined area where farmers are willing to work together in order to achieve a more sustainable way of farming (SWCB, 1994). This approach, although respecting the hydrologic link between one farm and the next, puts more emphasis on a social catchment rather than a hydrologic catchment.

Even the agencies commissioning conservation may have difficulty with the catchment concept. They are typically in the Ministry of Agriculture. For political or administrative reasons they may exclude some lands. For example, in the Jordan case, the Upper Zarqa Catchment was defined so as to exclude urban Amman (which nevertheless generated considerable deposits) and the desert areas used mainly by pastoralists — although the occasional rainfall there generated considerable sediment flows. In Kenya, soil conservation is planned on the basis of small catchments within the district, the normal administrative unit. Nevertheless, there are problems in obtaining the co-operation of those controlling public land such as roads and schools, although these can present a high proportion of current erosion problems in intensely cultivated areas. We have to recognise that not all erosion problems are caused by farmers.

THE PEOPLE WHO LIVE THERE

There are three important things we need to know about the people who live in a catchment — their numbers, their options for earning money, and the resources they own. People, profit, and property are bound up together, but we will discuss manpower first. Population density, and the related concept of the land:labor ratio, are critically important because they determine the type of farming and livestock care that is practiced, and the labor resources available for remedial measures. A low population density generally means that soil and water conservation are neither desired nor feasible. Generally speaking, in areas with 30 persons/km^2, land will be farmed for a few years until its fertility falls, when it will be allowed to revert to bush and new land will be opened. Unconfined and unsupervised animals will graze uncultivated land, including fallows and harvested land. The scarce resource is labor, and herding and fencing do not pay. Even if people are persuaded to put in trash lines or cut-offs in a special effort, animals will soon destroy them, as we saw in a sparsely populated area of coastal Kenya (Tiffen et al., 1996). As few as 4 years after the soil and water conservation campaign, the traces of its work were disappearing. The land users could not justify the labor, herding, and fencing requirements needed to maintain the soil conservation measures initially undertaken.

A scattered population, poor because of distance from markets, can make little impact on major problems of gullying, as we saw in parts of Baringo, a Kenyan district still notorious for its erosion problems despite soil conservation campaigns that began in the 1940s and continue today. It is no use recommending to the labor-scarce farmer measures that will increase yield, but which will take labor that the farmer has either not got or can more profitably use elsewhere. This leads on to the type of labor available. In most of these low-density areas, where farming is unprofitable, most of the men will be away working somewhere else. We must remember that one of their options is not to farm. The villages are inhabited mainly by old men, women, and children. What is feasible for them? Can any conservation project succeed? As population density increases, the fallow period shortens, initially increasing fertility problems. However, there may still not be enough cash generation to buy in fertilizer, or enough labor to make and transport manure and fertilizer, to make fences, or to pen or tie up livestock. Deterioration of the land may be apparent, but farming may still not be very profitable. At these intermediate densities, therefore, we need to ask, what are the necessary conditions to make conservation attractive? Very often, improving access to markets is a necessary first step to give people the incentive to put more capital and labor into farming. Farmers well know the value of a road. Women farmers in Uganda said the price of a piece of land by the road was three times that of a field farther up the hill (personal communication to Tiffen, 1994). Indeed a new main road may bring in additional settlers. This not only provides additional labor, but the competition for the land gives it added value and makes people more willing to invest that labor in its improvement.

By the time population densities reach 70 to 100/km^2, new inputs become both feasible and profitable, and the scarce factor becomes land rather than labor. At this point, farmers become interested in the return to land, the yield per hectare. The farmer will be experimenting with various methods to improve yield and farm income. In dry climates, people will be seeking to maximise the utility of rainfall and, in wet climates, to limit water damage. For example, in Kenya during the 1940s and 1950s, discharge of excess runoff from the farms to a roadside drain was a common practice. In the 1980s farmer realised the added benefits accruing if they instead trapped roadside runoff to improve the soil moisture regime, either to grow profitable crops that could not be grown without additional water or to reduce crop production risk and increase yield. Other common changes are the more careful control of livestock, to avoid damage to crops and to generate manure. At these densities, new ideas for soil improvement and water management will be welcomed and tested to see how they compare with existing techniques. If they are profitable, these new ideas will be adopted.

PROFITS, OPTIONS, AND RISKS

Profit as an Incentive

We often forget that farmers are businesspeople, whether operating on a small or large scale. If small-scale, they need to make their assets of labor, livestock, and land work hard to provide food and cash for the family. Pure subsistence farming is rare nowadays, but in some cases people farm mainly to provide food for their families, deriving their main cash income from other activities such as migratory labor, trading, or craft work. They may or may not wish to use cash from these activities to finance farming. If farming becomes more profitable due to good access to market, they will use their outside earnings to improve their land. Indeed, these outside earnings were of great importance in the improvements made in Machakos District, Kenya (Tiffen et al., 1994). Profit is the absolute essential for good, sustainable farming practices.

Profit Requires Flexibility

Even where there are good markets, farming only remains the best option if the farmer pays constant attention to changes in prices for different types of output (grains, vegetables, livestock products) and different types of input (fertilizer, hired labor, etc.). Any recommendations we make must allow for the necessary flexibility and change. Often, the chief benefit of soil and water conservation is that it allows the growing of new crops, including many that the original planners never thought of. Bunch and Lopez (1995) make an interesting study of crop change over time in Guatemala and Honduras. In Embu District in 1996 we observed that the additional high revenue earner was *qat*, a narcotic grown for the Somali market, but certainly not officially recommended (Tiffen et al., 1996).

Profit Is Related to Risk

The major risk to farm incomes in many areas is variable rainfall. A farmer can purchase in manure or fertilizer, and fail to get a return because of lack of rain. In such areas, measures which conserve water (and incidentally, soil) can be very attractive because they reduce the risk of crop failure. It is no good offering credit for the inputs if rainfall failure is going to turn credits into unrepayable debt. In some parts of Kenya farmers are very reluctant to take official loans for which their land would be security, for fear that their land will be auctioned if they fail to repay. In the semiarid areas of the Middle East there is a long-standing objection to loans with interest, confirmed by religion. Instead, dryland farmers prefer share-cropping arrangements. A common arrangement is that the provider of land, labor, and capital each get 33% of the crop — capital traditionally taking the form of plow teams and seed. Each provider shares the risk as well as the gain. Are we sure enough of our recommendations to take a financial risk on them, and to tie loan repayment to the success of the crops?

PROPERTY RIGHTS AND COMMUNITY RIGHTS

We need to know who owns our catchment, its land, and its animals, and how. In the light of this we can consult with the people concerned on what can be done to improve the land condition, by individual landowners or land-using households, or by co-operating neighbors. If neighborly co-operation is likely to be insufficient, but certain measures are desired by a majority, are there feasible and enforceable mechanisms to make certain actions compulsory, such as a bylaw of a local authority?

Land Ownership

The majority of the world's farmers are small landowners, although in some countries most farms are large. Some large farms may be farmed by small tenants. Tenants do not have the same long-term interests as owners, because they cannot pass the land on to their children. This can raise considerable problems for land improvement, particularly where the large land-owners are also politically powerful. Smaller owner-farmers generally take a long-term view, especially as land becomes more valuable and scarce as population density increases. Most parents want to pass on good land to their children. For most farmers the main question is whether they have the means to invest in the measures that will improve their farm, not whether they are unconcerned by its deterioration.

However, this does not apply in areas of very low population density. There, people may have very secure rights over the area regularly farmed near their house, but only temporary rights, covering the actual cultivation period, over the land that they farm and fallow farther afield. There may or may not be some kind of tribal or village authority governing access to this land. Very often there is not — it is free access. With shifting fields and wandering animals, there is no hope of long-term success in a soil conservation campaign. We must wait until the time is ripe.

Property rights can change flexibly over time and place. For example, Gombe Emirate in northern Nigeria experienced a rapidly growing population, partly because of an influx of immigrants. In 1967 village heads were asked what happened to the land of families who left the village permanently. In villages where cotton farming was profitable, indigenous farmers were enlarging their farms, and migrants were coming in, village heads responded that the family would, of course, sell the land before going. In areas attracting less immigration, some 50 to 100 km farther north, village heads said they would give it out to an incoming farmer, after 1, 3, or 7 years. In an isolated northern village with poor roads and poor soils, the village head said nobody bothered about abandoned land, and nobody consulted him before opening a new field. Land was available everywhere for people to farm. Formal law had also changed. In 1960 the senior local judge told village heads that land could be sold by some-one who had himself cleared it from the bush. By 1970 he said that land could be sold if it had been cleared, or inherited, or bought (Tiffen, 1976). When we look at property rights we need to look at the direction of change, as well as the present situation. The social forces driving toward tighter private control over more and more land as population grows are very powerful. It is better to go with them rather than to try to resurrect a past situation.

Once land has become sufficiently valuable to sell, it is generally also sufficiently valuable for farmers to wish to conserve or improve it. Their rights include decisions on how and what to plant. Any conservation campaign must be based on persuasion. In any case, compulsion has a long record of failure — it may last for a year or two, but the measures are not maintained or improved upon unless they are in the farmers' interest.

Community Authorities for Grazing Land?

Because we see the sense of dealing with the catchment as a whole, we would like to discover a community authority that has power to regulate its use. It would be convenient if the land was not owned by individuals, but was common property regulated by some kind of person or institution that we could influence to carry out our wishes, but is it actually so? It is still common to talk of land in Africa as being tribal or village property with individual households only having use rights. But as we have seen, that is unlikely to be the perception and practice of people living in those areas that have a population density of say, 50 persons/km^2. If we treat their land as common property, we are more likely to rouse their opposition than their co-operation. This also applies elsewhere. In Jordan, livestock specialists had assumed the large areas of uncultivated land they saw were common property that could be organised under grazing associations. Unfortunately, the local mayors said the only common land left were a few marketplaces and such. The uncultivated land was individually owned by men working away in oil-rich countries (Agrar, 1982). The Jordanian parliament would not be likely to pass a law taking away their rights given by legislation in the 1930s. If traditional authorities have lost power, we may find that a modern institution, such as the district, or even a sub-division of a district, has powers to make bylaws. Here we need to consider what bylaws it would be willing to make and enforce. Women farmers in a densely populated area of Uganda agreed it would be useful to have a bylaw to say animals should be tethered. However, they pointed out that the people elected as Parish Councillors were also the owners of the largest goat herds, and they doubted such a bylaw would be either passed or enforced (personal communication to Tiffen, 1994). This illustrates the importance of finding out who owns the animals, as well as who owns the land. Grazing associations for local owners of cattle and small stock would have limited effect in Jordan, even had their been communal land. Much of the damage was done not by local herds, but by herds owned by pastoral tribes who came in from the desert in times of drought and who had friends in very high places indeed. They also had guns.

Even where communal ownership of grazing land is acceptable, a problem arises when the members of the group question the inequity brought about by differences in the livestock numbers each pastoralist is grazing. This is particularly so if there is a wealthy and powerful pastoralist that own a large percentage of the total livestock herd. This creates a situation where other members start increasing their livestock herd. This leads to overstocking and associated vegetation and soil resources degradation. We have to consider whether any new institution that we may wish to recommend will be able to handle such internal tensions.

THE MACHAKOS EXAMPLE

Machakos District (pre-1992) is in Eastern Province of Kenya and borders Nairobi city to the east. According to the 1989 population census it had a population of 1.4 million out of which 8% were in urban areas. Machakos District has easy road and rail access to Mombasa and Nairobi, and this has provided a good market for meat, fruits, vegetables, charcoal, and building sand. About 10% of the district has a climate marginally suited for coffee (annual rainfall of 900 to 1200 mm), 40% is characterized as suitable for cotton, while the remaining is suitable for livestock and millet production (Jaetzold and Schmidt, 1983). There are two rainy seasons, with a high variation in rainfall amount and, in most of the district, there is less than 60% chance of enough rains for the preferred maize crop.

Machakos is a district in Kenya, whose northern boundary lies some 50 km from Nairobi, and which in 1990 stretched some 300 km south-east from there. There are two short seasons of rain, enabling, with luck, two crops a year. About 10% the district, its higher areas, has a climate marginally suited to coffee. About 40% is characterized by experts as suited to cotton, and the remaining 50% to livestock and millet (Jaetzold and Schmidt, 1983). There is enormous variation in the amount of rainfall experienced each season and, in most of the district, only a 60% chance of enough for the preferred food crop, maize.

FIGURE 19.1 Hillside near Machakos town, photograph taken in June 1937. Note the scarcity of trees, and the gullies, not only in the valley, but in the cultivated area above the road. Population density was about 75/km^2. (From Barnes, 1937.)

The Erosion Problem, 1930 and 1990, and Its Measurement

In the 1930s, soil conservation science was in its infancy, and soil erosion rates were not measured. The local agricultural officer (Hobbs) made estimates by location (subdistrict) of the amount of "uncultivated land subject to erosion" which ranged from 13 to 76%, most being in the range 50 to 60% (Thomas, 1991). The soil conservation specialist, Barnes, wrote in 1937 that

> I can say that there is really no part of the inhabited reserve that is free from erosion. Probably 75% suffers from severe erosion in various forms, parts of this almost amounting to complete destruction, 20% with less serious erosion, and there may be 5% that is protected by trees or natural conditions. (Barnes, 1937, quoted in Tiffen et al., 1994, p. 101, and Thomas, 1991)

Several other visiting experts formed similar opinions. Barnes' report is illustrated by photographs, many reproduced in Tiffen et al. (1994) and one example is shown as Figure 19.1. There were deep gullies in the valleys, small gullies developing out of cropped land, sheets of eroded land covered with a hard cap which rain could not penetrate. In one location, Kalama, which Hobbs had noted as almost 60% eroded in 1937, Thomas measured erosion by comparing 29 controlled microsites, using 1948 and 1972 air photographs (Thomas, 1974). He concluded that moderate erosion on nonarable land had increased from 22 to 26%, and severe erosion from

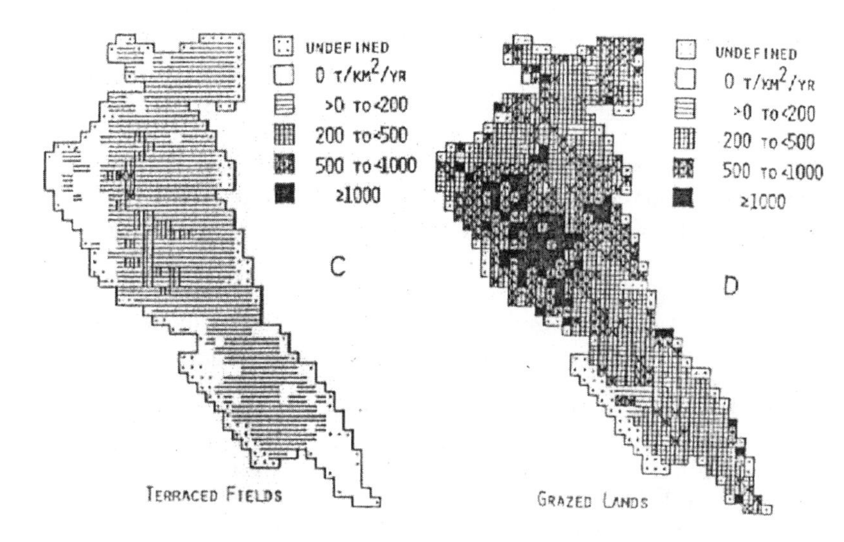

FIGURE 19.2 Sediment production rates from rill and sheet wash erosion on terraced fields and grazing land, Machakos District, 1982. (After Reid, 1982, maps 5B, 5D.)

48 to 63%. Cultivated land had increased from 18 to 26%. The cultivated land was much better protected against erosion in 1972. Taking both arable and nonarable land together, Thomas's measurements suggest 63% of the total land in the area suffered either moderate or severe erosion, not very dissimilar to Hobb's estimate, and that this was occurring mainly on pasture. It is worth noting that in five of the six seasons in both 1970–72 and 1934–36 rainfall was below average, as compared with only two of the six seasons 1946–48. Given Thomas's later conclusions on the importance of drought as a factor in erosion of grazing land, this suggest that while there had been an improvement in protection of cultivated land from 1937 to 1972, and an increase in the proportion of land cultivated, the position on grazing land had remained more or less static, when rainfall effects are discounted. An assessment of soil erosion in the district in 1981 was made by Reid (1982). Using low-level colour air photographs taken in a reconnaissance sample frame confirmed the difference between cultivated (now mainly terraced) and uncultivated land (Reid, 1982; quoted in Thomas, 1991, and Tiffen et al., 1994). This is shown in Figure 19.2. A second set of low-level photographs was taken 4.5 years later, in 1985, and showed an increase in the extent of erosion, especially in the areas of the district having least rainfall (Ecosystems, 1985–86). Between the two surveys, there had been three successive droughts in 1983 and 1984. Thomas concluded that the deterioration during this period was due to drought and that the results demonstrated the greater susceptibility of drier areas to drought-induced degradation and the greater vulnerability of soil conservation efforts in such areas (Thomas, 1991, p. 34).

From 1979 to 1990 there were various studies on soil loss, runoff, and sedimentation. Studies in the Iiuni catchment were linked to measurements of stream flow and sediment yield from a catchment of 11.3 km^2 and soil loss was estimated at 53.3 tons/ha/year on denuded grazing land (37% of the area), 16 tons on cultivated land (43% of the area), 1.1 tons on good grazing or woodland (20% of the area), giving an overall catchment soil loss of 5.4 tons/ha/year (Thomas et al., 1981). Reid's estimates for Kalama, taken as an example of a "cotton" area, and using the low-level colour photograph technique, provide an average erosion rate of 16 tons/ha/year off grazed land, 13 tons off terraced land, and 250 tons off roads (Reid, 1982). After making new field observations in 1990, and reviewing existing studies, Thomas concluded:

> ... in Machakos District ... erosion has been reduced since the 1930s, but has by no means ceased. Its intensity is highly variable over the District. ... The greater proportion of soil loss continues to occur on grazing land but the relative contribution from cultivated land varies according to the quality and maintenance of the terraces.... Erosion on grazing land has been reduced due to land demarcation and registration as a result of which ... communal grazing has almost disappeared. It has also been due to changing attitudes to land and livestock and greater awareness of the need to maintain cover. The higher rainfall in the last few years has aided the recovery of some areas which were formerly denuded.... Erosion on cropland has been reduced as a result of improvements in terracing....
>
> The rainfall record, taken in conjunction with the evidence on soil erosion ... suggests a strong link between periods of drought, denudation of grazing land, and intensified erosion.... Many of the older gullies, extremely active 50 to 60 years ago, have now become stabilised.

These observations show the great importance of taking into account rainfall conditions, changes in land use, land tenure, and population density. Most of the improvements on grazing land seem to have taken place since 1972, as farms became smaller, the proportion cultivated increased, and grazing land became increasingly privately controlled, and valued. Photographs taken in 1937 and 1990–91 provide the best evidence of change, in terms of revegetation of gullies and bare surfaces, increase in cultivated area at the expense of grazing land, increase in tree cover, and obvious increase in the value of output per unit area. Figures 19.1 and 19.3 are one of eight pairs contained in Tiffen et al. (1994). We can assume that Barnes chose some of the worst sites to illustrate his report, and that the 1990 photographs illustrate what has happened to once severely eroded land.

Land Rights

In 1937 there were about 80 people/km^2 in the better-watered hills, and about 50/km^2 on the lower slopes. Marketing was difficult due to lack of roads and the smallness of Nairobi town. Insofar as people made money from farming, it was from their livestock, which could be walked to a market. The experts blamed over-grazing for the problem, and in 1938 an attempt was made to confiscate "excess" livestock, and

FIGURE 19.3 The hillside near Machakos town, as transformed by January 1991. (The road takes a new route below the buildings on the left.) Note the additional trees, and terraces on the cultivated land. Population density was about 300/km^2. (Photograph courtesy of M. Mortimore.)

to sell them compulsorily at low prices. The Machakos people thwarted this by organising a protest march and sit-down in Nairobi, and getting questions asked at the Westminster Parliament. However, it left them deeply suspicious of the government.

These suspicions were heightened when experts talked of moving them off their sloping hills. This was interpreted by some Machakos inhabitants as another way of taking fertile land from the natives and giving it to the colonial settlers. Land tenure had already moved in local custom to the point where firm individual rights were acknowledged in cultivated land and where sales took place. Private grazing areas were already being marked out in the more densely populated areas. A new governor in late 1944 decided to take the erosion problem in hand and to make "tribal authorities" responsible for good management by its "tenant" farmers. The alarmed District Officer, who knew the state of local opinion, wrote, "We should, I think, not say too much about the evictions of bad tenants.... We should say we are respecting existing rights even when the land is closed and evacuated for reconditioning...." He followed this up 3 months later by reporting "widespread nervousness and a fear that the government is about to take away their land. When the Works Company went [to do some rehabilitation work] it was met by flat opposition, as they believed it a threat to their land rights.... They ... would prefer to do the work themselves." He went so far as to ask to be relieved of responsibility for the planned works. The perceived threat was because local custom gave land rights to the cultivator. Hence, anything resembling cultivation by government employees or tractors was strongly opposed (Tiffen, 1996).

By 1948 the colonial government had given way. People were to work two mornings on rehabilitation, but land rights were respected. Thus, a touring officer reported that a hill had been closed to grazing, *with the consent* of the grazing right owners, but that some young men had been fined for failing to turn up for work. When some grazing land was taken over for experiments in pasture rehabilitation, it was done after a clear agreement that the farmers concerned were only loaning the land, and would get it back when the experiment finished (Pereira and Beckley, 1952). As a consequence, suspicion diminished, and people worked more willingly, particularly when, later in the 1950s, they were allowed to choose their own group leaders and their own technologies.

Profit and Flexibility

The government during the period 1930 to 1955 was particularly concerned about grazing land. It was also conscious that people were poor and had few tools. Its recommendations at first focused on methods to get grass to take hold again on the bare pastures. On the land under maize and beans it initially enforced contour ditches, where the soil was thrown downhill. It thought most land was unsuited to cultivation and should be kept as pasture. However, when we look at Machakos today, we see, first, that much of the land once pasture is now cultivated; second, that there are few contour ditches, people having substituted bench terraces, formed by the harder initial effort of throwing the soil from the ditch uphill; and, third, that there are many new crops, including tree crops and vegetables. People do not necessarily do what the experts expect or recommend; they do what is profitable, as and when it becomes profitable. Figure 19.3 shows the 1990 view of the 1937 site shown in Figure 19.1.

How and why did the bench come to be utilised? Apart from a couple of experimental benches constructed near a school, in 1937, which do not seem to have spread, the first reference is in 1948, to "Chania of Kalama… who has made bench terraces (copying what he saw in India) and grows onions." In 1952 the Agricultural Officer reported "bench terraces were really catching on…. News was getting round of big money being made by some … cultivators. One … had sold Ksh 5000 worth of tomatoes from his smallish benched shamba!" (Tiffen et al., 1994). In both cases, we can see the strong profit motive, and that the first experiments were for high-value crops. By 1956 all the groups who could chose their own technology were choosing benches rather than contour ditches. The enthusiasm for the extra work this entailed was because the bench conserved water and made possible profitable crops that were becoming marketable as roads improved and Nairobi expanded. About the same time the government permitted Africans to grow coffee, under strict supervision and, in Machakos, on condition that they made benches. This was accepted without question, because coffee could only flourish with water conservation, and was very profitable. Once it was demonstrated that benches worked, they were applied also to less profitable crops, such as maize and beans. Indeed, they were so profitable that they continued to be built long after the government program stopped. From 1961 to 1978 there were no major government agricultural programs in Machakos, but this was the time when air photographs show that the bulk of the land was terraced (Figure 19.4).

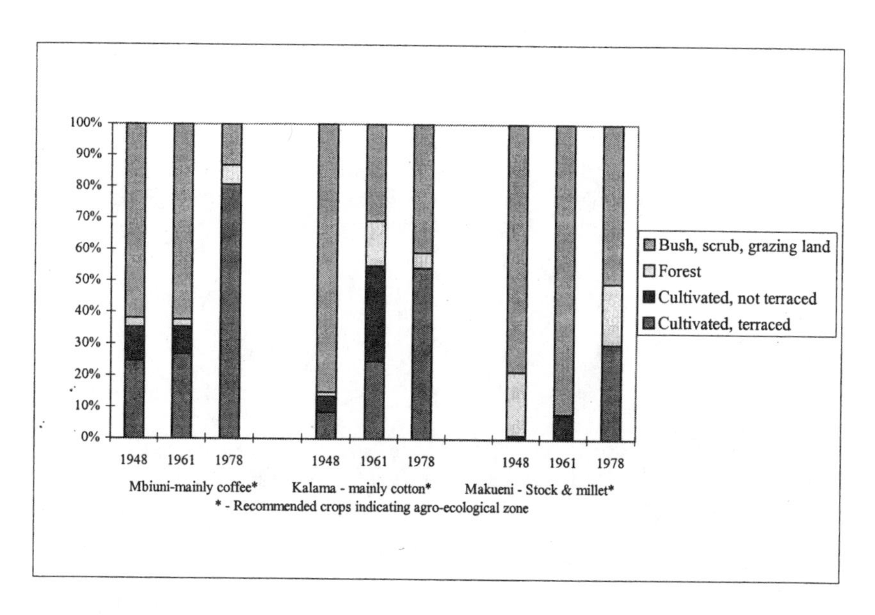

FIGURE 19.4 Progress in the conversion of pasture to cultivated terraced land, 1948 to 1978. (Courtesy of Tiffen et al., 1994.)

Figure 19.4 has been constructed from air photograph data. It also demonstrates that much of the former grazing land had by 1978 been converted to arable. The problem with grazing land is that intensive efforts to restore its productivity may well not be profitable. Indeed, Pereira et al. (1961) reported that the returns per acre on the pastoral experimental area were less than half those of crop production. Over time, the bare pasture lands restored by grass planting in the 1950s have been converted to arable. Thus, when we were admiring the productive terraced farm of a local pastor in 1990, his mother told us how she had been a member of a group that had planted grass on this once bare soil (Figure 19.5). Note again the link to tenure — the group had worked on land belonging to her and her husband, and inherited by her son. We met other farmers who were continuously carrying out small improvements to their pasture, and limiting their animal off-take from it, consciously because they wanted it to be in good condition for their sons, who would probably need to convert it to arable when they inherited a divided farm. Because of the lower rate of profit per unit area on grazing land, we can expect that farmers will devote resources first to their cultivated areas and, only after these have been improved, consider their grazing areas.

Profit and Productivity

After independence in 1963, many Machakos people moved their families into the dry Crown lands which they had previously only been permitted to use as pasture, establishing mixed farms, part cultivated, part for grazing. It might be expected that the utilization of this lower-grade land would reduce average productivity, when measured on the basis of the total land that they were allowed to use. However, if

FIGURE 19.5 Pastor Mutua on his farm. The terrace lip is protected by a local grass that is cut for fodder. There is a young mango tree in the background. In the 1950s this was sloping land with a hard crusted cap, and his mother was a member of a self-help group which planted grass to begin the restoration process. Since then, new terraces and cutoffs have been added, as resources allowed. (Photograph courtesy of D.B. Thomas, 1990.)

we convert all output to its 1957 value in terms of maize, we find that output per hectare had increased 11-fold during this period (Figure 19.6). This, as Figures 19.1, 19.3, and 19.4 illustrate, came partly from the conversion of pasture to arable, and partly increased cultivation of higher-value crops, such as the fruit trees visible in the photographs. Nevertheless, even in 1990, most cultivated land remained under maize and beans, and a higher average output of these crops was made possible by better soil water retention and better nutrient management, especially in the frequent seasons of below-average rainfall. In consequence, although population had risen five times, per head agricultural production in 1990 was three times above 1930 levels.

Potential production levels are still not attained, mainly due to water and fertility constraints. Low-input continuous cropping of maize has been reported to lead to a fertility decline as 1 t/ha crop of maize can remove as much as 20 to 40 kg of nitrogen, 2 to 4 kg of phosphorus, and 15 to 30 kg of potassium (Simpson et al., 1996). Farmers have recognised these constraints and in their efforts to boost production levels and reduce climatic production risks are undertaking various technological options. Water constraints are alleviated using *in situ* moisture conservation, water harvesting, runoff farming, and supplemental irrigation technologies. Due to the high cost of inorganic fertilizers, resource-poor farmers are using different

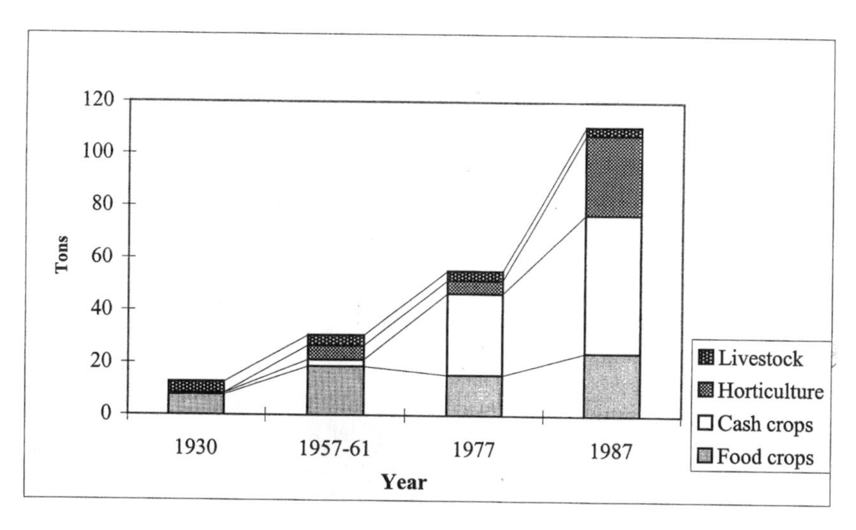

FIGURE 19.6 Value of output per square kilometer in maize terms, 1930 to 1987. (Courtesy of Tiffen et al., 1994.)

combinations of boma manure and compost, crop residue, green manure, grain legumes in crop rotation, agroforestry, and inorganic fertilizers to alleviate soil fertility constraints, but these are not always sufficient (Simpson et al., 1996).

Measuring Change in Productivity

Measuring change over time in productivity is subject to the same difficulties as the measurement of erosion. Production in Machakos varies greatly from year to year, according to seasonal rainfall. The type of output has varied over time, as farmers have responded to market changes and land:labor changes by changing their mix of crops and livestock. Further, the area settled and cultivated has changed over time, as people naturally settled first in the areas where soils and rainfall were most favourable, only spreading into the dryer half of the district after the 1960s. We examined rainfall, and found no downward trend over time. Rather, there were cyclic patterns of wetter and dryer sequences (Tiffen et al., 1994, chap. 3). Hence, we can rule out rainfall as an influence on *average* yields, despite its great influence in any given season.

There are two crops by which we can attempt to measure productivity, maize and coffee. The earliest data come from the agricultural census of 1930. This seems to have been based on rough estimates. Average yield was said to be about 500 kg/ha. A similar figure was given after a more careful estimate in 1957, a very good rainfall year. There was an agricultural census in 1960/61, which involved sampling. It reported only 200 kg/ha, but that was from very bad rains. From 1974 to 1988 we have a series, based on District Annual Reports, which in turn were based on yields in demonstration plots and estimates. The average for this 17-year period is 850 kg.

This type of average yield is confirmed by farm survey data. It is a low yield, which is not surprising given that seasonal rain in much of the district is frequently below the 250 mm, well distributed, thought to be the minimum for the crop. On average, it was coming from dryer land than in the 1930s, due to the spread of farming into the dry area. Nevertheless, the average appears to be substantially higher than it was

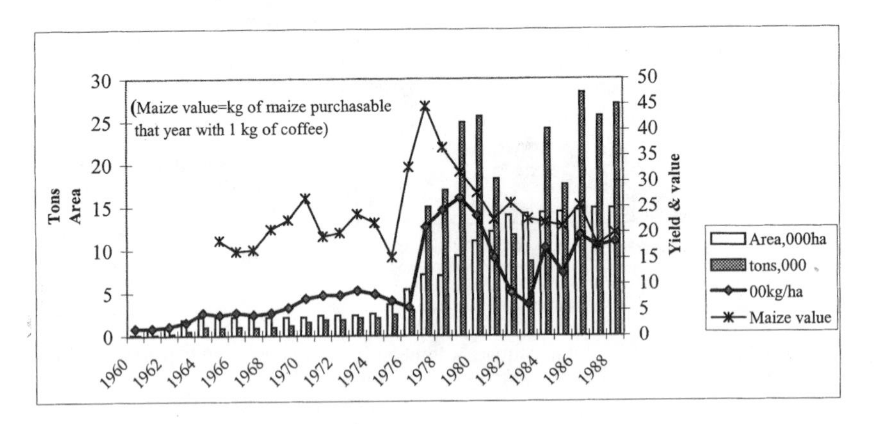

FIGURE 19.7 Output, maize value, yield, and area of coffee, 1960 to 1988. (Courtesy of Tiffen et al., 1994.)

in the 1930s and 1950s. Some of the addition will have come from using new drought-evading varieties, some to more manuring, and some to the water-retaining effect of the terraces. (The data are thoroughly discussed in Tiffen et al.,1996). It seems likely that the increase in terracing has contributed to a rise in yields over time, despite the use of dryer areas. Coffee was only permitted from the 1950s, and at first the area permitted to be planted was strictly limited, and farming methods highly regulated and supervised. Terracing was mandatory. New plantings were forbidden from 1964 to 1973, but after that control virtually ceased. Agricultural officers initially restricted coffee to the very best areas of what in Kenya are called agroecological zones 2 and 3. Zones 2 and 3 are the areas which were the most densely settled and cultivated in the 1930s, and by 1989 their population density had grown to more than 380/km^2. By 1990 a few farmers were also planting in the drier zone 4, recommended for cotton.

Coffee yield responds partly to rainfall, and the effect of the very low rainfall of 1983 can be seen clearly in Figure 19.7. However, as coffee is only planted on terraces in the higher rainfall areas, it responds also to the amount of inputs such as fertilizer, sprays, and labor. The farmer is much more assiduous when the price is high, as Figure 19.7 clearly shows. Price has been measured in terms of the number of kilograms of maize that a kilogram of coffee would buy. Higher prices stimulate increased plantings and higher yields. However, similar prices in the 1960s and 1980s produced higher yields (tons/ha) in the 1980s. Again, we can assume the increased terracing and other better farming methods played their role.

Thus, both with maize and coffee, yield per hectare has improved since the 1950s, despite increasing use of land receiving lower rainfall. Improved soil conservation has been one, but not the only, factor behind the increase.

Soil Fertility and Markets

Soil improvement has to be about the maintenance of soil fertility as well as the diminishment of erosion. Indeed, the two are closely linked, since well-nourished soil provides the more plentiful crop that gives the soil vegetative protection from the impact of rainfall. Businessmen farmers know that fertilizers can provide some of the necessary nutrients, but, to buy inputs as well as all their other necessities (clothing, school fees, medical care, etc.), they must have a market for their crops. They may have some manure and compost from their own resources, but these demand labor, labor that might better be spent looking for urban work if there is no market for farm crops. We noticed, when assessing the impact of the nation-wide soil and water conservation program in Kenya, that, in general, soil conservation and fertility management were of a higher standard in areas reasonably close to Kenya's large towns and much less good in equally well-watered areas in western Kenya, which are farther from international and urban markets (Tiffen et al., 1996).

Recently, the Organic Matter Management Network has developed very successful programs with farmer groups in western Kenya, in which the prime technology was deep digging and compost use in home vegetable gardens. This program has from the start been based on consultation with the farmers about their objectives, and a readiness to go where they wanted. For example, at a recent review workshop in Kakamega, 52% of the participants were farmers rather than extension advisers or non-governmental organization (NGO) officials. The organisers reported "a fierce pragmatism and ... concern about livelihood." "'We want more cash in pocket.... We support conservation because home gardens give money and food.'" They wanted help in finding markets, such as that for organically grown foods, and disturbed some official participants by their emphasis on cash, rather than food. Composting resulted in more vegetables; group effort and project guidance helped find a market; therefore, composting was worth-while. Unfortunately, as the authors note, there are few agricultural experts trained either to listen to farmers or to help with marketing problems (Cheatle et al., 1996). The Kakamega experience echoes the Machakos story, where bench terraces were first adopted to sell onions and tomatoes.

ROLE OF CATCHMENT MANAGEMENT: POTENTIAL AND CONSTRAINTS

Catchment management makes sense technically, but it can only work when it respects the aims, resources, and ownership rights of the people who live there. It cannot be based simply on the hydrological unit, but must consider socio-economic realities and the boundaries of farms and administrative units. The catchment belongs to people; it consists of their farms. However, there may also be some land that belongs to different government departments: those responsible for forests, roads, and some other public facilities. In very scantily populated areas there may also be land that belongs to no-one, which can be used, or abused, by any-one, or which is subject to some weak communal control. Here, the lack of people, and of markets, may be a constraint. Generally speaking, the catchment as a unit for soil conservation advice should take small units where people know each other, and are most likely to co-operate in any necessary group tasks. Respecting the aims of the inhabitants means listening to find out what they see as the constraints to successful farming, and helping them to overcome these, whatever they are. There are certain principles to remember:

1. People care most about their own farm and their own family's livelihood. Participation in a group takes time. Group activities should be restricted to the essential, and the reason for them should be well understood and accepted by the participants. This is particularly the case when the activity involved a long commitment for maintenance. We found, for example, in Kenya, that group nurseries for tree or grass planting material were almost invariably rather poorly maintained. Often, the necessity for them was not clear, since if farmers were convinced that certain types of grass or tree were useful, they could secure the material either from their own resources, or as gifts, or commercially (Tiffen et al., 1996).
2. People cannot be expected to work free of charge for the benefit of downstream farmers.
3. People will want to concentrate their efforts first on their best land, where their investment is most likely to pay off. If the techniques prove profitable, they will themselves want to use them over time on the rest of the farm. Pasture land is likely to be tackled last, as yielding least return to effort, unless the run-off from it is directly challenging a more profitable piece of land. Small farmers are likely to be able and willing to improve their farm over 10 or 20 years, and are highly unlikely to be able to do everything in the normal 1 to 3 year time horizon of a project.
4. Experts should contribute a realistic assessment of the causes of the sediment flows. A catchment-based program is only likely to succeed if the contributory factors to erosion are things which the people in the small catchment area itself have a realistic hope of tackling, with profit to themselves. A few people cannot tackle great geological forces. Over the decades we have wasted incredible sums of money and effort in places like Baringo in Kenya, or the Loess Plateau in China. Likewise, experts should be aware of the temporary effects of a series of dry seasons and the extent to which natural recuperation is likely when the rains improve. If the source of the sediment flows are in a town, or in the roads, they should be tackled by the appropriate authorities.
5. Concentrating skilled advice on one small catchment can show results, but is expensive in terms of government manpower and funds. If the work proves profitable, it will be copied. The catchment, therefore, should be planned from the start as a demonstration area, to which farmers from other nearby areas can be invited. The experts should plan to leave behind farmers who understand the principles, and who can therefore adapt prac-

tices as circumstances change, and teach others (Tiffen et al., 1996). It is wasteful to move from one catchment to the immediately adjacent one, since good ideas should automatically spread to the neighboring catchment.

6. If conservation is profitable, some farmers are likely to have already found good techniques. Outside advisers need to make a point of learning from the best existing practice, as well as bringing in new ideas which farmers can test out.

SUMMARY AND CONCLUSION

In planning soil conservation programs we have to remember:

1. People live in social units, not catchments. The smaller the catchment, the more likely that it coincides with some basic social units. The animals using a catchment do not necessarily belong to the owners of land there.
2. Population density affects farming systems, profits, and land tenure
   a. Under 30/km^2 scattered populations used shifting cultivation and free grazing. Most conservation work will fail. People have no rights in land they are not currently cultivating. The rest is free access.
   b. From 30 to 70 km^2 fallows shorten, and fertility deteriorates. People get firmer rights over their cultivated, harvested, and fallowed land. There may or may not be some community authority controlling access to the remainder, which is more and more taken up by new farmers. However, labor is still short and marketing expensive. Improvement in transport and marketing facilities may be a necessary prior condition before people can and will invest in land improvement measures.
   c. From 70/km^2 upward, land becomes increasingly scarce and valuable. Almost all land is claimed permanently by particular households and is bought, sold, rented, etc. Labour is more available and animals are controlled. Marketing is easier. Land improvement becomes necessary and profitable.
3. Farmers live by profits. If farming is not profitable, people pursue other options, reducing the labor available for land improvement. Profit-making requires flexibility in the face of new opportunities and constraints, and attention to risk.
4. In most places little land now comes under "traditional" or "tribal" control. However, there may be modern local authorities with powers to make bylaws. We need to understand local power structures to see what will be acceptable and enforceable rules. It is likely that wealthy owners of many animals have disproportionate power compared with many owners of one or two beasts.
5. There is need for decision support tools that would facilitate soil conservation experts in the assessments of the financial risks and gains resulting from a proposed soil and water management practice. (Farmers are already doing it by rough and ready methods.) On-going modeling initiatives in Kenya are expected to go a long way in generating data and information

that can be used by land users and experts alike to assess *a priori* the possible impacts of their future actions. The people who live in a place will always win in the long term, after the project has ceased. The successful project is the one that consults them and goes with, not against, social trends and needs.

ACKNOWLEDGMENTS

This chapter draws upon experience in Kenya and to a lesser extent in Jordan. Mary Tiffen and Francis Gichuki were members of a team from the Overseas Development Institute and the University of Nairobi, which traced environmental, social, and economic change in Machakos District, Kenya, 1930 to 1990, as reported in Tiffen et al., 1994. They also worked together on an evaluation for Sida and the Government of Kenya of the National Soil and Water Conservation Program in 1996, visiting six different districts. Mary Tiffen was also a member of a feasibility study team for the Upper Zarqa River Catchment Area in Jordan in 1979, work carried out for Agrar- und Hydrotechnik GmbH and the Government of Jordan.

REFERENCES

Agrar- und Hydrotechnik GMBH. 1982. Zarqa Soil Conservation Farming and Afforestation Program, Preliminary Report, Essen, July.

Barnes, R. O. 1937. Soil Erosion, Ukamba Reserve, Report to the Department of Agriculture. Memorandum, July, DC/MKS/10a/29/1.

Bunch R. and Lopez, G. 1995. Soil Recuperation in Central America: Sustaining Innovation after Intervention, Gatekeeper series no. 55, London, International Institute for Economic Development, London.

Cheatle, R.J., Nekesa P., and Nandwa, S. 1996. Farmers' Voice for Demand Led Services, Report 20/96, Association for Better Land Husbandry, Box 39042, Nairobi, Kenya.

Ecosystems. 1985–86. Baseline Survey of Machakos District, 1985. Report No. 2: Results of the 1985 photographic aerial survey. Sectoral analysis for the District Planning Unit. Report No. 3: Reinterpretation of the 1981 land use survey. Report No. 4: Land use changes in Machakos District 1981–1985. Nairobi: Ecosystems Ltd for Government of Kenya (Ministry of Finance and Planning, Machakos Integrated Development Programme).

Jaetzold, R. and Schmidt, H. 1983. Farm Management Handbook of Kenya, Vol. 2: Natural Conditions and Farm Management Information, Part IIc: East Kenya (Eastern and Coast Province), Nairobi: Ministry of Agriculture.

Pereira, H. C. and Beckley, V. R. S. 1952. Grass establishment on eroded soil in a semi-arid African reserve, *Emp. J. Exp. Agric.* 21(81):1–14.

Pereira, H. C., Hosegood, P., and Thomas, D. B. 1961. The productivity of tropical semi arid and thorn scrub country under intensive management, *Emp. J. Exp. Agric.* 29:269–286.

Reid, L. 1982. *Soil Erosion in Machakos District,* Nairobi: Ecosystems Ltd.

Simpson J. R., Okalebo, J. R., and Lubulwaa G. 1996. The Problem of Maintaining Soil Fertility in Eastern Kenya: A Review of Relevant Research, ACIAR Monograph No. 41, 60 pp.

SWCB (Soil and Water Conservation Branch), 1994. The National Soil and Water Conservation Program, Nairobi, Kenya.

Thomas, D. B. 1974. Air Photo Analysis of Trends in Soil Erosion and Land Use in Part of Machakos District, M.Sc. thesis, Reading University.

Thomas, D. B. 1991. Soil erosion, in Mortimore, M., Ed., Environmental Change and Dryland Management in Machakos District, Kenya 1930–90: Environmental Profile, ODI Working Paper No. 53, London: Overseas Development Institute.

Tiffen, M. 1976. The Enterprising Peasant: A Study of Economic Development in Gombe Emirate, North Eastern State, Nigeria, London: HMSO.

Tiffen, M. 1996. Land and capital: blind spots in the study of the resource-poor farmer, in Leach and Mearns, Eds., *The Lie of the Land: Challenging Received Wisdom in African Environmental Change and Policy,* London: International African Institute and James Currey.

Tiffen, M., Mortimore, M., and Gichuki, F. 1994. *More People, Less Erosion: Environmental Recovery in Kenya,* Wiley, New York.

Tiffen, M., Purcell, R., Gichuki, F., Gachene, C., and Gatheru, J. 1996. External Evaluation of the National Soil and Water Conservation Programme, Kenya, Final Report, London: Overseas Development Institute.

20 The Participatory Multipurpose Watershed Project: Nature's Salvation or Schumacher's Nightmare?

Robert E. Rhoades

CONTENTS

0-8493-0702-3/00/$0.00+$.50

INTRODUCTION

The participatory, integrated watershed project has become a major development thrust of local, national, and international agencies and governments in the past 5 years. Dozens of interdisciplinary, intersectoral, and interinstitutional projects — both large and small — have been funded and implemented in Asia, Africa, and North and South America. They address not only issues of natural resource management and sustainable agriculture, but also other concerns of governments and local populations (e.g., nutrition, health, community development). Through democratic, participatory planning, such projects aim to address the weakness of more conventional, top-down approaches which have not produced desired results and, in some cases, may have led to increased environmental degradation. This chapter makes an argument for both the watershed focus and the participatory method of engagement. However, evidence that this "new paradigm" works remains anecdotal and unspecified. Preliminary evaluations are clear that such projects are making both conceptual and implementation mistakes which are not the fault of the original conceptualization but of organizational, management, and transaction costs which have diverted such projects from their original goals. These "pitfalls" are listed and recommendations made of ways to overcome them.

GLOBAL ENTHUSIASM: PARTICIPATORY WATERSHED RESEARCH AND MANAGEMENT

The latest movement to hit the development world is "participatory multi-purpose watershed research and management."* Hailed as a reasonable strategy to satisfying the Agenda 21 megademand for sustainable natural resource management, sustainable agriculture, and sustainable livelihoods with one approach, everyone from the tiniest non-governmental organization (NGO) to the World Bank has jumped onto this "bandwagon" as development practitioners fondly call the periodic swings (or "jerks") in focus. Hundreds of communities located in Latin America, Africa, and Asia have been targeted to receive the participatory, integrated watershed approach and more are scheduled in the near future (UNCED, 1992).

The unabashed enthusiasm for "participatory watershed" solutions was highlighted recently in Kathmandu, Nepal, when World Bank President James D. Wolfensohn visited King Birendra (*Kathmandu Post*, 1996). During a press conference, the bank's chief executive officer hailed the signing of the Mahakali ratification treaty which would move forth another hydroelectric power dam and station in the Himalayas. Wolfensohn also said the Bank would throw in another $175 million to the Nepal Power Development Fund to get the derailed Arun Watershed hydroelectric project back on track. This was good news to interested bureaucrats and investors since a rather pesky local population, global environmentalists, and financial jitters

* The watershed scale has been selected as a focal critique in this chapter, but many of these points apply to other scale levels, such as the CGIAR ecoregional approach, which is promoted to be participatory, multisectoral, interdisciplinary, and holistic. Others include landscape, catchment, or river basin scales. The main requirement is that the project involves multiple zones, diverse stakeholders, and cross-ecosystem issues not found in components of farming systems research.

among investors had convinced the bank to cancel the project a year or so earlier. Wolfensohn assured everyone that he was still committed to alternative energy for Nepal as well "as the aspirations of the residents of the Arun Valley." Regarding the latter, Wolfensohn proposed the solution: "...the Bank is working very closely with the German Agency for Technical Cooperation (GTZ) at participatory approaches. I believe that by working with people in the region we will be able to come up with a plan to try and solve the frustration of the people and the acute sense of disappointment" (*Kathmandu Post*, 1996). This was refreshing news since the bank's own sociologist, Michael Cernea, has estimated that by the time the bank's current and projected schemes are finished, about 4 million people will be resettled (normally not voluntarily) and that 1 in 7 bank dollars is lent for projects that relocate people rather than address their aspirations (Chatterjee, 1994; Cernea 1996a). While Chambers (1994, 1995) bemoans the fact that the "last" remains the "last" in the development bureaucracies of the world, at least the participatory rhetoric has made it to the highest office of the world's biggest development bureaucracy.

But not only is the World Bank sold on the multipurpose participatory watershed projects or variants such as catchments, landscapes, river basins, and buffer zones; just about everyone else is as well. Agenda 21 asked committed governments to provide U.S.$13 billion alone for integrated watershed management between 1993 and 2000 (UNCED, 1992; Rhoades, 1997). With the hope for this level of funding, a mushrooming of natural resource management projects at the watershed scale has taken place at national, international, and bilateral levels over the past decade. Virtually every international development agency, most regional agencies, and all of the major bilateral donors have ongoing participatory watershed projects. There are scores of private foundations and NGOs, both large and small, involved as well. It is almost as easy to list the development agencies not pushing some kind of participatory watershed project as those multitudes which are. A few examples of specific projects include:

1. Food and Agriculture Organization (FAO)/United Nations Development Program (UNDP) "Support to Watershed Management in Asia," started in 1989 and covering 10 countries;
2. FAO/Netherlands Cooperative Project on Watershed Management Training;
3. The FAO Asian Watershed Management Network operating in 13 countries since 1994;
4. The U.S. Agency for International Development (USAID) Sustainable Agriculture and Natural Resource Management (SANREM) project operating in five countries;
5. The International Centre for Integrated Mountain Development (ICIMOD) Swiss-funded Watershed Development Project in four Hindu Kush Himalayan countries; and
6. Indo-German Watershed Development Program involving 50 NGOs in 74 watersheds. India, China, and Indonesia have large programs funded internally and with external support in the hundreds of millions of dollars. Both the Philippines and Vietnam have recently gotten multimillion dollar funding for the same end. In Australia, integrated catchment management

(ICM) is being promoted as a strategic framework to bring diverse stakeholders together for natural resource planning (Queensland Government, 1991; Shaw et. al 1995). In the U.S., participatory watershed research and development is making a comeback, as illustrated by the Kellogg-supported Ohio State University initiative (Grant et al., 1997). The trickle of publications on watershed management of a few years back has now turned into a flood, as has the convening of conferences and workshops on the theme (FAO, 1986; Farrington and Lobo, 1997; including the conference on which the present volume is based).

Despite this rush toward watersheds as a sustainable development focus, what is really known about how to do participatory research and development at the watershed scale? The truth is, it seems, that many agencies and specialists are venturing into unknown territory by combining multiple objectives with untried organizational approaches for the first time. Unless this process is given serious prior thought and continuous reflection, in a few years the reputation of both participation and watersheds might fall into disrepute and join the graveyard of other failed "development white elephants." A rather significant audience of critics, many representing the conventional production or component approach, is watching very carefully and very, very critically. If participatory watershed management falls flat, as it well might, this would mean a return of funding for the less participatory transfer-of-technology programs which have been so strongly criticized in recent years. Given that participatory watershed projects tend to be ambitious, wide ranging, and complex, they also possess a high probability of failure in numerous ways. Some early evidence, based on the author's observation in the field, suggests that participatory-justified projects are slipping back into a business-as-usual component approach (hydrology without the people) and emphasizing donor or researcher demands instead of local needs. The purpose of this chapter is to pause for a moment and think critically about both problems and needs in an effort to reflect on what participatory watershed research and management might need to succeed.

In doing so, the following set of questions will be addressed:

1. Why is there renewed international interest in watersheds?
2. Why is participation the preferred development approach?
3. What is the evidence so far that the participatory watershed approach is viable?
4. What challenges do participatory sustainable watershed management projects face?
5. How might we address these challenges?

WHY WATERSHEDS?

While the watershed approach is nothing new (e.g., in the U.S. or the Philippines), the recent intensity of interest and donor support can be readily justified on many grounds (Hamilton and Bruijnzeel, 1997). One good argument is that a rather large portion of the world population lives in highland-lowland watersheds. In Asia alone, approximately 25% of the world population lives along river systems in mountainous watersheds with another 25% living in the adjacent lowlands (Sharma, 1997). These natural systems function as both the water towers and pipelines for water supply to such important biomes as the Amazon Basin and the Gangetic Plains, to mention only two critical world areas in terms of global ecosystems. While the main reasons for watershed development are typically for irrigation to increase agriculture, hydroelectric generation, and flood control, multipurpose projects today are more ambitious and also involve conservation, health, education, and equity components. With global warming, deforestation, expansion of human populations up and down mountains, and resultant increases in drought, flooding, erosion, and crop loss, watershed degradation is reportedly widespread in all the major continents, both north and south (Hamilton and Bruijnzeel, 1997). Unless these vital systems are cared for and sustainability managed, the basic life-supporting systems on Earth may be permanently damaged. Certainly from a biophysical perspective, a focus on well-defined hydrological systems makes sense. Water and land use have reciprocal effects: land use is water dependent and water quality and quantity are impacted by land use. The two resources cannot be treated as separate development issues, as is often the case (Lundqvist et al., 1985). By studying interactions in the hydrological system instead of component research on crops or specific resources, the watershed (or landscape) unit broadens the analytical framework to encompass cross-ecosystem linkages, including upstream and downstream dynamics. Given that scientists can clearly delimit the study unit, input–output studies, decision-making models, and expert systems are practical and valuable (Yaalon, 1994). Compared with ecosystem, landscape, or catchment research or management frameworks, the "watershed" concept has the advantage of being understandable to laypersons, policy makers, and funders as an organizing and operational concept.

WHY PARTICIPATION?

Various authors over the years, however, have pointed to problems with focusing on watersheds solely as hydrological units if the objective is management and sustainable livelihoods (Jinapala et al., 1996). For starters, humans manage or degrade land and water. Beyond natural processes, land and water do not manage themselves. Second, people do not live or manage resources in the way that water flows, although this can sometimes influence their decisions (Fox, 1992; Maddock, 1996). Biophysical and sociocultural boundaries rarely coincide. In fact, watersheds are normally not management units at all for human populations (except supralocal institutions created by governments). Within, above, and across a typical watershed are numerous human boundaries (real and perceived) such as ethnic groupings, provincial governments, religious sacred grounds, villages, parks, or individual farms. Sometimes the function of a human community is to bridge two or more watersheds. In Nepal, for example, villages are frequently built along the ridges running between two watersheds. Although the watershed might be a clear unit around which to focus planning and convince donors, primary attention in development should logically still be given to the institutional landscape and human decision making since these constitute the causal factors underlying land or water degradation.

Given this messy overlay of human activity and naturally defined watersheds, a logical adjustment is to combine the watershed focus with another recent, popular development approach called "participation," that is, the full involvement of local populations in the identification of problems and solutions with teams of scientists, planners, and development specialists (Blackburn and Holland, 1998). Participation also dovetails well with the global objectives of decentralization and democratization, twin global objectives pursued by world powers in the post-Cold War era and aimed at integrating even the most marginal Third World communities into the world economy. Local people are also demanding more of a say in what goes on in their territory under the name of development. By giving them (or them reclaiming) a voice and by tapping their knowledge in making decisions on research and management questions, it is assumed more sustainable, locally relevant management systems can be designed and adopted (Hufschmidt, 1986). In promoting participation, evidence is marshaled that centrally controlled, government-run watershed projects in the past have suffered, since there have not been local ownership and management (Kerr et al., 1996; Farrington and Lobo, 1997).

By combining the two perspectives (watersheds and participation), the academic concerns of both social and biological scientists are also answered. Biological scientists complain that if watershed management is seen as purely a socioeconomic matter, then the powerful forces of nature are ignored. Conversely, the social scientists argue that if viewed as merely hydrology or natural processes, the powerful roles of institutions, culture, and economics are minimized. Also, the marriage of "participation" and "watershed" makes room for everyone: ecologists, agronomists, hydrologists, anthropologists, economists, planners, NGOs, national and local governments, farmers' organizations, and so on. Central to most projects today and guaranteed inclusion by most donors are NGOs, many of which base their development approach on the "small is beautiful" philosophy of Schumacher (1973). Some "organizational" variant of the appropriate technology philosophy must be included in every project design, mainly under the guise of "building local institutions" and "brokering or articulating local communities with scientists or government planners." The ideal is to carry out in a watershed, or more frequently these days a network of continental or globally linked watersheds, a kind of "Small is Beautiful" program ("economics [development] as if people mattered") which addresses complex "global" problems such as water quality/quantity, soil erosion, biodiversity loss, emigration, etc. The watershed approach makes the transfer-of-technology production focus of earlier decades look narrow, segmented, reductionist, top-down, and simplistic.

Participation at the watershed scale today means much more than the "participation" of farmers and agricultural scientists as operationalized in the farming systems programs of the 1980s. Few proposals for natural resource management or sustainable agriculture will be funded unless they guarantee broad-spectrum participation at many levels (interdisciplinary, intersectoral, interinstitutional), especially if the focus is on watershed development (Blackburn and Holland, 1998). To reflect reality, both biophysical and sociodemographic levels must be scaled and linked with each other. Logically, research and development must flow through these scales.

One of the major aspects of "involving local populations" is giving them a voice not only in the project design and execution under the control of academics and planners, but also in the up-front identification of the problems to be solved (Rhoades and Booth, 1982). These projects are, therefore, "demand-driven" (based on local perceived and articulated needs). Several "how to do it" manuals on participatory methodologies replete with interesting cartoons and exercises have been published, but these approaches do not fully capture the essence of the new paradigm (Mascarenhas et al., 1991; Paliniswamy et al., 1992). The new model of the multipurpose watershed project is far more ambitious; it is not only participatory with local people, but involves interaction between NGOs, governments, universities, international agencies, and the private sector in a "participatory brew." All of these stakeholders are seen as crucial for addressing complex problems and challenges in "improving the livelihoods of the people" while "saving nature" at the same time. In addition, such projects are expected to advise, catalyze, and network with policy makers to inform and perhaps change their decision making.

WHAT IS THE EVIDENCE SO FAR THAT PARTICIPATORY WATERSHED MANAGEMENT IS A VIABLE APPROACH?

While the participatory watershed rationale is appealing to donors (it answers several chapters of Agenda 21), operationalizing and executing such projects with field practitioners are proving to be far more difficult than realized. Some sympathetic observers with field experience hint that the participatory approach has not delivered the goods and should be reevaluated. Fisher (1995) even suggests doing away with the term "participation" since "token use of the terminology [participation] has devalued it; participation has come to mean so many things to different people that it means nothing." Instead, he prefers "collaboration" as a neutral term. Yet "collaboration" can also be a loaded term. In the Philippines, the term "collaborator" is a turncoat who supported the Japanese in WW II. In the halls of aid programs and development agencies, some critics of bottom-up development are starting to argue wistfully that the participatory rhetoric has outrun the ability to accomplish and that we should consider a return to component research, or — at a minimum — watershed management without the noise of participation. In fact, some participatory watershed projects end up as conventional hydrology studies despite the up-front courtesy to "people and participation" in the project paper justification.

Since the rationale for funding participatory approaches is redress of sins of the top-down, heavy subsidy approaches of the past which have alienated local populations and even contributed to further land and water degradation, the burden of proof has been — perhaps unrealistically — placed on the new "participatory" approach. Unfortunately, the latest boom in participatory watershed projects is fairly recent and the first assessments are only now starting to be made available. Most evidence of success or failure at this point is almost entirely anecdotal; if the participatory approach is the answer, then its proponents in a few years will have to prove beyond mere rhetoric that it actually works. One project, the New Horizons

Project, supported by IIED (International Institute for Environment and Development) and the Australian government, reviewed 22 participatory watershed development projects and concluded that participation is superior to top-down, coercive projects, but unfortunately the impact measurements are vague and based on the agencies' own findings instead of independent evaluation by external reviewers (Farrington and Lobo, 1997).

Since the late 1970s there has been a major thrust in getting scientists and planners in agriculture and natural resource management to accept and institutionalize the concept of participation (Rhoades and Booth, 1982). As late as 1990, however, it was still problematic for most agricultural scientists to accept that farmers could have an active role in technology design. Farmers were seen as recipients of science and technology, not as partners in the process. The groundbreaking work at CIP (Centro Internacional de La Papa) in developing and promoting the farmer-back-to-farmer model was one precursor to the farmer-first movement later promoted by Chambers (1994). The postharvest team that developed the farmer-back-to-farmer model was made up of four to five scientists. But the pre-Rio period was a much simpler development world made up of small interdisciplinary teams (maybe an anthropologist, an economist, and a couple of biological scientists) clearly focused on a single crop or technology. Today, participation has been elevated onto the stage of Big Science and Big Development and into new arenas such as social forestry, community forestry, joint forest management, participatory natural resource management (NRM), environmental stewardship, comanagement of protected areas, integrated conservation development projects, and so on. Since such interdisciplinary, multipurpose projects have to answer donor countries' commitments to the United Nations Commission on Environment and Development (UNCED), politicians and bureaucrats are also involved. Participatory research today refers to a radically different organizational structure than the small, clearly focused farming system teams of a decade ago. Donors now require participation at many levels, involving layers of stakeholder groups and communities, not just a target group of regional farmers planning a similar crop. A quantum leap from the early form of simple participation to today's complex interaction has taken place (see Schwitters, 1996, for a general account of the problems besetting "Big Science").

Unless practitioners think long and hard about how to do participatory watershed management, the movement may indeed fail. Failure, if it occurs, will not be caused by the critics, but because the proponents (both donors and implementers) of this exciting approach have not done their homework and come to grips with needed programmatic changes. In the absence of documented success case studies showing how participatory watershed management works, the attempt in this chapter is to alert the development community of problems being suggested by ongoing projects. Many of these observations come from the author's direct involvement in the U.S. AID-funded SANREM project in Ecuador and the Philippines.

THE ROAD TO PARTICIPATORY NATURAL RESOURCE RESEARCH IS LINED WITH PITFALLS

Two kinds of challenges are currently emerging in participatory watershed management: the conceptual demands and the implementation requirements. Although theory and praxis obviously crisscross and overlay, it is useful to separate them for heuristic purposes. If these challenges are left unattended, they can make many watershed projects more akin to Schumacher's nightmare than to nature's salvation as so many project proposals are promising to Agenda 21 donors.

THE CONCEPTUAL PITFALLS

Pitfall 1: Reinvent the Wheel

Field practitioners and project managers in watershed management need to be learning from other projects and creating mechanisms to foster cross-fertilization and cross-learning. Each project seemingly starts out in a vacuum with little attention to what has gone before, and — as a result — commits the same mistakes. Few honest published evaluations are available, and to date no international conference has been held specifically for the purpose of critically sharing experiences. The few available publications that evaluate successes and failures of participatory research have been published in-house with limited dissemination of the findings. For example, Sharma and Krosschell (no date) have written a paper called "Analysis of Lessons Learned from Case Studies in People's Participation in Watershed Management in Asia," which is really only available to the limited mailing list of the FAO/UNDP Participatory Watershed Project in Asia. In culling through the participatory experiences to see what worked and what did not in the FAO/UNDP network in Asia, they delineated three types of approaches to participatory watershed management:

1. Indigenous *in situ*, which by their nature are successful;
2. Building on local culture institutions, which affords some degree of success;
3. Facilitation by outsiders without heed to local culture, which are failure prone.

They found that *only* in situations where the project builds on the cosmology and indigenous institutions of the people will the project be long term and sustainable, a finding which seems highly relevant to many projects. This seems particularly relevant for projects trying to impose global concerns of climate change or biodiversity loss on local people. Their main finding, however, is that few guidelines or case studies are available to help make participatory watershed projects relevant to local populations.

Pitfall 2: Scale

Despite a large and thoughtful literature available in geography and ecology on scale and hierarchy theory, designers or implementors of participatory watershed or landscape projects have seemingly read very little (Stone, 1972; Allen and Starr, 1982;

Meentemeyer, 1989; Fox, 1992; Maddock, 1996). A great deal of confusion in watershed research comes from different kinds of researchers who study different scales without reference to their location in either the spatial or sociodemographic hierarchy. Planners typically plan with the same confusion. Scaling up and down between levels and across sites seems crucial, but this exercise is rarely carried out either in the planning stage or during project implementation. Another problem is that implementors of multiscale participatory projects have given very little thought to how spatial scales in physical environments are different from the spatial scales of human organization (Fresco, 1995). Similarly, scientific research teams and development/management teams typically operate on different spatial and temporal scales. Reductionist research operates on very fine spatial scales and for short time durations. Many scientists prefer to follow this pattern because of funding constraints, but watershed projects propose to expand spatial and temporal scales in reflection of a more holistic environment ethic. NGOs, on the other hand, prefer to operate at the level of community since this is an organizing unit, but they have few skills at the watershed or landscape scale. A great deal of pushing and shoving has taken place within projects to get funds and resources focused on whatever scale level is comfortable to each of the diverse stakeholders. These "scale wars" are often waged at the unconscious level, although with manifest conflicts between groups. NGOs insist on communities, provincial governors insist on province, the district agricultural officer the agricultural planning district, and the scientist tries to insist on the scientist's hydrological unit, while the donor insists on the regional or global scale. The challenge then becomes how to integrate results between disciplines/organizations and transfer the results from one scale to another. More theoretical and methodological attention needs to be given to the scale issue. One of the few projects that has concerned itself with the issue of scale and how scaling up from a specific study can be accomplished is that of the Indo-Germany Watershed Development Project (Farrington and Lobo, 1997), although it is basically confined to similar watersheds within India.

Pitfall 3: The Participatory Methodology Fetish

The 1990s has seen the growth of participatory research methodologies (participatory rural appraisal, rapid rural appraisal, etc.) best known through the writings of Chambers (1994). While these methods are a fresh counterpoint to the unimaginative questionnaire, the present reliance on and interpretation of such approaches may have become counterproductive and a violation of their original intent (IDS Workshop, 1998). Unfortunately, the emphasis on interaction and speed leads to superficiality in the way communities are approached and representativeness of data collected. Ironically, much of the participatory methodology becomes condescending and patronizing of local populations, just the opposite of the original intent of dispensing with the long questionnaire and formal research instruments which once alienated local people. Rather than treating local people with respect and as colleagues, participatory methods sometimes treat them more like school children by playing games, drawing exercises, and other 24-hour research remedies. Biological scientists who become exposed to these approaches often become more enthusiastic about participatory methods than seasoned social scientists. Worse yet, since NGOs position themselves as "facilitators" of participatory events, social scientists are sometimes marginalized or even pushed out of the project. All too often, social scientists are accused of being "top-down" by turf-guarding NGOs and overzealous biological scientists.

Another meaning of participation which is becoming more prevalent deals with the nonlocal project participants, that is, individual and institutional participation among scientists, NGOs, government officials, policy makers, and international agencies (Gaventa, 1998). These "participatory" interactions involve people who speak roughly the same "development" language, are paid by the project, and have professional objectives beyond the local setting. These interactions tend to be very costly in terms of travel, conferences, and salaries. The greater the number of nonlocal participants, the greater — in exponential increase — the "transaction" costs (see Pitfall 8, on Stakeholder Complexity, below).

Pitfall 4: Social Underdesign of Projects

In many participatory watershed projects, biological scientists are expected to take care of "science" while social scientists/NGOs presumably take care of "participation" (and an NGO can do participatory research just as well as a Ph.D sociologist). The very science we need most in watershed research, however — a solid social science — is the one seen as the most dispensable. Participatory methodologies facilitated by NGOs should not be confused with or thought of as a substitute for solid, professional social science research.* Although many NGO community facilitators have a first degree in an applied social science, they are rarely professional researchers knowledgeable about research design or data collection. Reliance on them, therefore, for social information may instead lead to the problem of the social underdesign of watershed projects (Kottack, 1991; 1995). Most serious professional social scientists have a great deal of problems accepting many of the fly-by-night participatory methods, although curiously many biological scientists, who often manage the projects, are ardent supporters. All too often, professional social science is confused with community organizing/social work as practiced by NGOs.

Serious social science questions which must be answered in watershed projects require as much careful research design and data collection as questions in the biological sciences. Critical questions involve:

What are the community boundaries?
Who are the legitimate authorities?
Who are the stakeholders?
What are inter- and intra-group conflicts?
What are the population livelihood dynamics?

---

* Another complaint expressed by developing country university professors is that it is no longer possible to place their students in development projects for practicums or thesis research because NGOs have cornered the market. Thus, it seems in some cases NGOs have a negative impact on both government and university involvement.

In the case of SANREM-Ecuador, only after conducting a complete household census was it realized that the participatory community methods had overlooked almost half of the local population in the problem definition exercise (43% of the population turned out to be landless or sharecroppers who did not attend the community meetings). Intensive social science research is *absolutely* necessary to make it clear that hidden agendas, internal conflicts, power struggles, shifting alliances, resource and territorial struggles within communities must be understood and accounted for in the project implementation. If these are not given credence along with "participation" and "technical solutions," the project will likely experience difficulties.

Implementation Pitfalls

Pitfall 5: Great Expectations

By promising to address multiple, often contradictory objectives, the participatory multipurpose watershed project can inadvertently become its own worst enemy. First, high expectations are created in the participatory process itself, when in meeting after meeting local people, NGOs, scientists, and government officials hope to attract attention to their vested interests. Demand-driven research and development means the various stakeholders talk about all local problems, not just water. The talk then gets confused about what can be realistically accomplished in the project time frame and budget. By having promised to involve all important stakeholders at the beginning ("participation"), each group creates its own set of expectations. Second, conflicting objectives embedded in participatory watershed projects create further ambivalence (food production and environment, development and environment, maximization of economic benefit and conservation, individual and societal costs). Is it really possible to accomplish all of these, or are some countervailing? Frequently, watershed projects that are primarily research (e.g., SANREM CRSP funded by U.S. AID) create confusion for both researchers and local populations by emphasizing "participation" as a way to set research priorities, but do not allow resources for development. The project *itself* may be evaluated on research output (publications; high-quality science) and not on attention to local people's needs. When project funds are cut, often local people are left with no tangible outputs, although they have expended a great deal of time in the participatory process. This is unfair to both local people and to researchers.

Pitfall 6: Tragedy of the Participatory Commons

Another problem in the implementation of such projects is a kind of organizational "tragedy of the commons" ("whatever belongs to everyone belongs to no one"). First, it is very hard to get a consensus if all stakeholders have the same weight in deciding what should be done (everyone has an agenda). Second, when people are not allowed to do what they do best (initiative and ideas are killed through group decision-making processes), they simply turn away from group interaction. If budgets are competitive, each stakeholder in the watershed project looks out for its needs. When budget cuts occur, each stakeholder group entrenches in terms of its own short-run goals, instead of focusing on what is best for the whole group. Like the tragedy of the commons, no one takes responsibility for the whole project. Organizationally, we pasture our animals (or fish) until the resource is depleted, always blaming others in the process. Unfortunately, rather than blaming organizational greed and poor management as the point of failure, the concepts of participation and watersheds may ultimately be the ones to catch the blame.

Business consultant Capozzi has written a small book of sayings for businessmen (Capozzi, 1997). One saying is appropriate: "In all my years in business, I have found that people in meetings tend to agree on decisions that, as individuals, they know are dumb." This is likewise true of participatory watershed projects which have true participatory decision making. The complexity of interdisciplinary, intersectoral, interinstitutional, participatory research, and management of natural resource and sustainable agriculture at the watershed level, may reflect both true need and real-world circumstances. However, unless carefully guided, full participation makes it very difficult to have agreed-upon assumptions, methodologies, goals, and operating procedures. The assumption of participatory research is that all parties should have a voice, but this often creates a lack of structure to accomplish the appropriate goals.

Pitfall 7: Duplicating Management Structures

Corollary to the "lack of focus" and "commons tragedy" is the tendency to create artificial, externally conceived committees/groups through which the watershed project management and workers can operate. Outsiders to a location (NGOs, foreign scientists, government agencies) strive for organized structure to work though. Locally, it is not always clear with whom you negotiate within watersheds (remember a watershed is not a social or political reality except to watershed scientists). Often no clear formal organizational structure is evident to outsiders, or there may be several competing structures. Frequently, in traditional societies the leaders rotate annually among households. The need for a formal structure is something that bureaucratic watershed projects need in the same way that colonial powers needed "chiefs," even when they did not exist. Most participatory watershed projects are also rich in on-the-ground gossip about local maneuvering between political rivals who are using the project as a stage upon which to build alliances, garner resources, and ultimately unseat the archenemy. Through local and external politics, project committees often evolve to have lives of their own beyond any local indigenous structure. A locally established project coordinating office can become another layer of bureaucracy, with its own vested interests, needs for resources, and control intrigues.

It is important to distinguish between *in situ* local organizations and those organizations set up or stimulated by outsiders to facilitate certain agendas (e.g., NGOs). A strategy that focuses on existing *in vitro* user-based institutions rather than on setting up new organizations or committees will likely be more successful (Sharma and Krosschell, no data). This does not rule out the possibilities that new structures might be needed, or new organizational features added, but any new structure is more likely to be effective if the project follows these steps recommended by Fisher (1995):

Step 1. Recognize existing systems and leave them alone (do not create new ones if not needed).
Step 2. Strengthen existing systems if and when they are inadequate, e.g., legal tenure, usufruct rights.
Step 3. Assist in establishing new institutions only if they are necessary, and pay attention to existing use and rights. Develop them on the basis of organizational parsimony (do not make it more complicated than it need be). Involve people with legitimate local interests.
Step 4. Where there are relatively separate populations with conflicting use rights, build new institutions capable of mediating between and communicating with diverse stakeholders.

Pitfall 8: Stakeholder Complexity and Competition

One of the needs for solid social science is to uncover differences and mutual interests among the many stakeholder groups in a watershed. There are conflicts within communities, among regional and national stakeholders, and between and within international groups such as development agencies, tourists, miners, and loggers. Due to the number of stakeholders, the probability for conflict increases. This reality runs counter to participatory rhetoric. For example, there seems to be an undercurrent of opinion that NGOs are essential if collaborative management is to succeed. Nonlocal NGOs are income-generating agencies beyond the immediate community of users that survive by linking with the international agencies/donors/governments to provide mediation and to act as gatekeepers. An associated view is that government agencies cannot implement collaborative resource management effectively, perhaps because of an unwillingness to give up entrenched power or because of an incompatibility with devoted management. Before dismissing the role of government in favor of NGOs, it must be remembered that NGOs have their own agendas, and they cannot perform all of the functions of government. Unfortunately, some NGOs are sometimes more interested in developing their own organizations and enhancing their payrolls than in helping local communities do the same.

Potential failures in the participatory watershed approach may also be related to our lack of understanding of how to do team work in multiobjective, multi-institutional settings. The truth is that large-scale team research and development has never been tested and proved at the natural resource/watershed level. The main experiences in interdisciplinary research have been in highly focused crop/component research programs (Rhoades and Booth, 1982). Even these are not easy, but they are workable. Participatory natural resource teams are of a totally different political order, with a complexity and intensity of interaction brought about by combining NGOs, local organizations, governments, universities (several disciplines) into a single watershed, none of which has ever really worked together in the past.

CONCLUSION

Is the participatory multipurpose watershed project nature's salvation (Agenda 21's answer), or will it become Schumacher's worst nightmare? Two conclusions can be drawn from the still inconclusive information now coming in regarding participatory watershed management. The first is that despite the enthusiasm for and the appeal of the approach, the ongoing projects have yielded more rhetoric than effective output (the "Green Revolution" had a couple of decades to prove itself; but participatory watershed management must do it in 3 to 5 years!). The second is that most agencies and observers still steadfastly remain convinced about the promise of such an approach. While learning to manage such projects should increase the likelihood that they will succeed, this is easier said than done. When confronted with the complex reality of actually carrying out such a project, faced with multiple local stakeholders with the sanctioned right to press for their needs (not that of Agenda 21 or scientists), practitioners need not only a "paradigm shift" but good science, appropriate methods, organizational skills, and donor patience. After all, many of these projects aim for nothing less than quality science and "impact-oriented" development, locally defined and globally relevant, and interdisciplinary/interinstitutional and intersectoral. All in one project! Without attention to organizational issues, the efficiency and clarity critics and donors are clamoring for these days may come at the cost of broad local participation, especially of local people (our hosts in the watershed, which makes us the guest, a fact often forgotten). It will probably be easier to reduce the number of local stakeholders than to dethrone a powerful university, government agency, or NGO. Is this the price to be paid? Will local communities, which all over the world are exerting for the first time their rights to determine what kind of activities go on in their territory, continue to buy the idea that scientists can run around their villages doing experiments, administering questionnaires or participatory rural appraisals? Will they continue to buy the idea that an urban-based NGO is necessary to link them to the outside world? Such communities are becoming "development" weary. We need to try and put ourselves in their place. If we do not convene soon to share experiences, learn from our mistakes, and provide hard-hitting assessments of the multipurpose participatory watershed project, the baby may indeed go out with the bathwater. Our hearts may be in the right place, but where are our heads?

REFERENCES

Allen, T. and T. Starr. 1982. *Hierarchy Perspectives for Ecological Complexity.* Chicago: The University of Chicago Press.

Blackburn, J. and J. Holland, Eds. 1998. *Who Changes? Institutionalizing Participation in Development.* London: Intermediate Technology Publications, Ltd.

Capozzi, J. 1997. *Excerpts from If You Want the Rainbow You Gotta Put Up with the Rain.* CT: JMC Industries, Inc., Villard Books, New York.

Cernea, M. 1996a. The Risks and Reconstruction Model for Resettling Displaced Populations. Keynote address at the International Conference on Reconstructing Livelihoods: Toward New Approaches to Resettlement, University of Oxford, September 9–13.

Cernea, M. 1996b. Public policy and responses to development-induced population displacements, *Econ. Political Wkly.* 31(24): 1515–1523.

Chambers, R. 1994. The origin and practice of participatory rural appraisal. *World dev.* 22(7): 953–969.

Chambers, R. 1995. Poverty and livelihoods: whose reality counts? *Environ. Urbanization* 7, April.

Chatterjee, P. 1994. Same old World Bank, same old mistakes, *Nation*, February 21.

FAO. 1986. Watershed Management in Asia and the Pacific: Needs and Opportunities for Action, RAS/85/017 Technical Report, Rome: FAO.

Farrington, J. and A. Bebbington. 1994. From Research to Innovation: Getting the Most from Interaction with NGOs in Farming Systems Research and Extension. IIED Gatekeeper Series No. 43.

Farrington, J. and C. Lobo. 1997. Scaling up participatory watershed development in India: lessons from the Indo-German Watershed Development Programme, *Nat. Resour. Perspect.* 17: 1–6.

Fisher, R. 1995. *Collaborative Management of Forests for Conservation and Development,* IUCN and World Wide Fund for Nature, Geneva, Switzerland

Fox, J. 1992. The problem of scale in community resource management. *Environ. Manage.* 16(3): 289–297.

Fresco, L. O. 1995. Agroecological knowledge at different scales, in *Eco-Regional Approaches for Sustainable Land Use and Food Production,* J. Bouman, A. Kuyvenhoven, B. Bouman, J. Luyten, and H. G. Zandstra, Eds., Dordrecht: Kluwer Academic, 133–141.

Gaventa, J. 1998. The scaling-up and institutionalization of PRA: lessons and challenges, in *Who Changes? Institutionalizing Participation in Development*, J. Blackburn and J. Holland, Eds., London: Intermediate Technology Publications, Ltd, 153–166.

Grant, L., T. Payne, and B. Stinner. 1997. Report to the Kellogg Foundation, Ohio Agricultural Research and Development Center, The Ohio State University, Columbus.

Grimble, R., M. Chan, J. Aglionby, and J. Quan. 1995. Trees and Trade-offs: A Stakeholder Approach to Natural Resource Management, IIED Gatekeeper Series No. 52.

Hamilton, L. and L. A. Bruijnzeel. 1997. Mountain watersheds — integrating water, soils, gravity, vegetation, and people, in *Mountains of the World*, B. Messerli and J. Ives, Eds., New York: Parthenon, 337–370.

Hinchcliffe, F., I. Guijt, J. Pretty, and P. Shah. 1995. New Horizons: The Economic, Social and Environmental Impacts of Participatory Watershed Development. IIED Gatekeeper Series No. 50.

Hufschmidt, M. M. 1986. A conceptual framework for watershed management, in *Watershed Resources Management*, Easter, K., J. A. Dixon, and M. Hufschmidt, Eds., Boulder, CO: Westview Press, 17–31.

IDS Workshop. 1998. Reflections and recommendations on scaling-up and organizational change, in *Who Changes? Institutionalizing Participation in Development*, J. Blackburn and J. Holland, Eds., London: Intermediate Technology Publications, 135–144.

Jinapala, K., J. Brewer, and R. Sakthivadivel. 1996. Multi-level Participatory Planning for Water Resources Development in Sri Lanka, IIED Gatekeeper Series No. 62.

Kathmandu Post. 1996. World Bank Chief Very Encouraged by Mahakali Ratification, Vol. IV, No. 237, October 13.

Kerr, J., N. Sanghi, and G. Sriramappa. 1996. Subsidies in Watershed Development in India: Distortions and Opportunities, IIED Gatekeeper Series No. 61.

Kottak, C. 1991. When people don't come first: some sociological lessons from completed projects, in *Putting People First*, M. Cernea, Ed., New York: Oxford University Press.

Kottak, C. 1995. Participatory development: rhetoric and reality, *Dev. Anthropol.* 13(1 and 2): 1–7.

Lundqvist, J., U. Lohm, and M. Falkenmark, Eds. 1985. *Strategies for River Basin Management Environmental Integration of Land and Water in a River Basin,* Boston: D. Reidel.

Maddock, T. 1996. Issues of scale in planning and resource management: the case of the Broad River Watershed, in A River Runs through It: Human Dimensions of Natural Resources Management in the Broad River Watershed, R. Rhoades and S. Talawar, Eds., Research paper 1, LANRA, Department of Anthropology, University of Georgia.

Mascarenhas, J., P. Shah, S. Joseph, R. Jayakaran, J. Devavaram, V. Ramachandran, A. Fernandez, R. Chambers, and J. Petty. 1991. Participatory Rural Appraisal, RRA Notes. No. 13, August. London: International Institute for Environment and Development and MYRADA, Bangalore, India.

Meentemeyer, V. 1989. Geographical perspectives of space, time and scale. *Landscape Ecol.* 3(3/4): 163–173.

Miller, R. 1994. Interactions and collaboration in global change across the social and natural sciences, *Ambio* 23(1): 19–24.

Paliniswamy, A., S. Subramanian, J. Pretty, and K. John. 1992. *Participatory Rural Appraisal for Agricultural Research at Paiyur, Tamil Nader,* London: International Institute for Environment and Development.

Queensland Government. 1991. Integrated Catchment Management: A Strategy for Achieving the Sustainable and Balanced Use of Land, Water, and Related Biological Resources, Brisbane, Queensland Department of Primary Industries.

Rhoades, R. 1987. *Pathways toward a Sustainable Mountain Agriculture: The Hindu Kush Himalayan Experience,* Kathmandu, International Centre for Integrated Mountain Development.

Rhoades, R. and R. Booth. 1982. Farmer-back-to-farmer: a model for generating acceptable agricultural technology, *Agric. Admin.* 11: 127–137.

Schumacher, E. F. 1973. *Small Is Beautiful: Economics as If People Mattered.* New York: Harper and Row.

Schwitters, R. 1996. The substance and style of "Big Science," *Chron. Higher Educ.*, February.

Sharma, P. 1992. An Overview of the Formal Watershed Management Related Education and Training Programs in Asia and Tentative Training/Workshop Work Plan (manuscript).

Sharma, P. 1992. The experience of Asian watershed management network (Asian WATMANET), paper presented at the regional workshop of the Asian Pacific Mountain Network, Kathmandu, Nepal, March 17–19.

Sharma, P., Ed. 1996. Recent Developments, Status and Gaps in Participatory Watershed Management in Education and Training in Asia, PWMTA-FARM Field Document No. 6, Kathmandu, Nepal.

Sharma, P, Ed. 1997. Participatory Processes for Integrated Watershed Management, PWMTA-FARM Field Document No. 7, Kathmandu, Nepal.

Sharma, P. and M. Wagley, Eds. 1995. The Status of Watershed Management in Asia, WMTUH/FARM RAS/93/063 Field Document No. 1 and FARM Field Document No. 3. Kathmandu, Nepal.

Shaw, R., J. Doherty, L. Brebber, L. Cogle, and R. Lait. 1995. The use of multi-objective decision making for resolution of resource use and environmental management conflicts at a catchment scale, *Am. Soc. Agric. Econ.*

Stone, K. 1972. A geographer's strength: the multiple-scale approach, *J. Geogr.* 71(6): 354–362.

United Nations Commission on Environment and Development (UNCED). 1992. *Agenda 21: Programme of Action for Sustainable Development,* New York: UN Publication.

Yaalon, D. H. 1994. On models, modelling, and process understanding, *Soil Sci. Soc. Am. J.* 58: 1276.

21 People's Involvement in Watershed Management: Lessons from Working among Resource-Poor Farmers

T. Francis Shaxson

CONTENTS

0-8493-0702-3/00/$0.00+$.50

INTRODUCTION

Watershed management projects imposed in areas peopled by small farmers have often been expensive failures. An altered approach to achieving conservation-effective production has changed agroecological and socioeconomic components. Good management of catchments ("watersheds") builds up from the aggregation of a multitude of microlevel improvements which develop over time and space from individuals' to whole communities' decisions and actions. People's own observations and perceptions provide good starting points for monitoring and evaluation of changes. Examples are given of the positive interactions between people and conservation-effective agricultural technologies which have resulted in large areas of farmland, on landscapes of catchments, showing sustainable and economic improvements in livelihoods, soil productivity, and stream flow hydrology.

Populations continue to increase in many tropical and subtropical countries, and tillage and grazing activities by resource-poor small farmers spread onto more fragile, steeper, and less-productive lands. Damage to land resources of soils, vegetation, water supplies, and microclimates are often associated with such expansions. "Watershed management" has often been projected as the logical way to avoid or repair such problems, although achieving successful and cost-effective results on a large scale has been problematic (Doolette and Magrath, 1990, p. vii). Is watershed management, as a relatively formal set of concepts and types of action, in every case the optimum approach from the people's perspective as well as from the technical perspective? Should it be instigated from the beginning, working from broad concept to required individual actions, or should the ideal end condition be reached by progressing from a multitude of small-scale, conservation-effective actions to catchment-wide coverage?

PEOPLE

FARMERS: ULTIMATE DECISION MAKERS

Irrespective of their relative lack of resources, small farmers are the ultimate managers and day-to-day decision makers about what happens to their lands, within the on-farm and off-farm constraints and potentials that surround them (Shaxson et al., 1989, p. 31). Thus, people are key and inseparable components of agroecosystems. From individuals' viewpoints there is a sequence and hierarchy of their self-interests, from the individual outwards, through family, common-interest group, local community, and beyond. Attachment to their land often transcends purely practical and commercial aspects of their lives and is often a powerful determinant in their decision making. But especially in conditions of poverty and stress, personal considerations generally override those of larger groupings of which the individual or family is a member. Maintaining productivity and outputs from a field or farm are primary personal and family concerns. If small farmers do not favor, cannot manage, or do not see acceptable benefit from some recommended practice, they are unlikely to implement or maintain it. If they feel they have been sidelined from decision making about what is going to be done in a particular situation — or have not even been consulted at all — they are unlikely to be very enthusiastic about what is proposed by others. But, on the other hand, as relevant and acceptable improvements are attained, individuals' interests begin to expand beyond the family into common-interest groups and ultimately involve whole communities.

EXAMPLE 1: PEOPLE'S CONCERNS ABOUT CATCHMENTS START SMALL

1. *A multilevel approach to participatory water-resources planning: Sri Lanka*

In the multi-level approach, meetings are held with farmers in local communities to get an initial information base and to introduce the planning approach and concepts. Then participants from different local communities within a watershed [catchment] meet at participatory planning sessions to exchange information about local conditions in different parts of the watershed. The participants used the enlarged information base to prepare water resources development plans for the whole watershed. This approach ensures that all local interests are reflected in the plans.

This approach was used in Sri Lanka to plan small tank(reservoir) rehabilitation activities. Preliminary studies found that farmers had little idea about the hydrology of parts of the watershed outside their village areas. The multi-level approach gave farmers the knowledge to prepare workable proposals for improving water distribution within the sub-watershed. Without the multilevel approach, farmers could only suggest fixing their tanks, an activity that would have little development effect since it would not increase irrigation water. The sub-watershed level plans, however, included means for augmenting tank water supplies and thus increasing irrigated areas.

The success of the approach was due partly to the constructive blend of scientists' knowledge of the watershed hydrology and the farmers' detailed knowledge of local hydrology, farming systems, and their own needs. A key point was that the farmers shared their local knowledge with farmers from other villages to produce useful water-shed level knowledge and plans. (Jinapala et al., 1996, p. 2)

2. *A sense of community action for catchment management grows gradually: North India*

Based on the field experiences learnt over the last 1½ years at different stages of participatory planning and program implementation with the villagers, this guideline has been prepared to further smoothen the process of joint implementation of programs as well as speeding up the transition from village to MMWS planning [MiniMicroWaterShed of up to 500 ha. in the Shiwalik Hills]. ...

The basic logic of approaching the problems of any MMWS is based on the assumption that the local people organise themselves to solve the problems of their homes first followed by those of their hamlets and the village and ultimately those of their MMWS. In reality, people always prefer to satisfy their individual needs first followed by taking care of common property resources. ... The sense of community action grows gradually in solving the problems within the village and then over a period of time it cuts across the village boundary. (Kar and Sharma, 1996)

Who Participates with Whom?

In working with such groups, the credibility of advisers is of crucial importance, for without it good dialogue cannot take place, nor can information flow easily in both directions to enrich the experience and knowledge of both the nonfarm agriculturists and the group members (Hudson and Cheatle, 1993).

If they do not find their advisers credible, they are unlikely to take the risks which may be associated with accepting what it is recommended should be changed. For improvements to occur, not only must advisers be credible in the farmers' eyes, but they must be able to suggest and facilitate the adaptation and adoption of appropriate and effective improvements which can produce quick perceptible benefits, and preferably increase income. In this context, much stress is currently given to "people's participation": for the purposes of this chapter, this is taken to mean the participation of outsiders — from governmental and nongovernmental organizations in particular — with farm families and groups in solving locally defined and prioritized problems, rather than the approach which expects farmers to participate in projects and programs conceived and designed by outsiders.

Common courtesy, let alone a desire for mutual enlightenment, dictates the need to approach supposed problems in catchments in a spirit of participation with the inhabitants both in the identification and prioritizing of key problems and potentials, in definition of appropriate solutions and strategies, in their implementation, and in the monitoring and evaluation of results (e.g., Douglas, 1996). "Designers of interventions in [not only] African NRM [Natural Resource Management] ignore indigenous knowledge and management systems at their peril.... Resource-oriented, technical initiatives which ignore the resource users are unlikely to survive the investment period" (Critchley and Turner, 1996).

WATERSHEDS AND CATCHMENTS: GEOGRAPHIC AND HYDROLOGIC RATIONALE

The land in many places has been shaped by water, forming topography that is composed of interpenetrating catchments (valley land units) and true watersheds ("whaleback" land units). Common usage of *watershed* to mean both types of land unit unfortunately has confounded the important and necessary distinction between them. A useful definition of a catchment is "a valley-shaped topographic unit of landscape, whose characterising output is water, and which has potential for satisfying some or all of the needs of its inhabitants" (IGCEDP, 1996). Its intrinsic significance relates specifically to movement and management of surface waters. Changes in the characteristics of flow at the lower end of a catchment (of whatever size) integrate the effects of upstream changes as they have affected the hydrology of the catchment.

A natural topographic catchment provides a discrete unit, which is independent of its neighbors with respect to surface water flows, in which to maximize the technical integrity and effectiveness of physical works, such as channels, banks, terraces, roads, etc., which are aimed at controlling flows of runoff.

Catchments may be of any size, in a nested hierarchy ranging from less than 1 m^2 to many hundreds of square kilometers (Shaxson et al., 1989). The smallest subunits may be man-made, such as tied ridges. The planning of catchments has generally proceeded from the larger to the smaller subdivisions, but the capacity of the land to detain and transmit water starts from the effectiveness of the aggregated small units.

Soils within larger catchment units may show patterns of catenary sequences from uplands to lowlands, but often the layout of land uses is unconformable with these technical patterns. Actual boundaries of fields and farms are often more dictated by local traditions, laws of tenure, and inheritance or by administrative fiat than they are by topographic or pedological boundaries (e.g., Shaxson, 1981).

But although there is much emphasis on watersheds/catchments as apparently logical units for many different activities, it is necessary to question the primacy of "watershed management" per se as the best means of tackling the problems of land and water decline which accompany poor land husbandry, because their roots may be as much in social, economic, institutional, or political factors as they are in technical difficulties alone.

TECHNICAL CONSIDERATIONS

RUNOFF AND EROSION: CAUSE OR CONSEQUENCE?

Water erosion is commonly cited as a prime direct cause of lowered productivity, as if erosion is an invisible force in its own right. Much effort has therefore been put into "erosion control", "combatting erosion" and similar actions, and this is commonly cited as the purpose of watershed management. There is a widespread assumption that the difference between pre- and posterosion yields can be closely correlated with the amount and quality of soil materials eroded from the area under consideration. However, this has been difficult to demonstrate convincingly either to extensionists or to farmers. In the tropics physical soil conservation works have been widely promoted on the basis that they would of themselves maintain or increase yields on "conserved" areas. This too has seldom proved to be the case, to the disillusionment of huge numbers of farmers. These works have in many cases added to costs without providing compensatory benefits to the people who are both poor and desperate to maintain soil productivity. In Lesotho, because of the areas of land taken out of preferred production, terraces and bunds are referred to as "field shorteners" (author's notes).

A different and more realistic interpretation of the facts is that accelerated runoff and erosion are foreseeable ecological consequences of disturbances to the complex interrelationships between the several features of land, in particular its geology, topography, water, soils, plants, animals, interacting under the dynamic influences of climate, gravity, and people. This indicates that changes in these interrelations precede the onset of the runoff and erosion, and that it is these prior "upstream" conditions which merit more attention than they have had to date. Productivity can fall because of changes in soil conditions even without consequent runoff and erosion (consider, e.g., waterlogging, loss of porosity, lowering of pH, soil pollution, etc.). If runoff and erosion occur from sloping lands, they are symptoms that damage has already occurred in the areas in which they arise. Of particular significance are changes in (1) soil exposure to high-energy rainfall (mediated by cover over the soil, as affecting rainfall erosivity) and (2) soil porosity at any level of the horizon, affected by the nature and stability of the soil architecture (mediated particularly by mechanical effects of tillage and the biophysical effects of organic matter and processes in the soil, as affecting soil erodibility). This is attested to by the technical advantages of minimum tillage/no-till systems, of mulches and other covers, and of other soil improvements following additions of organic materials. In this context, catchment topography is a secondary rather than overriding feature.

Yields and Rooting Environment

Posterosion yields are in most cases lower than before the erosion occurred. They are closely related to the volume and quality of the soil materials which remain behind, which may be the same, worse, or (sometimes) better as a rooting environment than before. This suggests the need for more emphasis on better management of soil as a rooting environment, as compared with the often overriding emphasis given to physical erosion control alone.

Organic Matter and Sustainability

The preeminent importance of organic materials above, on, and in the soil, together with the biological processes which transform them, now comes into focus. They can provide cover, humic structural gums, some nutrients, increased cation exchange capacity, and are key factors in the dynamics of the capacities of soils for both physical and chemical self-recuperation after damage and depletion. Through biotic processes in the soil they provide the only way to restore soil architecture on a repeatable basis after damage such as by tillage. They are therefore the key technical factors in considerations of sustainability of production into the future.

From Rainfall to Soil Moisture

In many situations, it is insufficiency of soil moisture that is more limiting to reaching potentials for high crop yields than lack of soil per se (FAO, 1995). Rainfall, in the vertical dimension, precedes runoff in the lateral dimension. Anything that minimizes the potentially damaging effects of high-energy rainfall at the outset — in particular compaction and sealing of the soil surface — will have important positive effects in limiting the volume and severity of effects of subsequent runoff, by favoring infiltration and absorption rather than loss. Water that has percolated the soil on the way to the water table and thence to springs and streams travels very much more slowly (days, weeks) than that which has been partitioned to runoff by an impermeable surface and travels over the surface to the same points (minutes, hours). The quality and regularity of flow from the catchment in the former situation is considerably greater than in the latter.

> It would seem to be much more efficient and effective to use the water when it is still a free good rather than let it run off, collect it in reservoirs, capitalize it with development loans, and encumber it with an irrigation bureaucracy before it is returned to the farmer's field. The precedent for this approach of conserving water where it falls has proven effective in parts of Australia and the Great Plains of the U.S. Studies have shown that of the increase in average wheat yields from 1936 to 1977 in the Central Great Plains of 750 kg/ha to 1800 kg/ha, 45% was due to water conservation technology. Other technologies contributing to this increase were wheat varieties (30%), harvesting equipment (12%), planting equipment (8%), and fertilizer practices (5%) (R. Meyer, quoted in Sanders, 1991).

Well-grown crops have more leaf area and more residues than poorly developed ones and provide more protective cover, which both contribute to maintaining infiltration rates at the soil surface. Soil in good physical, chemical, and biological condition for rooting is also (particularly through its porosity from the surface downward) good for reception, retention, and transmission of rainwater thus improving its hydrological conditions. Conservation of water and soil can therefore be *achieved*, by improvements in husbandry and raised production, more radically and effectively than "doing soil conservation" only to combat runoff and erosion. When cover and porosity are excellent (equivalent to good underforest conditions), runoff is usually minimal, and the need for physical works against runoff is much reduced, or even eliminated.

EXAMPLE 2: SUBSTITUTING A WELL-MANAGED TEA ESTATE FOR DENSE TROPICAL FOREST IN THE KENYA HIGHLANDS HAD MINIMAL EFFECTS ON WATER YIELD

(Note: After approximately the first 3 years in the field, a well-managed tea crop provides almost continuous cover to the soil at about 1 m above the surface and which increases in density over the years. A dense ground litter of prunings is maintained throughout the year.)

> Meticulous planning by soil conservation engineers and implementation by competent management have successfully replaced the hydrological controls of a dense tropical forest by a thriving plantation that includes a tea factory and housing and employs a large labor force. The main features are:
>
> - Contour planting
> - Hillside ditches (narrow-based terraces leading runoff to prepared waterways)
> - Careful road construction with grassed banks
> - Strips of original forest left to protect the steep stream banks
> - Construction of a small storage pond to supply water to the factory.
>
> The effects of the land-use change were monitored. During the forest felling, the peak flows increased, but control was rapidly regained.... The peaks remained sharper and

> were increased when the tea was pruned, but they were too small to be of practical significance. Soil erosion losses were rapidly reduced to a negligible level.
>
> Water yields were temporarily increased, by about 14 percent, for the first 3 years after clearing. The difference diminished as the tea developed, and when summed during the first 15 years, the extra yield was 9 percent. When fully mature, the tea estate matched the water yield of the forested control valley. Both yielded an average of 800 millimeters per year from an average rainfall of 2100 millimeters. (Pereira, 1989)

PEOPLE IN CATCHMENTS

FARMERS NOT ALWAYS CULPRITS

Farmers are often blamed for causing all the erosion which leads to sedimentation in and below catchments. In many situations they may in fact be responsible only for small proportions of it: most of the catchment surface may be well managed and well covered while the bulk of the eroded materials may come from poorly designed/implemented roadworks, irrigation works, mining activities, etc. (e.g., Doolette and Magrath, 1990; NSPWP, 1992). Also, in very steep catchments, geological earthquakes and other natural disturbances — such as rivers undercutting their banks — may result in much material coming down in debris slides, irrespective of people's actions. The situations over the majority of the surface even of steep catchments may be not nearly as bad as cursory observations might suggest, because the inhabitants may in fact have developed relatively sustainable ways of using the land safely, and which may also include management of erosion as the soil moves downslope rather then trying to stop it altogether (e.g., Tamang, 1993; Gardner and Jenkins, 1995; Fernandez, 1997).

PEOPLE AND SUSTAINABILITY

Sustainability of any agreed improvements depends on people wanting to continue to generate the benefits to which the improvement originally led. Therefore, no outside agency should antagonize rural people by adopting the attitude (implicit or explicit) that they are merely intransigent hindrances to the implementation of theoretically optimum technical solutions to catchment problems.

APPROPRIATE BOUNDARIES

Where the well-being of people (from whom improvements in resource management can flow) is the primary focus of assistance — rather than the technical optimization of land-use patterns in the catchment — it is appropriate at the outset to respect the boundaries with which people are familiar.

Coincidence of self-interest in the social sphere with the needs for improved management in the smallest productive units of a catchment — within a single field — in the technical sphere is likely to be the key to beginning to sustain preferred land uses indefinitely and at the same time to improve the hydrology of the catchment in which it is undertaken.

The boundaries of even the aggregated lands of a self-identifying community of people do not often accord wholly or even in parts with topographic catchment features. Rather, they drape unconformably over the topographic landscape, like a blanket thrown over an armchair. In addition, some people who live in the community may have plots of land outside the catchment boundary, and people living outside the catchment may have landholdings within it. The objectives and degree of interest of different landholders may not be homogeneous within a catchment, and their concerns about land management may vary accordingly.

The outsider's concept of "catchment" may not be important, or understood, within a rural community, as it may have little direct relevance to their daily activities. If it is conceived as representing a structural part of their environment, it may be limited by the margins that they can see, which may encompass an area less than the topographic catchment as a whole. Appealing to rural people to undertake "catchment management" may be meaningless because (1) they may not give significance to the concept and (2) if the phrase in reality is taken to mean (as it often is) only physical conservation works, they may be unconvinced of the benefits that might be obtained.

While the idea of a local "catchment management committee" may be important to outsiders considering the care of natural resources on a topographic basis, initial work with farmers in improving their livelihood may be much better centered within the physical and metaphorical boundaries of different common-interest groups within the community. Some of their concerns about land management may be unrelated directly to individual catchment areas, such as catching roof water for gardening and drinking water, better marketing, improved road networks, etc. (Kiara et al., 1997). Even if, initially, they do not acknowledge the validity of topographic boundaries, they will be familiar with farms and fields shown on administrative records of the local and national government

Insistence by a major project that land uses should conform with technical land-use capability classes can generate serious antagonisms not only to those who try to implement the work but also to any other outside person or institution presuming to give advice or assistance. The recommendation for reducing intensity of land use as capability class increases may not always be realistic or even valid. It was found in the hilly Machakos District of Kenya that, over a period of 60 years, while population of the area had increased fivefold, the environment was in much better condition in 1990 than in 1930. Increased pressure of population had resulted in greater care being taken of the land, in raised production per person and per hectare, and in less erosion, despite greater intensity of land use (Tiffen et al., 1994).

People's Priorities

People's primary concerns may not even be with natural resource management at the outset. Construction of a school or bridge, or solving of a disease problem, may be predominant concerns about which action would be needed first before people are willing to consider natural resource concerns in detail (Shaxson, 1989). Generation of cash by nonagricultural means, such as sewing, brick making, or off-farm working, may be a key starting point. Or the resolution of a problem — such as

"Who has run off with the money from the local coöperative?" — may be an important matter to be resolved before land-use concerns come to the top of the list (e.g., Vieira and Shaxson, 1995).

Rural people's awareness of topographic catchments and the need for improving their management may be reached through concerns about water supplies where these are limited or erratic in wells and streams. This may be of more immediate concern to rural families than even the increased production of crops which could be achieved through better soil conditions. The second concern can often be woven into the means of solving the water supply problem. Improving water supplies may be an important component for sustainability of chosen land uses in a particular area. If people are unable to sustain a desired level of livelihood because of water shortage, they may abandon the land, and sustainability is lost. Conversely, when agricultural production rises, runoff and soil losses decrease, and groundwater supplies improve, all due to soil improvements in catchments, people return to areas they had left. However, as happened in Parana, land prices may have risen so much as a result of land improvement that some who had earlier migrated away because of land degradation are unable to afford to buy back the areas they left earlier! (Author's field notes).

But even if at an early stage, the people do not pay much attention to catchment-wide aspects such as future road network, integrated water supplies, etc. It is incumbent on their advisers to keep such matters in mind and to provide at all times guidance which takes account of the fact that these matters are likely to come up in the future.

EXAMPLE 3: CATCHMENT IMPROVEMENT PLEASES FARM FAMILIES

1. In Sao Miguel, Rio Grande do Sul, Brazil, a farmer's wife said that she very much approved of the farming improvements in the catchment above the house. Tillage had been changed from use of disks to use of tines, and because the straw was no longer burned but remained as a protective mulch on the surface, much more rainfall soaked into the soil, and after big storms muddy water no longer flooded downhill through the back door and spoiled the rugs. (Author's field note)
2. At Linha Maneco near Toledo, Parana, Brazil, a farmer with a 5-ha farm was much afflicted by lack of water for most of the year in the river along one boundary. Irrigation was impossible, and only unirrigated crops could be grown. The state's program for catchment-based improvement reached the farmers in the agricultural catchment upstream of his farm, and similar improvements — in tillage and in soil-coverage by crop residues — ensured that much higher percentages of each rainstorm were absorbed by the soil. The duration of river flow then extended well into the dry weather, sufficient to allow the family to install irrigation of high-value crops like strawberries, and to create two feeder dams and seven fishponds. Pigs are housed on the banks of the fishponds, and a significant part of the family's increased income is now derived not only from irrigated crops

but also from fees charged to visitors who come to catch fish on the weekend. (Author's field note)

EXAMPLE 4: CATCHMENT IMPROVEMENTS RAISE CROP YIELDS AND CUT ROAD-MAINTENANCE COSTS: MUNICIPIO TUPANSSI, PARANA, BRAZIL

Improvements to land management, in predominantly agricultural catchments covering about 23,000 ha, included variously broad-based bunds, contour planting, direct planting through crop-residue mulch, liming, and cultivation with tines rather than disks. Water infiltration improved; soil erosion reduced. In one microcatchment of the Municipio (Santa Terezinha), such improvements resulted in increased yields of main crops (3-year means): Soya: 1700 ⇒ 2350 kg/ha; Wheat: 1200 ⇒ 2000 kg/ha; Maize: 2800 ⇒ 3500 kg/ha; Beans: 800 ⇒ 1200 kg/ha.

Mean turbidity in streams during the rains reduced 100-fold.

In the same Municipio, by realigning roads onto crests and controlled gradients, and linking drainage with cross-slope bunds, cross-road erosion and water ponding was markedly reduced. On account of this, costs of maintenance of 400 km of earth roads in the Municipio changed: formerly: three grader passes per year, reduced to one per year; formerly 3 h/km graded, reduced to 2 h/km; saving in costs of grader-fuel (diesel) equivalent to $U.S.26/km/year. (Parana, 1994)

INDICATORS FOR MONITORING AND EVALUATION

There is much literature at various levels of generalization about the use of indicators (e.g., Casley and Kumar, 1987, 1988; FAO, 1997), but not a lot of action to use them effectively. After assessing 133 projects which had a significant soil conservation component, Hudson stated, "It was surprising that there were so many projects which could not be evaluated because there was insufficient documentation available.... Every agency without fail recognised weaknesses in its monitoring, reporting and evaluation procedures" (Hudson, 1991). The World Bank assessed 35 of its watershed development, forestry, agricultural, and rural development projects, across 18 countries, and concluded that for 27 of them there was "no quantitative information" about project impacts/effects (Doolette and Magrath, 1990).

Huge numbers of possible indicators can be supposed by outsiders looking in on a situation of rural change of any sort. The difficulty is that of deciding which out of the many would be the "best" to show what is happening — whether as success or failure. Generally, a vast list is produced, with implications of large costs and amounts of time to observe and to measure them to a satisfactory statistical level of precision. This, in itself, is an inhibitor of effective monitoring.

INDICATORS OF IMPROVEMENTS IN PEOPLE'S LIVELIHOODS

Casley and Kumar (1987, 1988) emphasize the value of qualitative indicators, which are much underused but which can provide a wealth of useful information. Taking this further, it is apparent from talking to farmers that they notice changes taking place and have their own pragmatic indicators by which they judge the nature and directions of change resulting from various forces due to and/or in spite of project/program efforts. From these beginnings, it can be decided what further detail and what complementary work may be needed to complete the information provided and required (Shaxson, 1997).

Three broad indicators are important as parts of a framework for more-detailed monitoring, since they relate to rural people's satisfaction with improvements and, hence, to success as seen from their viewpoints:

- Rural people's comments about how much more/less satisfied they are with their livelihoods now than they were in the preproject times, and the several factors that they use in making this overall judgment;
- The degree of adaptation that farmers make to introduced recommendations/suggestions, as they fit what appears to them to be generally worthwhile techniques to their individual circumstances;
- The rate of farmer-to-farmer spread of new or improved methods which have been introduced, more or less independently of any extension service efforts.

Economic evaluations based on a sufficiency of recorded data about benefits of improved husbandry in catchments are sparse. However, a study of the results of improvements in agriculture in the context of catchments in the State of Parana, Brazil, provides important information.

EXAMPLE 5: CATCHMENT-BASED IMPROVEMENTS IN LAND HUSBANDRY RESULT IN SIGNIFICANT ECONOMIC BENEFITS IN STATES OF PARANA AND SANTA CATARINA, BRAZIL

1. Parana

Of the almost 20 million ha total area of the state, about 30% was under temporary and permanent crops in 1982 (the cropped area may have expanded somewhat since then). Between 1987 and 1990 the Programme for Integrated Management of Soils (PMISA) had already resulted in catchment-aligned improvements on 2.5 million ha. Further progress has been achieved since then under the expanded successor program ParanaRural, and up to 5 million ha has been improved to date.

The rural development programs in the State of Parana in southern Brazil are strongly based on farmers' implementation of a mix of conservation-effective farming and physical practices in the context of "micro-bacias." These are small catchments of around 500 to 3000 ha in which farmers and the local authorites collaborate with each other in physical planning and implementation of those actions — such as bunding and road alignments — which cross farm boundaries.

The recommendations are based on considerable in-state research in agronomy, tillage, and the physics, chemistry, and biology of soils, also on prior in-field observations of performance of physical and agronomic recommendations (Derpsch et al., 1991). Such research has shown, and farmers have demonstrated, that losses of water and soil can be reduced to minimal levels by using minimum-tillage and direct-drilling techniques, crop rotations, winter cover crops, contour planting (related to contour bund alignments), and that on account of the beneficial effects, bunds can be reduced in size and/or in number, or even eliminated altogether.

Economic analysis made in 1989 indicated that, at the level of the farm (and at the cost/price structures at the time of the analysis), improved husbandry increased net profit by at least 20% above that achieved by nonimproving farmers in the State. At the level of the state, cost–benefit analysis (including also the value of social benefits following from widespread adoption of conservation-effective production practices) show that investments of $19 million/year would yield an annual return of about 20%. (*Sources:* Parana, 1992; Sorrenson and Montoya, 1989)

2. Santa Catarina

> The overall concepts of improved management ([a] protect the soil from surface sealing by rainsplash, [b] maintain an open structure for infiltration of rainwater ..., [c] mechanical or vegetative barriers to dispose safely of such runoff as still occurs, from agricultural land or from roads, after particularly intense rain) not only appear to be well understood by land users, field staff, project management and local politicians alike: farmers have also been remarkably successful in picking on practices adapted to fulfilling these basic requirements in their own particular settings. Thus while the mechanised farmers of Parana [average farm size about 40 ha] have shifted dramatically toward stubble mulching and direct drilling to protect the soil surface from erosion by spring rains, the smaller-scale operators typical of Santa Catarina [average farm size about 25 ha] rely instead on winter green manuring and animal-drawn or manual punch planters ("matraca") to achieve the same basic aim. Similarly, mechanised farmers use broad-based contour ridges (on which they can still plant crops, so no space is lost) to check any remaining runoff, while non-mechanised farmers use vegetative barriers, spanning farm boundaries and planted as a group activity ("mutirao"). Distribution of pig slurry using communally owned spreaders bought with partial finance from soil conservation funds is giving savings on inorganic fertilizer, as well as reducing contamination of local watercourses. Visiting the field when many spring-sown crops had only recently emerged and following 60 mm of rain in the previous 24 hours, streams in "closed" [completely-treated] microcatchments were only lightly turbid.

In southern Brazil, such improvements are now widespread, as can be seen when traveling through many parts of the States of Sao Paulo, Parana, Santa Catarina and Rio Grande do Sul. (*Sources:* FAO, 1993; Busscher et al., 1996; de Freitas, 1997).

DISCUSSION

WHAT CONSTITUTES SUCCESS?

Whose Viewpoint?

Whether any agricultural development project — including "watershed management" projects — is deemed to have been "successful," and the nature of any such success, depends on whose viewpoint is considered. Views will differ among some or all those involved, e.g., a government department promoting export crops, the project manager, technical subject-matter specialists, extension staff, and farm families all concerned with the same area/watershed. Ultimately, watershed management projects in which there is supposed to be "people's participation" should be judged on whether people's livelihoods in the catchment itself have improved. In this light, their criteria and descriptions of success are clearly important, even if they are more qualitative than quantitative. These can be complemented by other indicators to provide explanatory detail.

Whose Objectives?

Assessments of success also depend on whose original objectives are used as the criterion or goal. As Woods has pointed out (personal comm.), "If the original objectives are set at realistic levels by the people in the villages themselves, and if they have been fully involved in all decision making, the chances of them reaching their own objectives successfully are high. By contrast, if the objectives were originally set by outsiders, they may be less realistic or appropriate, people's interest in reaching them may be less, and the chances of success within the same time are thereby reduced."

What Time Frame?

Was success judged before the project ended/immediately after formal completion/2/5/10/15 years after completion? The later the final assessment is made, the better the measure of the perpetuation of benefits which accrued from project/program actions and inputs. Farmers' own testing, adaptation, and possible adoption takes time, and at the beginning only the boldest may take up recommendations; others may join in as they see neighbors having some success with actions which can enhance their livelihoods. The graph of uptake rate is seldom a straight line; rather, it is an S-shaped curve over time, with barely perceptible change in the first 1 to 2 years, followed (we hope) by a more rapid expansion phase, to a later leveling-off at a higher plane when the majority have implemented various aspects of the improvements on offer. A project might still be judged successful after, say, 6 years even if little change had been perceived after only the first 3 years.

Bunch and Lopez (quoted in Hinchcliffe et al., 1995) point out that it is the process of innovation and experimentation (over a period of time) by farmers that leads to sustainability, not the sustained use of particular bits of technology, since these may necessarily change as circumstances vary.

Other Uncertainties

In the narrow terms of the catchment as a hydrologic unit, the chief indicators of changes within the topographic boundary relate primarily to the hydrologic regime at the outlet of specific catchments and subcatchments. They are, particularly, rainfall:runoff relationships in terms of volumes, flood peaks, duration of flows, and turbidity and sedimentation data — as long as the equipment and procedures are in place to collect the data.

Comparing "with and without" or "before and after" measurements in catchments requires that uniformity assessments be made before natural variations/differences can be subtracted from results to leave the true indications of net change due to treatment (see, e.g., Pereira, 1973; 1989). In large catchments, changes in sediment load due to changes in erosion upstream are notoriously difficult to partition between those attributable to natural events, to alterations in sediment delivery ratios, and to effects of human activities, and have to be considered most circumspectly before disentangling real effects due to treatment.

Whether any changes due to treatment are perceptible appears to depend on the size of the area considered. The smaller the plot, the more readily are positive changes discerned, but the less representative they may be of the overall effects on larger catchment areas. And the larger the area considered, the more difficult it is to take account of unpredictable severe climatic events such as produced by small rainstorm storm cells which affect only parts of the whole area.

It may often be difficult to determine, even with apparently adequate monitoring, whether observed changes are due (1) more to variations in exogenous weather conditions, markets, policies, infrastructure, etc. than they are to project efforts, or vice versa; and/or (2) to rural people taking up recommended improvements (such as terrace construction, tree planting and similar actions) more because of the chance to gain food or cash from employment or incentive payments than because they are persuaded of the technical benefits of doing them. Many kilometers of conservation bunds and terraces were constructed by gangs of labor in Lesotho under Food-for-Work schemes over more than 10 years, but there is very little evidence that those who did the work adopted and/or maintatined such measures on their own farms, without the incentive (Author's notes). What farmers say about such measures may depend on how they perceive the expectations or motives of those to whom they are speaking. Some Indian farmers praised technical effects of earlier-constructed conservation works when talking to a government researcher, but denigrated the works to others the next day when the officer was not present — but they admitted also that they would not mind doing a similar job again because it would provide employment in the lean season (Kerr et al., 1996).

Factors for Success and Failure

The imposition of "grand-design" catchment management plans, designed and directed by outsiders with whom the populace is expected to "participate," seems to have been ineffective in many situations. A World Bank study of 35 watershed management projects (20 in Asia) involving implementation of predetermined physical and vegetative erosion control technologies indicates that, despite the expenditure of more than U.S.$500 million, results were less than satisfactory in halting erosion or improving people's livelihoods, provoking regional staff to question the efficacy and benefits of the technologies and treatments used (Doolette and Smyle, 1990).

World Bank figures on rates of failure of agricultural and soil conservation projects between 1973 and 1984 were quoted as being between 26 and 40% (Hudson, 1991). Hudson implied that, within the conventional "mind-set" about soil conservation projects at that time, more attention to training, institution building, better accounting, improved evaluations, etc. would be sufficient to overcome this problem.

There are three additional features that now appear to be essential for rural people themselves to be able to undertake and sustain conservation-effective improvements within the catchments in which they live:

- Ensuring that rural people are the central pivot in identification and prioritizing of key problems and potentials, in definition of appropriate solutions and strategies, in their implementation, and in the monitoring and evaluation of results (e.g., Douglas, 1996);
- Great emphasis on organic materials and processes above, on, and in the soil as important (though not complete) means of achieving conservation within production processes, via improvements in both cover over the soil and the maintenance and regeneration of good architectural conditions within the soil;
- Optimization of the "envelope" of external "pulls" and pressures — economic, political, infrastructural, institutional, informational, etc. — within which rural people make their decisions that affect land use and husbandry (Shaxson et al., 1989).

It is not so much the act of implementing watershed/catchment management of itself that brings technical benefits, but rather the aggregate of the effects which people's improved husbandry actions have on the microlevel conditions of cover, and of the biological, chemical, and physical aspects of the soil as a rooting environment and as a store and conduit of rainwater. It is these effects, even without reference to topographic features, which provide the conservation-effective bases for sustainable land use, to which physical works are complementary when aligned relative to physical topography of the catchment (e.g., Vargas et al., 1986; Shaxson, 1988; Shone et al., 1991; Hinchcliffe et al., 1995; Pantanali, 1996; Hamilton et al., 1996).

Nevertheless, using the catchment broadly as the working unit area may be justified on the basis of organizational efficiency and convenience, as it allows concentration of, e.g., government advisory resources and efforts within a specified area for a limited time span. Advisory workers find it convenient to and effective to work with local common-interest groups, and on contiguous areas often formed by their members' farms. It is hoped that, if the participatory plans and programs are attractive and build on people's skills and enthusiasms during the initial period, the process of trials, adaptation, and consolidation by farm families will become self-replicating. Then the process should spread across and out of the original catchment, allowing advisory services to switch the bulk of their major attention to other as-yet-unimproved catchments nearby (e.g., Thompson and Pretty, 1996).

Ever-larger areas — from individuals' plots up to large catchments — will be covered by effects of better husbandry according to several parallel sequences of farmers' actions:

- Through increasing numbers: Person ⇒ Family ⇒ Common-Interest Group ⇒ Community;
- As successive attention is given to sequences of people's own prioritized problems until concern about maintaining soil productivity and water supplies comes to the top of the list;
- As improved practices enlarge in area from Field corner ⇒ Field ⇒ Farm ⇒ Neighbor's farm ⇒ Whole village ⇒ Grouped villages ⇒ Catchment;
- As improvements develop to include Farm/cultivated land ⇒ Pastureland ⇒ Woodland ⇒ Whole village system ⇒ Catchment.

Good husbandry all across catchments and watersheds is the desirable plateau condition to be achieved. That objective will more securely be reached where the journey follows the paths of people's well-informed and incremental decision making than where preformed plans are provided in whose fulfillment the inhabitants are expected to become involved.

REFERENCES

Busscher, W.J., D.W. Reeves, R.A. Kochhann, P.J. Bauer, G.L. Mullins, W.M. Clapham, W.D. Kemper, and P.R. Galerani. 1996. Conservation farming in southern Brazil: using cover crops to decrease erosion and increase infiltration, *J. Soil Water Conserv.* 51(3), 188–192.

Casley, D.J. and K. Kumar. 1987. *Project Monitoring and Evaluation in Agriculture,* Baltimore: Johns Hopkins University Press, 159 pp.

Casley, D.J. and K. Kumar. 1988. *The Collection, Analysis and Use of Monitoring and Evaluation Data,* Baltimore: Johns Hopkins University Press, 174 pp.

Critchley, W. and S. Turner. 1996. Introduction, in *Successful Natural Resource Management in Southern Africa,* Amsterdam: Vrije Universiteit, Centre for Devt. Coopn. Services. Windhoek (Namibia): Gamsberg Macmillan Publishers (Pty) Ltd., 1–17.

de Freitas, V. H. 1997. Transformations in the micro-catchments of Santa Catarina, Brazil, in Pretty, J. N. et al., Eds., *The Economic, Social and Environmental Impacts of Participatory Watershed Development,* London: International Institute for Environment and Development, IIED, London.

Derpsch, R., C. H. Roth, N. Sidiras, and U. Kopke. 1991. Controle da erosao no Parana, Brasil: Sistemas de Cobertura do Solo, Plantio Direto e Preparo Conservacionista do Solo, Sonderpublikation der GTZ, no. 245, Eschborn, Germany: GTZ, 268 pp.

Doolette, J. B. and W. B. Magrath, Eds. 1990. Watershed Development in Asia: Strategies and Technologies, World Bank Technical Paper no. 127, Washington, D.C., World Bank, 227 pp.

Doolette, J. B. and J. W. Smyle. 1990. Soil and moisture conservation technologies: review of literature, in Doolette, J. B. and Magrath, W. B., Eds., Watershed Development in Asia: Strategies and Technologies, World Bank Technical Paper no. 127. Washington, D.C.: World Bank, 227 pp.

Douglas, M. G. 1996. *Participatory Catchment Management,* Report OD/ITM 55. Wallingford (UK): HR Wallingford, 2 vols., 143 and 126 pp.

FAO. 1993. Brazil: World Bank Land Management Projects I (Parana) and II (Santa Catarina), Working paper — Synthesis of Supervision Mission Observations, Investment Centre Doc. 6/93 CP-BRA 52(WP) 15.1.93.

FAO. 1995. Agricultural Investment to Promote Improved Capture and Use of Rainfall in Dryland Farming, FAO Investment Centre Technical Paper 10. Rome: FAO, 46 pp.

FAO. 1997. Land Quality Indicators and Their Use in Sustainable Agriculture and Rural Development, Land and Water Bull. 5. Rome: FAO, 213 pp.

Fernandez, A. P. 1997. The impact of technology adaptation on productivity and sustainability: MYRADA's experiences in southern India, in Pretty, J. N. et al., Eds., *The Economic, Social and Environmental Impacts of Participatory Watershed Development,* London: International Institute for Environment and Development, IIED, London.

Gardner, R. and A. Jenkins. 1995. *Land Use, Soil Conservation and Water Resource Management in the Nepal Middle Hills,* London: Royal Geographical Society and Wallingford: Institute of Hydrology, 13 pp.

Hamilton, P., S. Bunyasi, C. Gichengo, J. Katua, L. Nkanata, J. Tum, and B. Meso. 1996. "Goodbye to Hunger!" The Adoption, Diffusion and Impact of Conservation Farming Practices in Rural Kenya, ABLH Report 12/97, Nairobi (P.O. Box 39042): Association for Better Land Husbandry, 206 pp.

Hinchcliffe, F., I. Guijt, J. N. Pretty, and P. Shah. 1995. *New Horizons: The Economic, Social and Environmental Impacts of Participatory Watershed Development,* Gatekeeper Series SA50, London: IIED, 22 pp.

Hudson, N.W. 1991. *A Study of the Reasons for Success or Failure of Soil Conservation Projects,* FAO Soils Bull. 64, Rome: FAO, 65 pp.

Hudson, N. W. and R. J. Cheatle, Eds. 1993. *Working with Farmers for Better Land Husbandry,* London: Intermediate Technology Publications, 272 pp.

IGCEDP. 1990. Global Challenges, Local Visions: Experiences in People-Centred Eco-Development in the Himalayan Foothills, Palampur (HP, India): Indo-German Changar Eco-Development Project, P.O. Box 25, 38 pp.

Jinapala, K., J. D. Brewer, and R. Sakthivadivel. 1996. Multi-Level Participatory Planning for Water Rescources Development in Sri Lanka, Gatekeeper Series SA62, London: IIED, 20 pp.

Kar, K. and M. Sharma. 1996. Participatory Planning for Integrated Mini-Microwatershed Development — Gaon se Pandol Tak, Internal Guideline Document, Palampur (HP, India): Indo-German Changar Project, 18 pp.

Kerr, J. M., N. K. Sanghi, and G. Sriramappa. 1996. Subsidies in Watershed Development Projects in India: Distortions and Opportunities, Gatekeeper Series SA61, London: IIED, 23 pp.

Kiara, J. K., L. S. Munyikombo, L. S. Mwarasomba, J. N. Pretty, and J. Thompson. 1997. The catchment approach to soil and water conservation: Ministry of Agriculture, Kenya, in Pretty, J. N. et al., Eds., *The Economic, Social and Environmental Impacts of Participatory Watershed Development,* London: International Institute for Environment and Development, IIED, London.

NSPWP. 1992. A Participatory Approach to Environmental Protection Measures for Hill Irrigation Schemes in Nepal, Nepal Special Public Works Prog. Manual no. 1, Kathmandu: HM Government of Nepal, Department of Irrigation, 150 pp.

Pantanali, R. 1996. Lesotho: A Note on the Machobane System, TCI Occasional Paper Series no. 7, Rome: FAO/Investment Centre Division, 26 pp.

Parana, 1992. Parana: A Forca da Campo, Brochure, Sec. de Estado da Agricultura e Abastecimento do Parana, 108 pp.

Parana, 1994. Descriptive technical handout for visitors, Realeza — PR, Comissao Municipal de Desenvolvimento Rural e Preservacao Ambiental, 20 pp.

Pereira, H. C. 1973. *Land Use and Water Resources in Temperate and Tropical Climates,* London: Cambridge University Press, 246 pp.

Pereira, H. C. 1989. *Policy and Practice in the Management of Tropical Watersheds,* London: Pinter Publishers/Belhaven Press, 237 pp.

Sanders, D. W. 1991. Soil Moisture Conservation, *FAO Soil Conserv. Notes,* 25, 1–2.

Shaxson, T. F. 1981. Reconciling social and technical needs in conservation work on village farmlands, in *Soil Conservation — Problems and Prospects,* R. P. C. Morgan, Ed., Chichester (UK): Wiley, 385–397.

Shaxson, T. F. 1988. Conserving soil by stealth, in *Conservation Farming on Steep Lands,* W. C. Moldenhauer, N. W. Hudson, Eds., Ankeny: Soil and Water Conservation Society, Ankeng, Iowa, U.S.A., 9–17.

Shaxson, T. F. 1989. Achieving sustainable productive land use: a framework for interdisciplinary field action, *Splash* (Maseru, Lesotho), 5(3/4), 4–13.

Shaxson, T. F. 1997. Land husbandry's fifth dimension,enriching our understanding of farmers' motivation, in L.S. Bhusham, I.P. Abrol, and M.S. Rama Mohan Rao, Eds., *Soil and Water Conservation: Challenges and Opportunities*, Proc. 8th ISCO Conf. ICAR, New Dehli, India, 2:1070–1076.

Shaxson, T. F. 1997b. Land quality indicators: ideas stimulated by work in Costa Rica, North India and Central Ecuador, in Land Quality Indicators and their Use in Sustainable Agriculture and Rural Development, Land and Water Bull. 5, Rome: FAO, 165–184.

Shaxson, T. F., N. W. Hudson, D. W. Sanders, E. Roose, and W. C. Moldenhauer. 1989. *Land Husbandry: A Framework for Soil and Water Conservation,* Ankeny: Soil and Water Conservation Society, 64 pp.

Shone, G., P. Evans, J.-E. Carlsson, Y. Khatiwada, P. Bergman, and G. Taylor. 1991(?). People and Their Land — The Production through Conservation Approach in Lesotho, Maseru: Min. Agri., Coops. & Marketing/Dept. Soil Cons., Forest. & Land Use Planning, Govt. Rinters, Lesotho, 33 pp.

Sorrenson, W. J. and L. J. Montoya. 1989. Implicacoes Economicas da Erosao do Solo e do Uso de Algumas Praticas Conservacionistas no Parana, IAPAR Boletim Tecnico no. 21, August 1989, Londrina, Brazil: Fundacao Instituto Agronomico do Parana, 110 pp.

Tamang, D. 1993. Living in a Fragile Ecosystem: Indigenous Soil Management in the Hills of Nepal, Gatekeeper Series SA41, London: International Institute for Environment and Development, 23 pp.

Thompson, J. and J. N. Pretty. 1996. Sustainability Indicators and Soil Conservation, *J. Soil Water Conserv.,* 51(4), 265–273.

Tiffen, M., M. Mortimore, and F. Gichuki. 1994. *More People, Less Erosion: Environmental Recovery in Kenya,* New York: John Wiley & Sons, 311 pp.

Vargas, M. Y., A. V. Villanueva, and J. A. Moreno. 1986. Impacto de la Conservacion de Suelos y Aguas in la Sierra Peruana, Lima: Min. Agri/Prog. Nac. de Cons. de Suelos y Aguas en Cuencas Hidrograficas, 19 pp.

Vieira, M. J. and T. F. Shaxson. 1995. Criterios para la Identificaion y Seleccion de Alternativas Tecnicas para el Uso, Manejo, Recuperacion y Conservacion de Suelos y Agua, San Jose, Costa Rica: Min. Agri. & Ganad./FAO Proj. AG:GCP/COS/012/NET Doc. de Campo no. 32, 37 pp.

Environment Quality and Watershed Management

22 Watershed Management for Mitigating the Greenhouse Effect

John M. Kimble and Rattan Lal

CONTENTS

INTRODUCTION

How we manage our natural resources to reduce the loss of carbon or conversely to increase the rate of its sequestration in the pedosphere is important in mitigating predicted greenhouse effects. An important strategy to address soil management issues is on a watershed basis. Processes that influence emissions of greenhouse gases from soil are (1) erosion and deposition; (2) leaching of dissolved organic carbon; and (3) mineralization of humus. In contrast, soil processes that sequester carbon at the watershed scale are (1) biomass production, (2) soil water and energy dynamics, and (3) aggregation. Erosion has a major influence on soil organic carbon (SOC) and SOC is related to greenhouse gases. If we are to control erosion we need to do this on a watershed basis not field by field. We need to understand the watershed as a unit to make reliable estimates of loss or sequestration rates of carbon. On the upper parts of many watersheds there can be large soil losses but much of this material may be deposited on lower or flatter slopes. Just saying a field lost "X" tons of soil is not enough, as it may really not be moved out of the watershed. However, erosion is not the only process of concern within a watershed. The use of buffer strips near streams may greatly reduce the soil loss and consequently reduce SOC losses and increase the rates of sequestration. A watershed is an integrated unit

0-8493-0702-3/00/$0.00+$.50

and what is done on one part may have major effects on others, which may even be under different ownership. High levels of fertilization on one area may have effects on other fields lower in the landscape. This may change the C:N ratio again affecting SOC content and its dynamics. In the past much of our efforts have been focused on fields, not on watersheds or more diverse ecosystems to be effective in mitigating SOC losses. Management of other soil properties is needed at broader scale for sustainable use of these resources. The interactive effects through a watershed need to be addressed.

The idea that the Earth's climate is being changed by what is called the "greenhouse effect" has become a hot issue of the 1990s. The greenhouse effect refers to the process whereby is where the Earth's atmosphere act in the same manner as the glass in a greenhouse. The incoming solar radiation is trapped within the system and global warming occurs. There are several radiatively active or greenhouse gases (GHGs) which contribute to this, they are CO_2, N_2O, NO_X, CH_4, and CFCs. Many agricultural practices are involved in the sequestration or release of these gases. Intensive tillage operations mix the soil and increase the decomposition of soil organic carbon (SOC) where as conservation tillage and better residue management may increase SOC. These process and effects have been extensively addressed in recent books (Lal et al., 1995a; b; c; 1997; 1998 a; b) and will not be covered in detail in this chapter.

In recent years the idea of managing a single resource has become obsolete and the idea of total ecosystem management has become the vogue. In the book *Ecosystem Geography* (Bailey, 1996), Jack Ward Thomas, Chief USDA Forest Service, states, "Land management is presently undergoing enormous change: away from managing single resources to managing ecosystems. From forest to tundra, to desert, to steppe, the world's ecosystems vary vastly." Watersheds can vary in size and contain many different ecosystems (the Missouri drainage for example starts in the Rocky Mountains of Wyoming, Colorado, and Montana and extends though much of the Great Plains. It can be subdivided into many small watersheds. The climatic conditions vary greatly in terms of the USDA Soil Taxonomy from frigid to thermic in soil temperature, and from aridic to udic and even aquic in some areas in its soil moisture regime. This is not a single watershed but a complex of many smaller watersheds all of which need to be managed differently so as the system functions as a complex of many integral parts. From the standpoint of managing watersheds for mitigating the greenhouse effect, we need to deal with much smaller watersheds, ranging in size from a few hundred hectares to several thousand hectares, but in all of this we must keep in mind what we do in one component may well affect many others. No part of a watershed is or should be considered alone; all parts are linked. Some have compared ecosystems to a spiderweb (Bailey, 1996). In watershed management, as in ecosystem management, we can use the spiderweb analogy. If we do something to a spiderweb (cut one or two strands) there are changes to the whole web and the structural integrity of the web is broken and the spider rushes to fix it. Likewise, when something happens to one part of a watershed all parts are affected and in the past these changes have been ignored. In watershed management, if we accelerate or reduce erosion, change the land use, or add fertilizers, etc. on one part, it has multiple effects on other parts.

An example of a change in a watershed having multiple effects is the following. In large watersheds we have built many dams; this changes the flow of water to the oceans. At a meeting in Nanjing, China in 1995, this effect was discussed with several Chinese scientists. They are finding a decrease in the fish catch in parts of the South China Sea. The first reaction is that there is overfishing. Yet, ongoing research shows some of the changes may be a result of the reduced movement of sediments (nutrients) into the sea. The dams now trap the sediments which before provided nutrients for the plankton which was the start of the food chain. The overall watershed was affected in a negative way to give positive effects for other reason (hydroelectric, reduced flooding, irrigation, etc.). But we have also had unexpected and unplanned changes in a broader area. To manage a system we need to consider all parts as equals and to look at the relationships of one field to another, and not at fields or even parts of fields as single units.

SOIL PROCESSES AND THE GREENHOUSE EFFECT

Processes that influence emissions of GHGs from soils are (1) erosion and soil fertility depletion; (2) leaching of dissolved organic carbon; (3) mineralization of carbon, (4) methanogenesis, and (5) nitrification in uplands of N_2O. When erosion in a field is examined, the idea that all of the material has moved to some vast waste land or the oceans and out of the system is really not the case. Erosion (geological) has developed most of the fertile parts of the world (river plains and deltas). All too often, we look at erosion and not the corresponding deposition. The areas of deposition may greatly affect the rates of leaching of dissolved organic carbon (DOC) and even the rates of mineralization. In contrast to soil erosion, soil processes that sequester carbon at the watershed scale are (1) biomass production, (2) soil, water, and energy balance, and (3) aggregation. Figure 22.1 shows both the processes that influence GHG emissions and/or the sequestration of soil organic matter. It must be kept in mind that there is an interaction among these processes; they are all linked and should not be considered separately.

EROSION AND DEPOSITION

Erosion is a naturally occurring geological process. Human intervention, however, has changed the dynamics of the process. The natural grass cover of the extensive grasslands in the great plains has been removed and replaced by row crops. This has accelerated both wind and water erosion. The natural functioning of a watershed has been altered. To overcome some of these effects, terraces have been constructed, crops are planted on the contour, and extensive use of minimum tillage has developed. These practices have helped to reduce erosion and will continue to be evolved. But in doing so we have also transformed the hydrology of the system. Many times practices were on a field-by-field basis without consideration to other users either upstream or downstream. Water flow may be channeled through grass diversion which reduces erosion but the downstream flow may be accelerated. This can reduce the water moving through the soil profile and the downward leaching of DOC. Carbon in solution is now moved off the field and into downstream water bodies.

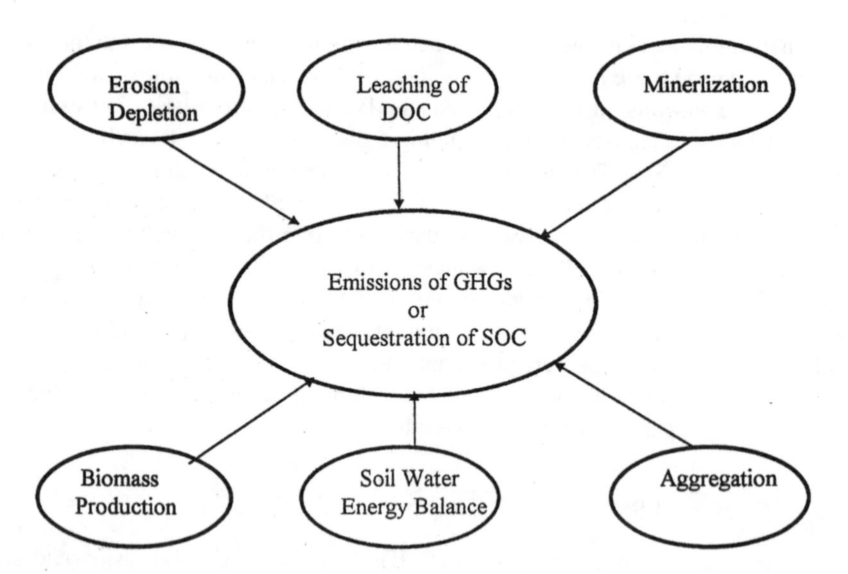

FIGURE 22.1 Factors which influence greenhouse emissions or sequestration of soil organic matter.

SOC may be buried in the terraces and this is not considered in the overall balance of the system.

In areas prone to accelerated soil erosion not only is the carbon removed by the erosion, but more is exposed to mineralization where it had once been protected, lying deep in the profile. In the areas of deposition, carbon is now protected from mineralization. Preliminary data from a study in southeastern Nebraska show varied amounts of total SOC in the top 100 cm of the soil profile for different land uses (Table 22.1). In a continuously cropped upland area where there is accelerated

TABLE 22.1
SOC in the Top 100 cm of a Sharpsburg Soil under Different Managements and Thickness of Mollic Epipedon

Treatment	SOC, kg/m³	Thickness of Mollic Epipedon, cm
Continuously cropped		
Eroded phase	12.04	39
Depositional phase	17.28	122
Continuous grassland	18.75	76

erosion because of tillage operations, the SOC was 12.04 kg/m³, whereas in a downslope continuously cropped area where there is deposition of the material

TABLE 22.2
Examples of the Effects of the Degree of Erosion on Grain Yields

	Erosion Phase			
	Slight	Moderate	Severe	Deposition
		Corn		
Yield grain (Mg ha^{-1})	7.0	6.6	5.6	4.9
		Soybeans		
Yield grain (Mg ha^{-1})	3.3	2.9	2.4	4.2

Source: Fahnestock, P. et al., *J. Sustainable Agric.*, 7(2/3):85–100, 1995. With permission.

eroded for the upslope areas, the SOC here was 17.28 kg/m^3. The SOC content of the downslope position is almost equal to that of an area of grassland (18.75 kg/m^3). All of these sites were in the same map unit and in undisturbed (non-human-accelerated erosion) situation would have been expected to have had a similar SOC content. So we face a quandary — What is the real level of SOC in the system? The eroded uplands have lost SOC; the area of deposition on the lower slope has gained SOC. Soil maps many times consider such areas as a single unit, and it may be assigned a single number for computational purposes. The question that still remains unanswered is whether there is a net gain or loss of SOC within the system. There are analytical data for three points within the unit. How can this information be aggregated to larger and larger areas for making estimations of the system SOC? Is the SOC in the lower depositional areas protected against mineralization or not? Our purpose here is not to answer these questions but to raise the issue of how we deal with contradicting point data in a large unit. By raising these questions, in time, we hope to find answers to such questions on a watershed basis.

There also may be a loss of productivity on the upper slopes and a net gain on the lower slopes, so what is the balance or net effect? Table 22.2 contains selected data from Fahnestock et al. (1995), which show the effects of erosional phases on corn and soybean yields. Only 1993 data yield from one site for each crop are shown; however, similar results were reported for other years and other sites. Over all, as the degree of erosion increases, the grain yield decreases. In areas of deposition, the yields tend to be only slightly less in some cases than slight erosion and actually higher than areas of slight erosion. This may be because of the deposition of the SOC-rich, higher-fertility material from the eroded sites at the depositional sites. This also shows the importance of SOC to soil fertility and yield. If the eroded material moves off the field into a nearby water body, it has further confounded the effects. The sediment may be deposited in the water bed, accentuating eutrophication but in the long run removing the SOC which is attached to the soil particles from the effects of oxidization and mineralization.

Leaching of Dissolved Organic Carbon

Factors that increase the rate of DOC leaching deep into the soil profile increase the rate of carbon sequestration, thereby enhancing SOC storage. With climate change, there may be an increase in DOC if there is an increase in net primary production (Moore, 1998). Other antropgenic factors also can influence the amount of DOC in the pedosphere. But no major changes would be expected, an increase in runoff may reduce leaching, and may increase the export of DOC from the pedosphere (Moore, 1998). Tillage operations which prevent runoff and increase infiltration (conservation tillage) increase biological activity, improve soil quality, increase the rate of carbon sequestration, reduce mineralization (lower soil temperatures, improved infiltration), and allow more water to move down into the profile moving the DOC to depths where it is sequestered and less susceptible to mineralization. Carried to the extreme, this may also be a problem as along with the DOC there are other chemicals (nitrates, pesticides, and herbicides) associated with the dissolved material, and these can have a negative effect on water quality and downstream users. Again, we need to look at both "on-site" and "off-site" effects of our actions. What is good on the corn field (reduced erosion, more infiltration) may have a negative impact on the groundwater and in surface water bodies. We must look at the whole watershed as a complete unit and not just as individual parts. Most of the DOC is in the oceans with an estimated 700 Pg and only 1.1 Pg in the pedosphere. There is a lot of internal movement of DOC in soils without any export, and much of the DOC is consumed by microbes within the soil system (Moore, 1998). Humans can influence what happens to the DOC by changing the drainage of the soil, increasing erosion, changing the net primary production. Understanding the dynamics of DOC must be done on a watershed basis as the DOC moves in all parts of the systems or catchment.

Mineralization of Carbon

Many workers have presented data on the loss of SOC by cultivation (Mann, 1986; Rasmussen and Parton, 1994). Cihacek and Unger (1995) summarized data for soils in the great plains and the data presented showed losses ranging form 7 to 51%. The amounts lost depend on the soil type, the cultivation method, and the length of the cultivation period. Recent data for Canada (Dumanski et al. 1994) shows average losses of SOC from mineral soils due to cultivation of 13 to 34%, and of about 35% for Histosols (Table 22.3).

The SOC can easily be mineralized by intensive farming systems (stirring the soils is like stirring a fire, the rate of oxidation is increased and the carbon is lost). In cultivated areas up to 50% of the native carbon may be lost in a few years after cultivation. We expose the soil to higher temperatures, and reduce the input of material to the soil carbon pool. After the initial loss due to clearing, subsequent changes are a function of soil management (Rasmussen and Collins, 1991). To reduce the emission of GHGs we can increase the use of minimum tillage (this tends to lower soil temperature) giving lower rates of mineralization. If we use practices that increase the movement of carbon deep into the soil profile we also reduce the rates of mineraliza-

TABLE 22.3
Mineralization Rate of Carbon Loss from Cultivation of 43.8 M ha of Land in Canada

	Precultivation, Gt	Postcultivation, Gt	Range, %
Surface carbon	3.81	3.04	
Total carbon	7.61	5.93	
Loss by cultivation			13–34

Source: Dumanski, et al. (1994).

tion. Increases in runoff and its rate of flow to lower fields can increase erosion, reduce leaching of DOC, and at the same time increase the rate of mineralization.

The problem of managing on a field-to-field basis is that it does not allow for the understanding of the whole watershed and the interaction of processes which take place at a scale larger than a single field. The movement of surface and underground water does not recognize the field boundaries of private ownership. Yet for years we have tried to manage on a field-by-field basis. Many times ownerships change between fields in a watershed, and there has not been an integrated effort to look at the overall watershed or ecosystem. As we realize that we must manage watersheds, we will see that what happens on one field will affect other fields.

We are moving into an era of what many call "site-specific management" (SSM; to many optimization of yield). But if SSM is really considered, it is not just yield optimization but also improving water quality, reducing use of chemicals on areas where production will not increase no matter how much NPK is added. We need to address the interactive effects of SSM and practices to mitigate SOC losses. Increased rates of fertilization may increase biomass production and also increase SOC storage. But yield is not the only indicator that needs to be considered in SSM. The SSM concept needs to be expanded to a watershed basis not just used to consider fertilization inputs to a specific area of the field. The technology is now available to manage very small areas, but all these small areas must be integrated into a complete system of management, not just considering yield as the only parameter of interest.

Methanogenesis from Wetlands

Methanogenesis is the process where methanogens use CO_2 or SOC and produce CH_4 which may then be released into the atmosphere. Details of the rates and fluxes of CH_4 emission are give in Mitsch and Wu (1995). We must manage wetlands very carefully because they are a part of an overall system and they function in many ways within a watershed.

Mitsch and Gosselink (1993) discussed the role of wetlands in the overall landscape. The wetlands are important preserves of biodiversity; they have a role in preventing flooding and droughts and in improving water quality. Mitsch and Wu (1995) feel wetlands may play a major role in the stabilization of the climate. They feel that there may have been a net shift in wetlands from a net sink to a source of carbon because of anthropogenic changes in the last 100 years. There are large areas of drained soils (former wetlands) which are being farmed. In other areas we have created anthropogenic wetlands by the accumulation of drainage waters from irrigated fields, and even in drained fields there are still periods of seasonal wetness which can lead to methanogenesis. Many peat soils are farmed and the dynamics of these systems are changed by different farming practices. When we drain or irrigate an area we affect other parts of the watershed, and this has ramifications not only on the field that is being managed but also on other areas within the watershed. If we remove water from an area by drainage, it can change the overall dynamics within the watershed. Whereas once the water moved slowly through the system, drainage may greatly speed the removal and may affect fields other than the one being managed.

Wetlands are considered to be the largest natural source of methane emissions (Moore and Roulet, 1995). Many peat lands are being drained for a variety of reasons, which has greatly changed the rates of methanogenesis. Armentano and Mengeus (1986) and Gorham (1991) have estimated that as many as 12 to 30 km^2 of peat lands have been drained in temperate and subarctic regions. What effects this has had on the rates of methanogenesis and also on overall carbon dynamics of the other parts of the watersheds are not known. Wetlands are very complex systems and the rates of fluxes are results of the interactions of the soil temperature and the water table (Moore and Roulet, 1995). As we change the hydrology of a watershed, we can affect both of these properties and therefore the flux rates.

Nitrification in Uplands

The largest single application of an agrochemical on upland soils is fertilization of nitrogen (Kitchen et al., 1997). From uplands we can expect nitrification and leaching of the applied N fertilizer. The nitrification can put N_2O into the atmosphere. The N_2O concentration is increasing in the atmosphere at the rate of about 0.8 ppbv/year (EPA, 1995). The primary source of the N_2O is applied nitrogen fertilizers. Agricultural practices can also enhance the emission of N_2O from soils (Mosier et al., 1991). In soils with high SOC content (>1.5%) and warmer temperatures, conventional agricultural practices enhance N_2O emissions. The reverse is true in soils with low SOC content (Li, 1995). If we apply equal rates of nitrogen to all parts of a watershed we can expect different N_2O fluxes. This leads to the need for the adoption of SSM. In upland eroded areas there may well be a need for lower rates of fertilization because of the eroded nature of the soil and many interacting factors. If high rates are applied, there can be a movement of the fertilizer downslope, which will lead to excess levels in other parts of the system and then more nitrification and N_2O emission.

With the application of nitrogen fertilizers in a watershed, we need to consider all parts of the system. We also need to consider the interaction of N, P, and S and how they are related to the overall carbon cycle The interactive nature of N, S, and P in the carbon cycle and the sequestration of carbon are discussed by Himes (1998).

If any one element is missing, the application of N, for example, will result in a net loss through nitrification. The time, as well as the amount of applied nitrogen fertilizers, needs to be closely watched so that we do not increase the N_2O emission. In recent years there has been a major concern about the movement of nitrates into the groundwater, indicating an overapplication of nitrogen to the system. Along with the movement of the nitrate we can also expect higher rates of nitrification and gaseous losses from the watershed.

CONCLUSIONS

Aldo Leopold stated, "A thing is right when it tends to preserve the integrity, stability, and beauty of the biotic community" (Leopold, 1949). This idea can be put on a watershed scale. For the whole watershed to be right, all of its parts must be right. If we are to mitigate the greenhouse effect we must do it on more than a field-to-field basis. All parts of the system must be brought together and considered equally. We must look at the interaction of the cropped field with the buffer strips, windbreaks, grass waterways, and the relationship of one field to another field. We must manage the whole ecosystem in a sustainable manner, and this is best done by managing on a watershed basis and looking at the impacts of both on-site and off-site effects.

How we manage our natural resources to reduce the loss of carbon or, conversely, to increase the rate of its sequestration in the pedosphere is important in mitigating possible greenhouse effect. One of the best ways to deal with soil management is on a watershed basis. Processes that influence emissions of GHGs from the pedosphere are (1) erosion and deposition, (2) leaching of DOC, (3) mineralization of humus, (4) methanogenesis, and (5) denitrification. In contrast soil processes that sequester carbon at the watershed scale are (1) biomass production, (2) soil water and energy balance, and (3) accretion of organic matter in microaggregates. Erosion has a major influence on SOC and is related to the emissions of GHGs. If we are to control erosion, we need to do this on a watershed basis not field by field. We need to understand the watershed as a unit to make reliable estimates of losses or sequestration rates. On the upper parts of many watersheds there can be large soil losses, but much of this material may be deposited on lower or flatter slopes. Just saying a field lost "X" tons of soil is not enough, as it may really not be moved out of the watershed. However, erosion is not the only process of concern within a watershed. The use of buffer strips near streams may greatly reduce the soil loss and consequently reduce SOC losses and increase the rates of sequestration. A watershed is an integrated unit and what is done on one part may have major effects on other parts, which may even be under different ownership. High levels of fertilization on one area may have effects on other fields lower in the landscape. This may change the C:N ratios, again affecting SOC. In the past, much of our efforts have been focused on fields, not on watersheds or more diverse ecosystems. To be effective in mitigating SOC losses and other soil properties management is needed at broader scale. The interactive effects through a watershed need to be addressed.

REFERENCES

Armentano, T.V. and E.S. Menges. 1986. Patterns of change in the carbon balance of organic-soil wetlands of the temperate zone, *J. Ecol.* 74:755–774.

Bailey, R.G. 1996. *Ecosystem Geography,* Springer-Verlag, New York, 204.

Cihacek, L.J. and M.G. Ulmer. 1995. Estimated soil organic carbon losses from long-term crop-fallow in the Northern Great Plains of the USA, in Lal, R., J. Kimble, E. Levine, and B.A. Stewart, Eds., *Soil Management and Greenhouse Effect,* CRC Press, Baco Raton, FL, 385 pp.

Dumanski, J., L. J. Gregorich, V. Kirkwood, M.A. Cann, I.L.B. Culley, and D.R. Coote. 1994. The status of land management practices on agricultural land in Canada. Centre for Land and Biological Resources Research, Agriculture and Agri-Food Canada, *Tech. Bull.*, 1994-3E, 46pp.

EPA, 1995. Inventory of U.S. Greenhouse Gas Emissions and Sinks, 1990–94, U.S. EPA, Washington, D.C.

Fahnestock, P., R. Lal, and G.F. Hall. 1995. Land use and erosional effects on two Ohio Alfisols: II. Crop yields, *J. Sustainable Agric.,* 7(2/3):85–100.

Gorham, E. 1991. Northern peatlands: role in the carbon cycle and probable responses to climatic warming, *Ecol. Appl.,* 1:182–195.

Himes, F.L. 1998. Nitrogen, sulfur, and phosphorus and the sequestration of carbon, in Lal, R., J.M. Kimble, R. Follett, and B.A. Stewart, Eds., *Soil Processes and the Carbon Cycle,* CRC Press, Boca Raton, FL, 609 pp.

Kitchen, N.R., P.E. Blanchard, D. F. Huges, and R.N. Lerch. 1997. Impact of historical and current farming systems on groundwater nitrate in northern Missouri, *J. Soil Water Conserv.* 52(4):272–277.

Lal, R., J. Kimble, E. Levine, and B.A. Stewart, Eds. 1995a. *Soils and Global Change,* CRC Press, Boca Raton, FL, 440 pp.

Lal, R., J. Kimble, E. Levine, and B.A. Stewart, Eds. 1995b. *Soil Management and Greenhouse Effect,* CRC Press, Boca Raton, FL, 385 pp.

Lal, R., J. Kimble, and E. Levine, Eds. 1995c. *Soil Processes and Greenhouse Effect,* NRCS, Lincoln, NE, 178 pp.

Lal, R., J.M. Kimble, and R. Follett, Eds. 1997. *Soil Properties and Their Management for Carbon Sequestration,* Natural Soil Survey Center, NRCS, Lincoln, NE, 150 pp.

Lal, R., J.M. Kimble, R. Follett, and B.A. Stewart, Eds. 1998a. *Management of Carbon Sequestration in Soil,* CRC Press, Boca Raton, FL, 457 pp.

Lal, R., J.M. Kimble, R. Follett, and B.A. Stewart, Eds. 1998b. *Soil Processes and the Carbon Cycle,* CRC Press, Boca Raton, FL, 609 pp.

Leopold, A. 1949. *A Sand County Almanac: And Sketches Here and There,* Oxford University Press, New York.

Li, C. 1995. Modeling the impact of agricultural practices on soil C and N_2O emissions, in Lal, R., J. Kimble, E. Levine, and B.A. Stewart, Eds., *Soil Management and Greenhouse Effect,* CRC Press, Boca Raton, FL, 385 pp.

Mann, L.K. 1986. Changes in soil C storage after cultivation, *Soil Sci.* 142:279–288.

Mitsch, W.J. and J.G. Gosselink. 1993. *Wetlands,* 2nd ed., Van Nostrand Reinhold, New York, 722 pp.

Mitsch, W.J. and W. Wu. 1995. Wetlands and global change, in R. Lal, J. Kimble, E. Levine, and B.A. Stewart, Eds., *Soil Management and Greenhouse Effect,* CRC Press, Boca Raton, FL, 385 pp.

Moore, T.R. 1998. Dissolved organic carbon: sources, sinks, and fluxes and role in the carbon cycle, in Lal, R., J.M. Kimble, R. Follett, and B.A. Stewart, Eds., *Soil Processes and the Carbon Cycle,* CRC Press, Boca Raton, FL, 609 pp.

Moore, T.R. and N.T. Roulet. 1995. Methane emissions form Canadian peatlands, in R. Lal, J. Kimble, E. Levine, and B.A. Stewart, Eds., *Soils and Global Change,* CRC Press, Boca Raton, FL, 153–164.

Mosier, A., D. Schimel, D. Valentine, K Bronson, and W. Parton. 1991. Methane and nitrous oxide fluxes in native, fertilized and cultivated grasslands, *Nature,* 350:320–332.

Rasmusen, P.E. and H.P. Collins. 1991. Long-term impacts of tillage, fertilizer, and crop residue on soil organic matter in temperate and semi-arid regions, *Adv. Agron.,* 45:93–134.

Rasmussen, P.E. and W.J. Parton. 1994. Long-term effects of residue management in wheat-fallow: I. Inputs, yield, and soil organic matter, *Soil Sci. Soc. Am. J.,* 58:523–530.

Schimel, D.S. 1995. Terrestrial ecosystems and the carbon cycle, *Global Changes Biol.,* 1:77–91.

23 Managing Watershed for Food Security and Environmental Quality: Challenges for the 21st Century

Rattan Lal

CONTENTS

INTRODUCTION

Soil degradation is a serious issue because of its global adverse impacts. Both economic and environmental impacts are related to on-site and off-site effects of soil degradation (Figure 23.1). On-site impacts are due to decline in soil quality with attendant reduction in productivity and additional costs of input needed to grow plants under suboptimal conditions (Lal, 1998). Decline in soil quality has also adverse environmental effects related to emissions of greenhouse gases, and reduction in the ability of soil to denature pollutants. The latter results in pollution and eutrophication of natural waters. The off-site adverse effects of soil degradation may be agronomic due to reduction in plant growth, economic due to siltation of waterways and reservoirs, and environmental due to pollution and eutrophication of water. An important off-site adverse economic effect of soil erosion by water is the damage to recreational facilities and aquaculture (e.g., the fish and shrimp industry). In view of these concerns, several important issues have been raised among the scientific community:

- Relative magnitude of on-site vs. off-site impacts of soil degradation,
- Success of watershed management and participatory strategies for global food security and environment quality, and
- Knowledge gaps and research and development priorities.

0-8493-0702-3/00/$0.00+$.50

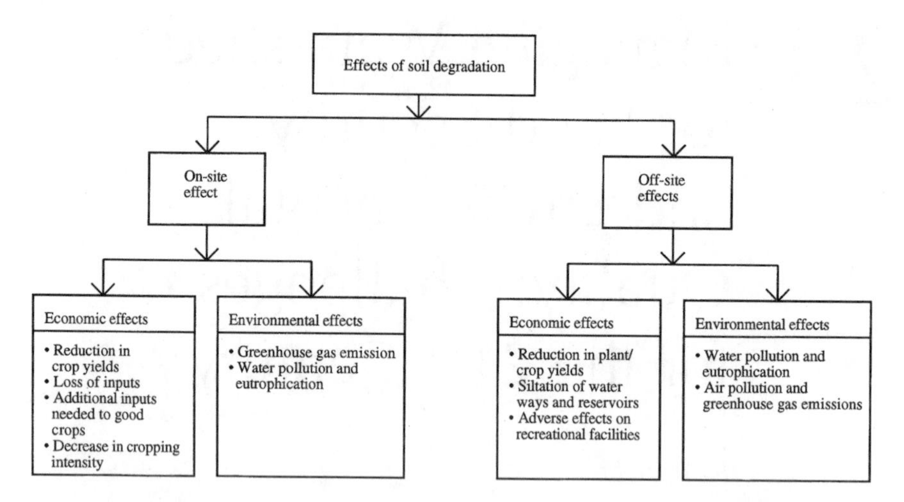

FIGURE 23.1 The economic and environmental effects of soil degradation.

ON-SITE VS. OFF-SITE IMPACTS OF SOIL DEGRADATION

The relative magnitude of on-site vs. off-site effects has become a debatable issue. There are two opposing schools. One argues that off-site impacts due to reduction in soil quality (and attendant decline in productivity) are rather small compared with the off-site economic and environmental impacts (Crosson, 1997). The other believes that both off-site and on-site impacts of soil degradation are strong, and appropriate policies need to be identified and implemented to mitigate the adverse effects and restore degraded soils and environment (Pimentel et al., 1995). The debate is perpetuated by the paucity of research information, especially from the developing countries where resource-poor farmers cannot invest in soil ameliorative measures and soil degradation is extremely severe. An objective approach to resolving the debate requires site-specific information on agronomic productivity in relation to the severity of soil degradation for major soils in principal ecoregions. These data are needed from long-term experiments especially designed to establish the cause–effect relationship.

On-site effects of soil degradation are less severe in farming systems of the temperate (Figure 23.2) than the tropical regions (Figure 23.3); in intensive and commercial farming systems (Figure 23.4) than in extensive and subsistence land-use systems (Figure 23.5); in deep soils of high inherent fertility (Figure 23.6) than in impoverished and highly weathered soils of low productive capacity (Figure 23.7); and in developed economies where farmers can invest in agriculture (Figure 23.8) than in developing economies where off-farm inputs are prohibitively expensive for the resource-poor farmers (Figure 23.9). The on-site impacts of severe soil degradation are easily masked by high inherent soil fertility, deep surface soil horizon with absence of a root-restrictive layer at shallow depth, intensive off-farm input,

FIGURE 23.2 Farming systems in temperate environments are less prone to adverse effects of soil degradation.

FIGURE 23.3 Farming systems of the tropics are strongly impacted by and lead to soil degradation.

and liberal use of soil restorative measures (e.g., fertilizer, manure, improved varieties, water table management, irrigation, deep plowing, etc.). Larson et al. (1983)

FIGURE 23.4 Commercial farming systems can absorb the impact of soil degradation.

observed that reduction in crop yield due to current level of soil erosion in U.S. for the next 100 years was merely 4%. In contrast, Lal (1995) and Dregne (1990; 1992; 1995) have reported severe on-site effects of soil erosion in Africa, Asia, and Australia, often in the range of 20 to 50% and even within a short time frame of a human generation after conversion from natural to agricultural ecosystems. Soil degradation can grossly accentuate the production cost for the same level of output.

In addition to the magnitude of the on-site vs. off-site effects, there is also the question of sustainability. Although the reduction in crop yield may be low over the human time frame (one or two generations), a degradative system may not be sustainable over a long time frame (several generations or centuries). The issue of sustainability is complex. It involves economic, environmental, social, political, and cultural issues. Some of these issues can be quantified and assessed with known indexes and analytical procedures and others cannot.

Important arguments by both schools (low vs. high on-site impacts of soil degradation) are outlined in Table 23.1. The main issue is the magnitude of yield

FIGURE 23.5 Extensive systems of the tropics are highly prone to soil degradation.

reduction, incentives and approaches that facilitate adoption of improved technology, and ability of soil/land to restore itself under agricultural intensification. Answers to these questions lie in a multidisciplinary and holistic research approach to sustainable use of natural resources.

SUCCESS OF WATERSHED MANAGEMENT AND PARTICIPATORY STRATEGIES

The watershed management approach is not a new concept, and several development organizations have adopted and implemented it since the 1970s, if not earlier. The success of this strategy is a debatable issue (Table 23.2). Some argue that watershed management based on participatory approach has met with only a limited success. The proponents of this school argue that there are numerous examples of failure, and the returns on heavy investments are rather meager. Watershed-based programs have been particularly unsuccessful in the case of small landholders of tropical regions.

Under appropriate circumstances, the success of the participatory approach has primarily been due to enhanced communication between extension workers and the farmers. The principal reason of the failure is a top-down approach whereby farmer involvement is sought in activities planned by outsiders.

FIGURE 23.6 Productivity of deep soils of high inherent fertility is less prone to soil degradation.

KNOWLEDGE GAPS AND RESEARCHABLE PRIORITIES

The watershed management strategy is constrained by the lack of appropriate baseline data on natural resources in relation to all aspects, i.e., biophysical (soil, terrain, vegetation, climate, drainage), socioeconomic (farm size, land tenure, education, health, access to market, credit facilities, availability of inputs), and cultural (customs, religious preferences). Once the baseline is established, the impact of land use on natural resources needs to be quantified. There are obvious knowledge gaps with regard to the extent and severity of soil degradation within a watershed, the cause–effect relationship between land use and soil degradation, the absolute and relative magnitude of on-site vs. off-site economic loss due to degradative processes, and the resilience characteristics of landscape units and their response to restorative measures.

There are numerous socioeconomic issues that need to be addressed. Success of the watershed management strategy depends on a clear identification of both

FIGURE 23.7 Improverished soils of low inherent productivity are highly prone to soil degradation.

FIGURE 23.8 Productivity losses are less severe where farmers can invest on land.

agricultural and nonagricultural products and services of the landscape. The value of nonagricultural services of the watershed program are likely to increase with an increase in income. A clear link needs to be established between on-site and off-site

FIGURE 23.9 Degradation-caused reduction in productivity is greater on soils where off-farm inputs are prohibitively expensive.

TABLE 23.1
Debatable Issues on Soil Degradation and Productivity

On-Site Effects Are Severe	On-Site Effects On Crop Yields Are Not Severe
• Yield losses are high • Farmers cannot afford ameliorative measures • Property rights are not secure or well defined, and do not provide incentive for farmers to invest in land • Agronomic productivity is a limited measure of soil quality; there are other issues (C emissions, wildlife habitat, aesthetic values) which are also important • Soil is a nonrenewable resource over the human time frame	• Yield losses are low • Farmers are rational and would adopt measures to control soil degradation • Market economics will take care of the problem • Soil has resilience and capacity to restore • Indigenous knowledge and local skills can play an important role in soil restoration • Use of participatory and other modern approaches can faciltate adoption of improved technology • Awareness of linking socioeconomic and biophysical approaches can solve problems that limit technology transfer

effects of soil degradation. Involvement of nonfarm residents of the watershed in a development program is essential.

Future programs in watershed management need to explore other strategies. It is pertinent to identify and promote win–win scenarios at the watershed level.

TABLE 23.2
Success or Failure of the Watershed Management Approach

Limited Success or Failure	High Success
• As a hydrologic unit, it does not correspond to social and political boundaries • Residents are not clear about the benefits and costs involved in participatory activities • Nonfarmer residents are not well represented • There are too many small farmers within a watershed with often conflicting interests	• Problem diagnosis or identification with participatory approach • Facilitates a holistic approach to natural resources management • Applicable to large landholders with know-how and resources to invest

Further, positive aspects of the strategy should be identified and appropriately promoted. Examples of positive aspects include enhanced productivity, nonagricultural products and services (e.g., recreational opportunities, wildlife, clean water). Some do's and don'ts of watershed management strategy are outlined in Table 23.3.

TABLE 23.3
Some Do's and Don'ts of Watershed Management Strategy

Do's	Don'ts
• Stress positive aspects • Promote win–win solutions • Help participants in generating lasting success so as to justify political decisions • Adopt a holistic approach to natural resource management linking biophysical and socioeconomic issues • Encourage two-way information flow • Ensure long-term continuity • Improve marketing system • Generate nonfarm income • Strengthen institutional support • Involve nonagricultural resident	• Avoid reductionist and disciplinary-based solutions to the complex problem • Don't provide unnecessary financial incentives with hidden agendas and excess baggage • Don't ignore the relevance to farmers' real needs • Avoid excessive instrumentation and analyses of water, soil, and biota for the sake of analyses

Although watershed management and associated participatory approaches had limited success, it is important to learn from the experience rather than give up. Adopting a positive approach highlighting the bright spots of past successes and promoting win–win scenarios are bound to be successful. Disciplinary-based experiments are needed to provide specific answers. However, overinstrumentation (Figure 23.10), detailed analyses of soil and water merely for the sake of analyses, has often been counterproductive and wasteful of limited resources.

FIGURE 23.10 Watershed management research is expensive because of equipment needed for monitoring and evaluation.

REFERENCES

Crosson, P. 1997. The on-farm economic costs of erosion, in R. Lal, W.H. Blum, C. Valentine, and B.A. Stewart, Eds., *Methods for Assessment of Soil Degradation,* CRC Press, Boca Raton, FL, 495–511.

Dregne, H.E. 1990. Erosion and soil productivity in Africa, *J. Soil Water Conserv.,* 45: 431–737.

Dregne, H.E. 1992. Erosion and soil productivity in Asia, *J. Soil Water Conserv.,* 47: 8–13.

Dregne, H.E. 1995. Erosion and soil productivity in Asia and New Zealand, *Land Degrad. Rehab.,* 6: 71–78.

Lal, R. 1995. Erosion-crop productivity relationships for soils of Africa, *Soil Sci. Soc. Am. J.,* 59: 661–667.

Lal, R. 1998. Soil erosion impact on agronomic productivity and enviroment quality, *Crit. Rev. Plant Sci.,* 17: 319–464.

Larson, W.E., F.J. Pierce, and R.H. Dowdy. 1983. The threat of soil erosion to long-term crop production in the USA, *Science,* 219: 458–465.

Pimentel, D., C. Harvey, P. Resosudarmo, K. Sinclair, D. Kurz, M. McNair, S. Crist, L. Shpritz, L. Fitton, R. Saffouri, and R. Blair. 1995. Environmental and economic costs of soil erosion and conservation benefits, *Science,* 267: 1117–1123.

Index

0-8493-0702-3/00/$0.00+$.50

D

E

F

G

H

I

L

M

N

T

U

V

W

Y

Z